D1389874

009351165

In today's increasingly uncertain, competitive and fast-moving world, companies must rely more and more on individuals to come up with new ideas, to develop creative responses and push for changes before opportunities disappear or minor irritants turn into catastrophes. Innovations, whether in products, market strategies, technological processes or work practices, are designed not by machines but by people.

This book shows how people can exert more leverage in organisations and initiate innovation, thus contributing to their companies' and their own successes. But it is also about the circumstances that make it possible for people to contribute their ideas. The development of "participation management" skills and environments, Rosabeth Kanter argues, provides the key to a corporate renaissance generated from within the organisation itself. The author shows the contrasts between "segmentalist" companies, which are anti-change and tend to have a narrow, compartmentalised perspective on corporate problems, and "integrative" companies, which provide a wide-open, team-oriented environment, in which innovation occurs throughout the organisation, not merely at its top or bottom. Rather than resting on the laurels of past accomplishments, integrative companies look ahead to the challenges of the future.

Here are detailed analyses of both the pathways and pitfalls in the search for innovation at some of the most important companies in America, including Hewlett-Packard, General Electric, Polaroid, General Motors, Wang Laboratories, and Honeywell. Rosabeth Kanter has looked behind the scenes in these and other firms, and describes their organisational structures, their corporate cultures, and their specific strategies—precisely what they are doing to free the powers of innovation and the entrepreneurial spirit within their own people.

The Change Masters vividly demonstrates that when environments and structures are hospitable to innovation, people's natural inventiveness and power skills can make almost anything happen. Professor Kanter's fascinating book is an indispensable guide for individuals who seek to realise their entrepreneurial potential, for executives who want to see their companies grow, and for all those concerned with business ideas and management education.

Rosabeth Moss Kanter is Class of 1960 Professor of Business Administration at the Harvard Business School. Active as an adviser to blue-chip companies and other major organisations, she was a founder of the Boston management-consulting firm Goodmeasure, Inc., and is currently Chairman of its Board of Directors. She is the author of eight other books, including the best-selling *Men and Women of the Corporation*, winner of the 1977 C. Wright Mills Award for the year's best book on social issues and *When Giants Learn to Dance, Mastering the Challenges of Strategy, Management and Careers in the 1990s*, also available from Unwin Paperbacks.

Also by Rosabeth Moss Kanter

WHEN GIANTS LEARN TO DANCE

THE
CHANGE
MASTERS

CORPORATE ENTREPRENEURS AT WORK

Rosabeth Moss Kanter

London and New York

First published in Great Britain by George Allen & Unwin 1984
First published in paperback by Unwin Hyman Limited 1985
Fourth impression 1988
Reprinted 1988, 1989, 1990, 1991

Reprinted 1992
by Routledge
11 New Fetter Lane, London EC4P 4EE

Simultaneously published in the USA and Canada
by Routledge
a division of Routledge, Chapman and Hall, Inc.
29 West 35th Street, New York, NY 10001

© 1983 Rosabeth Moss Kanter

Printed in Great Britain by
Cox and Wyman Ltd, Reading

British Library Cataloguing in Publication Data

A catalogue record for this book is available from the British Library.

ISBN 0–415–08467–9

Library of Congress Cataloging-in-Publication Data

A catalogue record for this book is also available from the Library of Congress.

Acknowledgments

Special thanks are due to the many people who assisted me at various stages of the research and manuscript preparation. This book itself benefited from the efforts of many working teams who deserve public credit.

Important members of my research team, active in both data collection and analysis and as authors of background case studies, include David Summers, Barry Stein, and Ken Farbstein. Karen Belinky, Janis Bowersox, Allan Cohen, Karen Handmaker, Myron Kellner-Rogers, and Mary Vogel also played key roles in the innovation research; Daniel Isenberg and Marcy Murninghan were central members of the research/consulting teams involved in the "Chipco" and "Petrocorp" projects, respectively, under Barry Stein's direction. Others who contributed to interviewing or data compilation through the years include Derick Brinkerhoff, Martha Cox, Ora Fant, Henry Foley, William Fonvielle, Ann Glickman, Cynthia Ingolls, Gary Jusela, Irene Schneller, and Meg Wheatley. Professors Allan Cohen and Richard Beckhard graciously involved their management classes in carrying out some of the pilot interviews for the middle-manager portion of the research.

Manuscript preparation was a particularly time-consuming task, as the book went through multiple drafts and was worked and reworked. Particularly meritorious contributions were made by Debra Boots, Pamela Colesworthy, Mona Kotch, Linda Rossman, Deborah Stockton, and Fritz Winegardner.

There are many corporate people who gave generously of their time and insights, most of whom must remain anonymous, as I promised them anonymity in return for their candor. But I can at least thank publicly a few who were helpful in making official arrangements for research or were key collaborators in testing the ideas in action projects. Among them are Richard Boyle, John Buck, and Charlaine Hobson of Honeywell; Larry Sly of General Electric; Jean Barron of Hewlett-Packard; David Bertolli of Wang Laboratories; and John Harlour of Polaroid. I also owe a special debt to Frances Grigsby.

Then there were the professional colleagues who read early versions (many of them barely resembling the final product) and gave me the benefit of their critical reactions. Many thanks to Howard Aldrich, By Barnes, Richard Beckhard, Warren Bennis, David Brown, Jane Brown, Allan Cohen,

Martha Cox, David Gleicher, Richard Hackman, Walter Powell, Paula Rayman, Jonathon Rieder, Barry Stein, Karl Weick, and Eleanor Westney.

Early reports of some of the research and conclusions appeared in various places under these titles: "Building the Parallel Organization: Toward Mechanisms for Permanent Quality of Work Life," in the *Journal of Applied Behavioral Science*, Summer 1980; "The Middle Manager as Innovator," in *Harvard Business Review*, July-August 1982; "Dilemmas of Managing Participation," in *Organizational Dynamics*, Summer 1982; and "Power and Entrepreneurship: Corporate Middle Managers," in *Varieties of Work*, edited by Phyllis Stewart and Muriel Cantor, Sage Publications, 1982. The book also builds on themes in my 1980 Plenary Address to The American Sociological Association Annual Meeting in New York, "Power and Change: Setting New Intellectual Directions for Organizational Analysis." In addition, a project on a related topic funded by the William H. Donner Foundation of New York gave me the opportunity to formulate some of the ideas here, and I am grateful to the Donner Foundation.

Then there are the most important thanks of all. First, to my editor and friend, Erwin Glikes, appreciation for his encouragement, patience, creative insights, and outstanding skills. Next, to my son, Matthew Moss Kanter Stein, gratitude for constant love and warmth even when I spent endless weekends having to close my office door to get on with the writing—an abstraction one could hardly expect a three-year-old to understand, but he did. And then, to my husband and colleague, Barry Stein, overdue public credit for all the private support and sacrifices during the years of this effort, partnership in action-research projects (where he was indeed the "change architect"), and many major intellectual contributions to the ideas in this book. Barry, I have truly learned from you.

New Haven, Connecticut
December 1982

To Matthew,
my own quality-of-work-life program

And in loving memory of
Helen Moss,
whose disposition Matthew inherited

Contents

Change Masters: Those people and organizations adept at the art of anticipating the need for, and of leading, productive change.

PART I

THE NEED FOR AN AMERICAN CORPORATE RENAISSANCE

CHAPTER

1

Introduction

> Arnold Toynbee once described the rise and
> fall of nations in terms of challenge and re-
> sponse. A young nation, he said, is confronted
> with a challenge for which it finds a success-
> ful response. It then grows and prospers. But
> as time passes, the nature of the challenge
> changes. And if a nation continues to make
> the same, once-successful response to the
> new challenge, it inevitably suffers a decline
> and eventual failure. As we begin the last two
> decades of the 20th century, the United States
> faces such a challenge.
>
> —William S. Anderson,
>   Chairman, NCR Corporation

IN MY TRAVELS around corporate America, in settings as distant geo-
graphically and culturally as a Los Angeles bank's high-rise headquarters, a snow-bordered Minneapolis electronics factory, a dingy
Detroit engine plant, a Seattle instrumentation lab sitting among
fishing boats, and a New York garment district sportswear show-
room, I have been struck by an ever-louder echo of the same question: how to stimulate more innovation, enterprise, and initiative
from their people.

In every sector, old and new, I hear a renewed recognition of
the importance of people, and of the talents and contributions of
individuals, to a company's success. People seem to matter in direct
proportion to an awareness of corporate crisis.

Indeed, only when an organization exists in stable circumstances, when its operations resemble clockwork, unvarying in their
practices, can individuals be taken for granted or ignored without
peril. Not long ago, when American companies seemed to control
the world in which they operated, it was easier to ask the people

within them to just fit into the "system" and to assume that the "system," not the people, was responsible for success. In turn-of-the-century organization theory and its "scientific management" legacy, individuals constituted not assets but sources of error. The ideal organization was designed to free itself from human error or human intervention, running automatically to turn out predictable products and predictable profits. Management was there to handle the few unexpected events that could occur.

But as world events disturb the smooth workings of corporate machines and threaten to overwhelm us—from OPEC and foreign competition to inflation and regulation—the number of "exceptions" and change requirements go up, and companies must rely on more and more of their people to make decisions on matters for which a routine response may not exist. Thus, individuals actually need to count for more, because it is people within the organization who come up with new ideas, who develop creative responses, and who push for change before opportunities disappear or minor irritants turn into catastrophes. Innovations, whether in products, market strategies, technological processes, or work practices, are designed not by machines but by people.

And so, after years of telling corporate citizens to "trust the system," many companies must relearn instead to trust their people —and encourage their people to use neglected creative capacities in order to tap the most potent economic stimulus of all: *idea power*.

This book shows *how* people can exert more leverage in organizations and thus contribute to their company's, and their own, success. It describes how individuals can help corporations stay ahead of a changing environment by moving their organizations beyond what they already know, into the more uncertain realm of innovation. Even small-scale innovations initiated by individuals— "micro-changes"—can eventually add up to "macro-changes"—the company's ability to be responsive and adaptive as circumstances demand.

But this book is also about the circumstances that make it possible for people to contribute their ideas. It describes the environments that stimulate people to act and give them the power to do so: how some companies systematically encourage innovation by the design of their systems and the treatment of their people, while others stifle or ignore it. *The degree to which the opportunity to use power effectively is granted to or withheld from individuals is one operative difference between those companies which stagnate and those which innovate.* The difference begins with a company's approach to solving problems and extends throughout its culture and structure.

Let me cite one striking finding about the relationship between a company's willingness to invest in its people and its ultimate success. In 1981 I asked an expert panel of sixty-five vice-presidents of human resources in major companies to identify those corporations who had been the most progressive and forward-thinking in their systems and practices with respect to people—the ones who had pioneered in designing effective human-resource systems, participatory workplace alternatives, and affirmative-action programs. Forty-seven "most progressive" companies were identified, led by IBM, General Electric, and Xerox. (The complete list is in the Appendix.) My research team developed a "match" for each progressive company—a nonnominated firm in the same industry and as close as possible in assets, sales, and number of employees in 1980. Then the twenty-year financial performance of the progressive companies was compared with that of the matches. *The companies with reputations for progressive human-resource practices were significantly higher in long-term profitability and financial growth than their counterparts.*[1]

Of course, given the kind of data to which I had access, it is impossible to untangle causality or be entirely sure that reputation reflects reality. Perhaps richer companies could afford to spend more in the human-resource area than poorer ones. But I doubt that this is the case. I think instead that both investment in people and long-term profitability were related to another vital factor: the willingness to take the lead in innovation—be it a good affirmative-action plan, a venture into labor–management cooperation, or use of a new production technology—in response to changing social and economic circumstances. And in a certain sense, when a company innovates in practices which ensure that more kinds of people, at all levels, have the skills and opportunity to contribute to solving problems and suggesting new ideas, then it is establishing the context for further innovations. Improved organization designs and human-resource practices can be a company's *innovation-producing innovations.*

THE INNOVATION ISSUE
AND THE AMERICAN ENTREPRENEURIAL SPIRIT

As America's economy slips further into the doldrums, innovation is beginning to be recognized as a national priority. But there is a clear and pressing need for more innovation, for we face social and economic changes of unprecedented magnitude and variety, which past practices cannot accommodate and which instead require innovative responses.

While economists argue about whether the first letter in the economy of the early 1980s is R (recession) or D (depression), the critical R&D (research and development) for our economic future is still too often neglected, as ventures are milked instead for short-term profits. Certainly, as Senator Paul Tsongas of Massachusetts has argued persuasively, the government has a role to play in stimulating investment in innovation, by underwriting risky research, removing antitrust barriers to some joint ventures, funding technical training and higher education in general, and encouraging savings to permit investments. But so do the policies and practices of large corporations themselves—and the extent to which they encourage their people to solve problems, to seek new ideas, to challenge established wisdom, to experiment, to innovate is crucial.

Indeed, where America leads, it leads because of innovation. In certain vital, innovative industries such as computers, pharmaceuticals, medical electronics, and telecommunications, the United States is ahead even of Japan; though Japan is the world's number two exporter of large computers, a sizable share of them is made by IBM of Japan.[2] Thus, playing to national strengths means continuing to be innovative.

The term "innovation" makes most people think first about technology: new products and new methods for making them. Typically the word "innovation" creates an image of an invention, a new piece of technical apparatus, or perhaps something of conventionally scientific character. If most people were asked to list some of the major innovations of the last few years, microprocessors and computer-related devices would be mentioned frequently. Fewer people would mention new tax laws or the creation of enterprise zones, even though those are innovations too. Fewer still, if any, would be likely to mention such innovations as quality circles or problem-solving task forces. This is unfortunate, for our emerging world requires more social and organizational innovation. Indeed, it is by now a virtual truism that if technical innovation runs far ahead of complementary social and organizational innovation, its use in practice can be either dysfunctional or negligible. The advanced technology incorporated in nuclear plants clearly needs more *organizational* innovation to prevent the frequent breakdowns of both components and human controls. Even many "productivity improvements" rest, at root, on innovations that determine how jobs are designed or how departments are composed.

Innovation refers to the process of bringing any new, problem-solving idea into use. Ideas for reorganizing, cutting costs, putting in new budgeting systems, improving communication, or assembling products in teams are also innovations. Innovation is the generation, acceptance, and implementation of new ideas, processes, products, or services. It can thus occur in any part of a corporation, and it can involve creative use as well as original invention. Application and implementation are central to this definition; it involves the capacity to change or adapt. And there can be many different kinds of innovations, brought about by many different kinds of people: the corporate equivalent of entrepreneurs.[3]

"Corporate entrepreneurs" can find opportunities for innovation in nearly any setting, but opportunities are most abundant in particular domains that depend on a company and its industry.[4] Much of this difference comes from the stage of development of the industry or of the products and services within it.

Early in an industry's growth, there will typically be many innovations in the products themselves and much time spent figuring out how to tailor them to particular customer preferences. Most entrepreneurs will be concerned with the design of new products and services or with incorporating rapid technological advances into current products—as in both financial services and personal computers today.

As an industry "matures" and both its products and its technologies become older and more stable, innovations are more likely to center around saving costs and improving performance, thus encouraging creative effort in manufacturing efficiencies, work methods, or quality control—a very different realm for innovation, one that "smokestack" industries like steel and autos are currently emphasizing. Look at Procter and Gamble, for example, producer of a traditional array of household products, from soap to toothpaste. Since the 1960s its commitment to new, people-sensitive plant designs and modifications of older facilities has been as serious as that of any firm in the world. And now that the company faces stiffer competition from cheaper "generic" (unlabeled) goods, its cost savings through innovations in manufacturing are a part of its competitive advantage.

Innovations in management methods and organizational practices thus constitute a broad range of opportunities for internal entrepreneurs. In resource-rich environments, when working capital, expert staff, and hungry customers are all abundant, emphasis is likely to be placed on potential breakthroughs in technology and extensive research and development activities because the company can afford them. But even when resources are in short supply, potential innovators need not—and do not—give up. In resource-lean times, the domain for innovation simply shifts to managerial procedure and organizational practice—as in the design of new ways to engage employees in solving problems.

The *form* that innovation takes may not be the same in an insurance company or a bank as in a computer manufacturer or a telecommunications firm, but the need for innovation—and thus, for people with the power to act on their ideas for change—exists in both kinds of companies.

A variety of innovations are essential for our economic health. In the corporate environment that has been emerging since the 1960s, business *must* be innovative, learning to operate in an unfamiliar context and within new constraints stemming from worldwide trends, changes in the labor force, the ups and downs of government regulation, and technological developments, as I show in the next chapter. Much attention has been focused on this problem, commonly under the rubric of productivity. But productivity, a word of many different meanings, is often treated as a mechanical issue of input and output, reflected in accounting costs of the products turned out. This sort of productivity, though relatively easy to measure—indeed, that is probably its main source of popularity—is the result of the confusion of efficiency with effectiveness. As Peter Drucker has long pointed out, efficiency is doing things right; effectiveness is doing the right things. And doing the wrong things less expensively is not much help.

More than that, excessive concern with short-term productivity can lead to lower effectiveness for several fundamental reasons. For one thing, this emphasis implicitly makes more difficult the diversion of resources from current applications (or their withdrawal as "profit") to the longer-term and more systematic development that is needed to make the organization more genuinely viable in the long run. The quarterly financial statement is not a good measure of economic health—as a growing chorus of critics is pointing out.

Moreover, the aspect of productivity that needs serious attention is *not* the mechanical output of a production facility; it is, rather, the capacity of the organization to satisfy customer needs most fully with whatever resources it has at its disposal. This may require *modification* of the product, development of entirely new products, or changes in the ways they are delivered to customers. But mechanical notions of productivity lead often to products that meet ever more refined minimum standards, frequently resulting in a decline in customer satisfaction with them. The former thrust calls out for innovation—indeed, for innovative thinking on every level of the organization's affairs—while the latter confines innovation to a marginal and unexciting role.

In the present environment, with scarce resources, low profits, reduced growth, and pressure to increase current profitability, the capacity to invest resources in the future and promote broad innovation is even more sharply reduced than would normally be the case. Given the standard equation of innovation with risk, the usual propensity for large firms to be risk-averse may under present circumstances become exaggerated. But think about the opposite problem: under present circumstances, maintaining the status quo is itself risky.

If America is to build on its past competitive strengths and to secure a better future for itself, innovation—and the risk and change that it implies—is a necessity not because it produces more profits, but because it alone ensures our survival (and ultimately, therefore, profit). To get more innovation, we need to reinfuse more American organizations with the entrepreneurial spirit responsible for America's success in the past.

Keep in mind, however, that in the past entrepreneurs and the entrepreneurial spirit have for the most part been seen as existing outside of and apart from the corporation; technological invention, for example, has largely derived from the small enterprise.[5] Thus, it is still an open question whether large organizations can accommodate, let alone take advantage of, individuals with an entrepreneurial spirit, especially where there has been a tradition of subordinating individuals to the routines and rules of bureaucratic systems.

Nevertheless, we need to create conditions, even inside large organizations, that make it possible for individuals to get the power to experiment, to create, to develop, to test—to innovate. Whereas short-term productivity can be affected by purely mechanical systems, innovation requires intellectual effort. And that, in turn, means *people*. All people. On all fronts. In the finance department, the purchasing department, and the secretarial pool as well as the R&D group. People at all levels, including ordinary people at the grass roots and middle managers at the heads of departments, can contribute to solving organizational problems, to inventing new methods or pieces of strategies.

These "corporate entrepreneurs" can help their organizations to experiment on uncharted territories and to move beyond what is known into the realm of innovation—if the power to do this is available, and if the organization knows how to take advantage of their enterprise. Those "if's" are the subject of this book.

TRAVELS THROUGH CORPORATE AMERICA

I started examining the conditions for corporate innovation five years ago. Over these years, I was fortunate enough to have firsthand contact with several thousand people in well over fifty companies across America, and I used those opportunities to try to understand the differences in culture, structure, and performance apparent across companies and between units within companies. It soon became clear that some organizations seemed to represent a "new wave," happily out ahead by every trend indicator, the first to exploit new technologies, both physical and social, and thus progressive in their treatment of people and workplace issues as well as leaders in product development.

These were the innovating companies. As I saw them, they reflected a clear departure from the standard industrial model of the past. Their success challenged much received wisdom about the orderly nature of organizational structure and the inevitability of burdensome bureaucracy in large organizations, because many of them were more complex and less orderly than the organization-theory texts suggested. Furthermore, they had grown to large size (in two cases, to more than $3 billion in sales and more than 50,000 employees) without replicating every aspect of the machine bureaucracy of traditional industrial firms. And even more telling, the people in them seemed to be enjoying themselves, and they seemed proud of their companies' successes. There were clearly problems in these companies, but the problems could be discussed, and there seemed to be someone always ready to convert any problem into a project.

A pervasive feeling of powerlessness did not seem to characterize this cluster of companies. Certainly people seemed fascinated by the idea of power and could identify the powerless among them, but there were also larger numbers of people at all levels who could grab and use power. And with this power to act came the chance to innovate. So in these places, I thought, the entrepreneurial spirit was allowed to flourish even in a large corporation.

Perhaps I was seeing glimmers of models that could possibly replace the traditional corporate bureaucracy, weighted down as it was with too much hierarchy, too little opportunity, overly concentrated power, a variety of nonmeritocratic inequities in the treatment of women and minorities—and the difficulty of changing itself. The "new style" innovative company, in contrast, seemed not only different in organization but also better suited to the demands of our times for more responsiveness to the needs of educated, rights-conscious employees who come from both sexes and many races. Then, as the economy worsened, it became clear that the adaptive, innovative postures of these newer-model corporations were not just luxuries of good times but necessities for survival in bad times.[6]

I set out to refine and deepen this perspective by a series of systematic research studies and change projects, described in detail in the Appendix. I looked both at the organizational conditions supporting innovation and productive change, and at the experiences and activities of the innovators who bring this about. I wanted to see how people acquire and use power in empowering organizations

and how this contributes to innovation and the mastery of change.
And I wanted to pinpoint the specific factors in a company's structure and culture that made innovation flourish. Armed with hypotheses derived from my informal observations, consulting
experiences, and extensive reading of the research of my colleagues,[7] I looked for a set of companies to examine in depth, in
addition to the data I was gathering on a large set of "progressive
companies."

Ten major companies with which I had intensive contact over a
period of time as consultant or researcher served as the focus for the
more rigorous and detailed investigations that would eventually
translate my general ideas into concrete form. Six were "new style"
firms or major sections of firms in industries where America was still
a world leader. These included Hewlett-Packard, a highly respected
manufacturer of computers and electronic test equipment, known
for quality products; Wang Laboratories, a leader in the competitive
word-processing-equipment field; Polaroid, technological leader in
photographic equipment; and a leading manufacturer of computers
that I have given the pseudonym of "Chipco," because the company
prefers anonymity.* Also among them were two high-technology
sectors of older industrial firms—Honeywell's group of defense and
marine systems operations, an industry profit leader; and General
Electric's medical-electronics unit, manufacturer of X-ray equipment, CT (computer tomography) scanners, and other diagnostic devices with the leading market share. These six constitute the cluster
of companies from which I derived the positive models in this book
—examples of how American companies can encourage innovation
and prosper from it.

I also examined four companies in older, more traditional industries, industries racked with change and thus with a similar need for
innovation. General Motors was chosen not only as the automotive-
industry leader, representing the struggles to be innovative in a
difficult—some would say, declining—sector, but also because of its
importance in American life and in the history of American business.
Finally, three companies represented the old style of less innovative
organization, known here as "Meridian Telephone," a communications company; "Southern Insurance," a large insurance and financial-services company; and "Petrocorp," a raw-materials refiner. I
promised anonymity to these companies, but I would have used
pseudonyms in any case, because I want to present the typical American corporate problems these firms revealed to me without singling

---

* Throughout this book, the use of pseudonyms for companies or individuals is signified by the use of quotation marks surrounding the name at its first mention in a
chapter.

them out for "criticism." (And, in addition, changes that would make them different places today are already in motion in some of them, stimulated in part by my research findings.)

Depending on the material and people available, I developed one or more of several kinds of case studies on these ten core companies: holistic accounts of organization, culture, and strategy; and/or detailed documentation of workplace-changes efforts. In total these cases built on hundreds of interviews and thousands of hours of direct involvement and observation, as well as review of documents and published material. Eight of the ten companies were also involved in a structured, quantitative study of the degree (and method) of innovative accomplishments led by their managers—six as full participants in the main study ("Chipco," General Electric Medical Systems, Honeywell Defense and Marine Systems Group, Polaroid, "Meridian," and "Southern") and two later, in a more informal way, to check the findings (Wang and parts of Hewlett-Packard). More than 250 managers were interviewed in depth (26 in a pilot study, 208 in the six main companies, and 30 in the "checkup"), and each of the six main companies was scored on an "innovation index"—the proportion of effective managers initiating or contributing to new products or market opportunities, new work or production methods, new structures, and new policies. This study confirmed my distinction between the highly innovating and the innovation-smothering companies; "Chipco" (at 71), GE Med Systems (at 64), Polaroid (at 62), and Honeywell DMSG (at 61) all scored much higher than "Southern" (at 45) or "Meridian" (at 36) on the "innovation index." Both Wang and H-P had high innovation profiles as well.

Overall I examined more than 115 innovations in detail, including many with significant financial, strategic, or organizational implications. I explored developments as diverse as futuristic X-ray machines, computerized data libraries, underwater sensing devices, and projects to improve office productivity—and many others listed in the Appendix. And I saw how the skills of corporate entrepreneurs were allowed to flourish in the companies getting more innovations of all kinds.

I kept my ten core research companies and the complicated web of intersecting projects in which they participated in perspective by continually checking my emerging findings against the experience of other companies with which I had contact (including other divisions of GE and Honeywell) and available financial and social indicators of the state of American industry. My own studies of the forty-seven "progressive companies" and a related survey of the changing personnel department in thirty-two companies were an important part of this. (The Appendix describes the series of research projects

I conducted and which companies were involved in each.) I am thus able to draw on examples from a wide range of companies in this book.

While a breadth of view, including statistical associations, is important to show that my findings relate to corporate America more generally, the depth that my focus on ten companies permitted enabled me to learn in detail about exactly *how* people came to be connected to productive and innovative actions. I felt that since change was my theme, I needed to take movies rather than snapshots, and so I relate more of my findings in terms of human drama than in terms of statistical associations, although my studies elicited both. I could thus understand in rather fine-grained detail exactly why some circumstances produced more enterprise and achievement than others, and I could watch a flow of events in change sequences unfold over time. I could see not only what held things in place but also how they could be made to change over time. More importantly, I could understand the human reality of how work life was lived in different kinds of settings—and thus, how the abstraction of organizational design related to the concreteness of individual human action.

What emerged from these travels and investigations was a set of conclusions about the conditions supporting the mastery of change through encouraging the ability to innovate. Underlying the difference between innovation-stimulating and innovation-smothering situations is an important difference in the way problem solving occurs.

INTEGRATIVE ACTION VERSUS SEGMENTALISM: KEYS TO INNOVATION

I found that the entrepreneurial spirit producing innovation is associated with a particular way of approaching problems that I call "integrative": the willingness to move beyond received wisdom, to combine ideas from unconnected sources, to embrace change as an opportunity to test limits. To see problems integratively is to see them as wholes, related to larger wholes, and thus challenging established practices—rather than walling off a piece of experience and preventing it from being touched or affected by any new experiences.

Entrepreneurs—and entrepreneurial organizations—always operate at the edge of their competence, focusing more of their resources and attention on what they do not yet know (e.g., investment in R&D) than on controlling what they already know. They measure themselves not by the standards of the past (how far they have come) but by visions of the future (how far they have yet to go). And they do not allow the past to serve as a restraint on the future; the mere fact that something *has not* worked in the past does not mean that it cannot be made to work in the future. And the mere fact that something *has* worked in the past does not mean that it should remain.

Integrative thinking that actively embraces change is more likely in companies whose cultures and structures are also integrative, encouraging the treatment of problems as "wholes," considering the wider implications of actions. Such organizations reduce rancorous conflict and isolation between organizational units; create mechanisms for exchange of information and new ideas across organizational boundaries; ensure that multiple perspectives will be taken into account in decisions; and provide coherence and direction to the whole organization. In these team-oriented cooperative environments, innovation flourishes. There may be differences recognized and even encouraged—an array of different specialties, a diversity of people—but the mechanisms exist for transcending differences and finding common ground. Units may be temporarily isolated in order to do an important piece of work without interference—as in a portion of the innovation cycle reported in Chapter 8 —but the isolation is only temporary.

The contrasting style of thought is anti-change–oriented and prevents innovation. I call it "segmentalism" because it is concerned with compartmentalizing actions, events, and problems and keeping each piece isolated from the others. Segmentalist approaches see problems as narrowly as possible, independently of their context, independently of their connections to any other problems. Companies with segmentalist cultures are likely to have segmented structures: a large number of compartments walled off from one another—department from department, level above from level below, field office from headquarters, labor from management, or men from women.[8] Only the minimum number of exchanges takes place at the boundaries of segments; each slice is assumed to stand or fall rather independently of any other anyway, so why *should* they need to cooperate? Segmentalism assumes that problems can be solved when they are carved into pieces and the pieces assigned to specialists who work in isolation.[9] Even innovation itself can become a specialty in segmentalist systems—something given to the R&D department to take care of so that no one else has to worry about it.

Companies where segmentalist approaches dominate find it difficult to innovate or to handle change. Change threatens to disturb the neat array of segments, and so changes are isolated in one segment and not allowed to touch any others. In searching for the right compartment in which to isolate a problem, those operating segmentally are letting the past—the existing structure—dominate the future. The system is designed to protect against change, to protect against deviation from a predetermined central thrust, and to ensure that individuals have sufficient awe and respect for this course to maintain their role in it without question—though they may fight over their share of the proceeds.

As soon as a problem is identified, it is surrounded and isolated. Each person, each department, each level has only a part of any problem and no assumed need to worry about any other part. "Overspecification" of resources leaves little slack for experimentation anyway. The instructions are assumed to be clear, and so are the expected outputs. "Success" means that each segment works independently, with minimum need for communication; it would imply "failure" if some other segment (a neighboring function, the next level up) had to start worrying about *your* problem. Praise and blame attach to jurisdictions.[10] Diverse and deviant activities may occur in a segmented system, and individual segments may develop good, innovative ideas, but there is little impetus or mechanism for the transfer of this knowledge from one segment to another.

Much managerial and academic wisdom about organizational problem solving stems from the study of segmented structures and segmentalist cultures, and so it is not very helpful for understanding innovation and change. Segmentalism as problem solving is close to what has been called "local rationality" in theories of decision making: the idea that any decision problem is best factored into subproblems, with each assigned to a different subunit, so that each has only one goal and one piece of the problem.[11] But note that this fragmentation of problems is only one possible approach to good decisions —a segmentalist approach that fits with segmented structures and cultures. In the integrative mode, people do the *opposite*. They aggregate subproblems into larger problems, so as to re-create a unity that provides more insight into required action. This helps make possible the creative leap of insight that redefines a problem so that novel solutions can emerge. Similarly, instead of isolating the fragments in subunits, they assign the larger problem to supraunits— integrative bodies that can consider the whole before taking action: a multi-unit team, a task force, a project center.

Conditions favoring the integrative mode broaden the search for solutions further beyond what the organization already "knows" (or, to be more accurate, beyond what its *leaders think* they know). But segmentalism inhibits innovation at every step of the solution-search process.[12]

First, the motivation to solve problems declines in segmented systems. Segmentalism discourages people from seeing problems—

or if they do see them, from revealing this discovery to anyone else. In the traditional mechanistic bureaucracy, the isolation of departments and levels means that people will see only local manifestations of problems, and they will perhaps appear puzzling or idiosyncratic. Furthermore, the message to the troops is clear: keep the lid on; the messenger with bad news will be shot. Because each segment is expected to do its work without troubling any other segment, communicating primarily in prespecified ways and mostly to transfer good news (results or output), then identification of a problem is a sign of failure likely to get the identifier in trouble. (And if the job is designed around security and routine, then identification of a problem is also likely to create more work for the identifier—a clear disincentive for saying anything.)[13]

A second source of motivation for activating a search for solutions to problems is also not present in a segmented organization: the desire to apply a pet idea or carry out a pet project. One of Stanford professor James March's great contributions to the understanding of organizational behavior was his insight that not all change is triggered by problems; people with solutions can go out looking for places to use them.[14] But if people's activities are confined to the letter of their job, if they are required to stay within the fences organizations erect between tasks, then it is much less likely that people will ever think beyond what they are given to do or dream about things they might do if only the right problem came along.

But if a problem is identified and a search undertaken, other aspects of segmentalist culture come into play to inhibit innovation. Organizational cultures that favor sorting issues into preexisting categories—e.g., the array of boxes on the organization chart—and stress precedent and procedures are not likely to encourage anyone to look beyond what already exists to find a novel solution. The principle of psychic economy means that if people find a solution near at hand, why expend the energy to look any further?[15] If a precedent will do, why create novelty? At the same time, the isolation of organizational segments also means that few alternatives will be known to the searchers; their choices will be severely constrained.

Finally, specialist biases and political conflicts are more likely to inhibit innovation in a segmentalist organization. Problems are, after all, being left with specialists who have little or no career incentive to consult with other, different specialists. The segment-specific nature of career paths makes it unlikely to find many people with integrative experiences bridging specialties. And in a segmented organization, where is the inducement for cooperation anyway?

Segmentalism, then, is what keeps an organization steady, on course, changing as little as possible, making only minimal adjustments. This style, this mode of organizing protects the successful organization against unnecessary change, ensures that it will repeat what it already "knows." For those activities which *should* be repeated—the areas of high certainty where routine, habitual action is efficient and desirable—segmentalism works.

But segmentalism also makes it harder for the organization to move beyond its existing capacity in order to innovate and improve. When people in a segmented organization do use an integrative mode to carry out important innovations, segmentalism makes it harder for the organization to incorporate and use these innovations. Segmentalism inhibits the entrepreneurial spirit and makes the organization a slave of its past—a victim, not a master, of change.

A tradition of success via the efforts of individuals is found in American entrepreneurial and corporate history; it is part of the American heritage. But many companies have "forgotten" this critical part of their own past success, stifling much individual initiative by oversegmenting aspects of corporate life: assigning people to fragments rather than larger pieces; emphasizing uncrossable boundaries between functions, between hierarchical levels, between central staffs and field operations, and even between kinds of people.

The entrepreneurial spirit is stifled in such an atmosphere, except for a few hardy souls who manage to innovate *despite* the organization because of special status as an outsider, an unanticipated gap in the system, or location at the periphery and support from a dissident subculture, as I will show in Chapter 3. But even when innovators do emerge in segmented environments, it is much less likely that their innovations will be used more broadly or, in some cases, survive beyond their initial trials. Indeed, innovations themselves may suffer from segmentalist treatment: isolation as a cultural island, confinement to a few special circumstances, like the examples in Chapter 4.

The task of stimulating more innovation is a difficult one for a large proportion of American organizations. Think about how many segments are walled off from one another in the traditional industrial firm: corporate staffs from divisions, management from labor, levels below from levels above, functions of one kind from functions of another kind, clericals from professionals, hourly workers from salaried workers, men from women, blacks from whites, office employees from factory employees. The structural barriers to communication, to exchange of ideas, to joint effort to solve problems are matched by attitudes that confine people to the category in which they have been placed, that assume they are *defined* by that category, and that fail to allow them to show what they can contribute beyond it.

The organization that produces a great deal of innovation must, by definition, be less category-conscious. It must, by definition, give more people the opportunity to reach for power despite the box on the organization chart they occupy. It must continually create teams that represent new and different configurations, offering the potential for many more people, in theory, to find a connection with nearly everyone else. Thus, while the high-innovation companies I studied have not eradicated all problems of race, sex, or class (as I point out in Chapter 7), they do tend to go further toward an egalitarian, meritocratic ideal than their counterparts. They are considered by professionals progressive and exemplary in human-resource policies,[16] and they are reputed to have generally higher proportions of women, minorities, and former blue-collar workers in management than the rest of U.S. industry.

If innovation is inhibited by segmentalism at every step of the problem-identification and solution-search process, it is stimulated by the integrative cultures and structures described in Chapters 5 and 6, with encouragement for individuals to identify problems in the first place, incentives for defining new problems to work on, and sufficient looseness of activity boundaries that people can envision new sets of personally intriguing projects. Such organizations generally approach "disturbing" events as *unique* occurrences, so that problems are less likely to be stuffed into preexisting slots than to be approached as new challenges; indeed, "making the familiar strange" is a central tenet of creativity-enhancement techniques. Further, there is virtually unrestricted communication about the projects, experiments, and perspectives of other areas and about the alternatives that already exist in the organization; there are team mechanisms that pull people together across specialties to solve problems, and individuals with integrative backgrounds bringing together several dimensions of organizational experience.

Organizations that are change-oriented, then, will have a large number of integrative mechanisms encouraging fluidity of boundaries, the free flow of ideas, and the empowerment of people to act on new information.

A sense of unity and identification with the whole is more likely to exist in innovating, integrative companies than in highly segmented ones because of the structure and the culture that grows out of it. A "shared philosophy" and "family feeling" cannot be simulated or imposed artificially because top management wants to "create a Japanese-style organization"; it has to derive from the way work is done. People go to work to do a job for *pay*, and if they also find a "community," that is because the work is done in an environment of mutual respect, participating teams, multiple ties, and relationships that crisscross the organization chart. Furthermore, the large amount of off-the-job socializing that takes place around innovating organizations is not merely an "enlightened fringe benefit"; it serves an important task-related purpose: building a foundation of cross-cutting relationships to make integrative team formation that much easier.

One other aspect of the legacy of segmentalism may make it difficult for many American firms to increase their levels of innovation: a preference for being guided by the past rather than the future, by what is already known rather than what is not yet known. Maintenance-oriented operations should, of course, be guided by the past, measured and compared against standards reflecting accumulated organizational experience. But innovation will not be produced very easily. Innovation requires a trust in the future that is difficult to arouse or sustain in organizations constantly looking to the past. I argue in Chapter 10 that the change master is partly a historian who knows which pieces of the past to honor and preserve while moving toward a different future, but that is not at all the same thing as letting the past *define* the future. But in a segmentalist environment, that is indeed what happens.

For example, the reward systems in segmentalist systems tend to look backward, paying people after they have achieved. They are *payoff-centered* rather than *investment-centered*. This idea is captured nicely by the "Peter Principle" of promoting a person to his/her level of incompetence. The promotion is a reward for doing the last job, not a bet on capacity for doing the next one. The bonus is available to take home after all the results are in, not a pool of funds to use to get results in the first place. The prize money is awarded after the experiment has borne fruit and money is readily available, not at conception time when the money is desperately needed. *Afterward*, looking back, it is clear where rewards should go. But unwillingness to bet on potential means that much potential innovation is lost.

The reward systems in innovating companies tend to be much more investment-oriented, more future-oriented. Seed capital is available for projects. People are promoted before they are ready, so that they stretch to do the next job. The most important reward to many corporate entrepreneurs occurs at the time they get the go-ahead for their project, not when "bennies" are handed out at its end. And then after completion of the accomplishment, the most important reward is to be in line for a bigger and better project.

For an organization to increase the extent to which it is future- rather than past-directed requires two other forms of integration. First, it requires conviction that everyone in the organization is at least facing in the same direction, since "forward" is not a precise trajectory. Here integrated long-range plans play a role, as at GE Medical Systems, shown in Chapter 6—plans to which all major units have contributed. Even though the plan is not a guarantee, and deviations are an important part of realizing it, a plan helps provide unity that allows letting go of the past. The second form of unity is cultural. I have already alluded to this: a sense that there *is* a whole and that there are clear principles guiding it.

The other requirement for empowering people to reach for a future different from the past is respect for the individuals in the organization. For people to trust one another in areas of uncertainty, where outcomes are not yet known, they need to respect the competence of the others. In segmentalist companies, "the system" is trusted more than the individuals. Indeed, "the system" is often designed to protect *against* individual actions. The lack of respect people have for one another in segmented environments will be described in Chapters 3 and 4: the infighting among staffs, the feeling that only outsiders could handle new challenges. Then contrast this with the culture of pride at Hewlett-Packard, "Chipco," or General Electric in Chapter 5: the feeling that anyone who works here must be superior. Respect for the individual, then, is not only a matter of basic human dignity; it is also a necessity for the leap of faith required to let entrepreneurs innovate.

"Individual" and "team" are not contradictory concepts in the innovating organization. Teams—whether in formal incarnations or as an implied emphasis on coalition formation and peer cooperation —are one of the integrative vehicles that keep power tools (information, resources, and support) accessible. The use of participative mechanisms helps ensure that segmentalism will not prevail, because individuals must constantly seek additional viewpoints, more pieces of the puzzle, solutions that have payoffs for others as well. At the same time, *more* individuals with ideas to contribute above and beyond their own job have a way to do this.

"Participation" has been touted to U.S. companies more for its "quality of work life" and motivational benefits than for its contributions to innovation. It is true that human dignity and motivation for work effort are increased when more participation is built into organizations. But there is so much more. Participation in team projects above and beyond the rote requirements of the job is a device for tapping unexpected *individual* contributions. It helps ready people for change by giving them a broader outlook and more skills. And it ensures that people have information beyond their limited purview.

In the model innovating companies, participative teams are not equivalent to "groupthink," or inaction without consensus, or management by committee—three negatives to many American managers. They are action bodies that develop better systems, methods, products, or policies than would result from unilateral action by one responsible segment, or even from each of the team members working in isolation from the others. The results are likely to be more innovative and more easily used. And the individuals involved are more empowered by the access to the additional power tools the team offers than they would be even if exercising their clear and unquestioned authority within one segment.

Throughout this book I will stress the link between individual entrepreneurs and their coalitions or teams. Individuals initiate— sometimes set into motion by a segment-transcending assignment— and then work through teams to bring idea to innovation. Prime movers push—by getting more and more people involved in action vehicles that express the change being promoted. At the same time, more formal teams at the grass-roots level represent ways that more individuals can become involved in taking initiative in collaboration with others.

Here is my argument, which unfolds in the pages that follow.

American corporations are at a critical watershed because they face a transforming economic and social environment which has emerged since the 1960s. This new context for corporate America makes past responses less effective; it changes the tasks for management at all levels and encourages the search for better ways to involve the entire work force in innovative problem solving.

But where segmentalism prevails, companies are likely to stifle their own potential for greater innovation, making it all but impossible for any but a few hardy individuals to contribute to solving problems—and highly unlikely that the company can even take advantage of those innovations in structure and practice that do occur.

There is hope, however, in the experience of several leading, innovating American firms that already operate in an integrative mode, thereby encouraging entrepreneurial behavior and employee involvement leading to productive, responsive changes. The practices of these companies—their structures, cultures, and reward systems—provide indigenous American models for ways in which innovation can be stimulated. Their experiences represent important lessons for American organizations.

Three new sets of skills are required to manage effectively in such integrative, innovation-stimulating environments, where individual "corporate entrepreneurs" work through participative teams to produce small changes that can later add up to big ones. First are "power skills"—skills in persuading others to invest information, support, and resources in new initiatives driven by an "entrepreneur." Second is the ability to manage the problems associated with the greater use of teams and employee participation. And third is an understanding of how change is designed and constructed in an organization—how the microchanges introduced by individual innovators relate to macrochanges or strategic reorientations.

Of course, identifying the models—the organizational practices and skills that encourage change mastery—does not guarantee their use; and many people wonder whether ailing, older American companies can indeed transform themselves. The experience of an archetypal American industrial corporation, General Motors, is revealing, because it did succeed in producing major innovative responses to its changing environment in the 1970s, using the methods and techniques I have identified. But then a legacy of segmentalism may be preventing full use of the company's own innovation-stimulating practices. The tragedy for many older industrial giants is how close they may have come to major productive changes, stopping short before their potential was realized.

Thus, American companies must go further than making piecemeal responses to problems treated in isolation. They must begin at the top to build more integrative responses to problems, as well as more flexible, integrative organizations to support this. Public policies can help encourage these changes, but ultimately the responsibility lies with corporate leaders. There are already active examples of major companies' beginning to set into motion the kinds of organizational changes I recommend.

American companies *can* return to economic leadership through a better use of people and the encouragement of innovation—if their leaders appreciate both the magnitude of the transformations that have taken place and the magnitude of the response that is required.

CHAPTER

Transformations in the American Corporate Environment, 1960s–1980s

> But I'm coming to believe that all of us are ghosts. . . . It's not just what we inherit from our mothers and fathers. It's also the shadows of dead ideas and opinions and convictions. They're no longer alive, but they grip us all the same, and hold on to us against our will. All I have to do is open a newspaper to see ghosts hovering between the lines. They are haunting the whole country, those stubborn phantoms—so many of them, so thick, they're like an impenetrable dark mist. And here we are, all of us, so abjectly terrified of the light.
>
> —Henrik Ibsen, *Ghosts*

> We thought they could never catch up, but they tried harder, and here we are—to paraphrase an old American slogan, a Sony in every house and two Toyotas in every garage.
>
> —George A. Keyworth II,
>   Director of the Federal Office
>   of Science and Technology Policy

BUSINESS ORGANIZATIONS are facing a change more extensive, more far-reaching in its implications, and more fundamental in its transforming quality than anything since the "modern" industrial system took shape in the years between roughly 1890 and 1920.[1] These changes in the American business environment come from several sources: the labor force, patterns of world trade, technology, and political sensibilities. Each of these by itself has changed significantly at other times. The present situation is unusual not only in that each is undergoing transforming changes, but that the changes are profound, and that they are occurring together.

In this second transforming era, between 1960 and 1990, American business organizations will need to learn to operate in a wholly new mode. Those organizations which recognize the immensity and the scope of these forces, and carry out the required organizational changes, will probably survive; many, indeed, will prosper enormously. Those organizations which either fail to understand the need for the change or are inept in their ability to deal with it will fade and fall behind, if they survive at all. It is already clear, from the record of the past decade, that a few farsighted organizations have begun to make the necessary changes; these few are signposts to the future. Others have already failed; they are memorials to the past.

Recent business history is filled with the skeletons of companies that failed to innovate or even to recognize the need to adapt to obvious change. International Harvester's financial woes—in 1981, it predicted a $302-million profit for 1982; by the middle of 1982, the prediction was for a $518-million *loss*, and at the end of the 1982 fiscal year the total loss was $1.64 *billion*—are often traced to its inability to control its environment, as interest rates skyrocketed and product demand slumped. But Caterpillar Tractor, almost a line-by-line competitor, faced exactly the same situation. Unlike Harvester, however, it has not only been holding its own, but even improving a bit. Its strategy involves concentrating on quality and service, building and maintaining long-term relationships with customers, innovative products and services (such as a willingness to support dealers by buying back unsold parts), and, perhaps most important, a continued and long-standing tradition of investment in manufacturing technology and development of people at all levels.

AM International (previously Addressograph-Multigraph) seems to be rapidly disappearing as it sells off divisions, lurches from one "strategy" to another without clear or consistent direction, and demonstrates an evident inability to adapt to the changing environment.[2] In 1981, it sold its U.S. operation of the Addressograph Division, the business that started the company in 1893, along with its credit-card recorders—perhaps its one "modern" product line. By that point, it had already sold off four other businesses, and two more were up for sale. Since 1967, AM had had three CEOs; its headquarters had been relocated twice in four years. Roy Ash, who took over in 1976 after a legendary career building Litton Industries, tried to mesmerize Wall Street by creating an effective image, a strategy that worked well in the go-go years of the early 1960s, but was out of touch with the new realities of later times.

There are also examples of American companies' almost literally handing a market over to the Japanese by failing to respond quickly or adapt to change. By 1981, Kyocera International, Inc., had 70 percent of the U.S. market for ceramic semiconductor housings, and its parent, Kyocera Company, had a similar worldwide market share, having taken the market completely away from the U.S. companies that pioneered this critical high-technology component.[3] Kyocera opened its first U.S. sales office in 1968, bought a Fairchild Camera (another loser) plant in 1971, and acquired Honeywell's San Diego ceramics plant shortly thereafter. Finally, in 1975, Du Pont stopped its ceramics production because Kyocera was the overwhelming leader.

Customers who switched from American producers to Kyocera faulted American companies for noninnovativeness, resistance to change, inability to adapt to fluctuations in demand, and slow response time compared with the Japanese producer's. The manager of raw-materials purchasing for Signetics Corporation complained: "We would go to American Lava with a request to quote a price on packages for us, and typically they would come back with two sheets of exceptions—all the things we required that they said we couldn't do." American Microsystem's materials manager echoed this: "If you called the president of an American company, you wouldn't hear from that company for three months. A salesman might finally come by and say, 'I understand you have a problem.'"

It was not just the legendary Japanese quality consciousness that paid off for Kyocera, but also a greater flexibility in the use of its people compared with its American rivals', and thus a greater ability to regroup to handle change. Kyocera's American marketing vice-president, who left a U.S. competitor to come to Kyocera, commented about his previous employer: "When they hit a deep valley, they simply disband things—like most of the U.S. companies. They shut off machinery, tear it up, and send people away because there is no work." Kyocera, in contrast, finds something else for its people to do to get ready for the next set of changes: executives might go on the road to market; production people might do forward planning.[4]

Over the last two decades, we have also witnessed a dramatic change in America's world position. Foreign competition, which used to be brushed off lightly, like dandruff, is now overtaking many of our major industries. (And the success of countries like Japan lies in some measure, ironically, in their use of models of "workplace democracy" or "team organization" well known in the United

States, models that we have never fully implemented.) Not only does America have a different position in the world society than twenty years ago, but foreign nations also control critical supplies that our country needs—petroleum, and other raw materials. This too changes the context in which our organizations operate.

For example, during the period from 1960 to 1980, sales of Japanese autos in the United States went from one-quarter of 1 percent to 22 percent of the market—a hundredfold increase. The results for the American automobile industry are well known. During the same period, consumer prices went up by 180 percent; that is, nearly tripled. In manufacturing, American output per labor hour rose 3.4 percent per year between 1970 and 1975; the equivalent figure for 1975 to 1980 was 1.6 percent. Compare these figures with Japan's: 6.7 percent and 7.9 percent—output increasing while America's dropped precipitously. Japan, of course, is a well-known economic miracle; but look at France (4.6 percent and 5.1 percent), or Italy (4.6 percent and 4.9 percent), or the Netherlands (6.2 percent and 6.6 percent). Overall, the United States lost 23 percent of its share of world markets in the 1970s, according to Commerce Department calculations.[5]

Even in innovation—America's classic strength—there are signs of decline. After leading the world in percentage of Gross National Product spent on R&D, American companies' average yearly expenditure on industrial R&D (excluding our high military R&D expense) fell to 1.5 percent of GNP, trailing both West Germany (at an average of 2.0 percent) and Japan (at an average of 1.9 percent). Meanwhile, West Germany had been doubling its proportional spending, and Japan had increased its proportion by 20 percent. Furthermore, the American edge in invention was also declining. In the 1950s, according to Stanford Research Institute figures, the United States initiated more than 80 percent of the world's major innovations; today it is close to 50 percent, and foreigners are acquiring a larger share of U.S. patents. American firms' share of U.S.-issued patents dropped from 78 percent to 63 percent between 1967 and 1977, whereas Japan's share increased from 2 percent to 10 percent and West Germany's from 6 percent to 8 percent.[6]

Business failures have also been going up steadily. By June of 1982, U.S. business failures had reached the highest level since the Great Depression of the 1930s.[7] This is simply one indication of a set of wholesale shifts in industrial adaptation—or lack of it. The financial-services industries are, of course, in turmoil. Some organizations, however, are taking great advantage of the potential opportunities by innovative and even revolutionary changes in products, services, and market orientation. Brokerage firms, for example, have been branching out into insurance, broader financial services, investment alternatives and—except in name—banking. Banks themselves are reciprocating. Those which are innovative and are able to transcend their traditions are doing very well; for example, Bank One of Columbus, Ohio, an otherwise obscure regional bank, has leapfrogged its competitors by sewing on a vehicle for credit-card services.

Even though big companies can still ride out downturns better than smaller ones, size is no longer guaranteed protection against decline and either closing or—more likely—acquisition by another firm.

The total scope of what needs to be done is, of course, highly variable, in large part because it depends on the particular organization and industry. What *is* clear, however, is the need for innovation at every level—innovation not merely in the traditional sense of new products and services, but in the very ways that organizations operate, in their view of themselves, and in the mechanisms that can develop and engage their resources to the maximum extent possible. Most important, organizations need innovation to shift from the present tendency to deal with their tasks in a relatively single-minded, top-directed way and to a capacity to respond innovatively, locally, and promptly to a whole variety of organizational contingencies—to change shape, so to speak.

The organizations now emerging as successful will be, above all, flexible; they will need to be able to bring particular resources together quickly, on the basis of short-term recognition of new requirements and the necessary capacities to deal with them. They will be organizations with more "surface" exposed to the environment and with a whole host of sensing mechanisms for recognizing emerging changes and their implications. In such an organization, more people with greater skills than ever before will link the organization to its environment.

Until now, most organizations have attempted to deal with forthcoming change and with environmental contingencies by ever-more-elaborate mechanisms for strategic planning—essentially designed to help organizations feel in control of their futures. There will always be a need for this, of course, but the balance between planning—which reduces the need for effective reaction—and structural flexibility—which increases the capacity for effective reaction—needs to shift toward the latter. The era of strategic planning (control) may be over; we are entering an era of tactical planning (response).

And just as the economic challenges facing American business seem to demand more flexible, responsive people and sensitive practices, so do the accumulating social challenges indicate that a new era is at hand.

SIGNALS OF A TRANSFORMING ERA: FACTORS IN ORGANIZATION DESIGN

There are a few periods in history that deserve the label of "transforming eras," when circumstances change sufficiently to warrant a major shift of assumptions. Thomas Kuhn, the historian of science, has pointed out that major change takes place only occasionally, in what he called paradigm shifts, when the working assumptions on which people have depended become so inappropriate that they break down, to be replaced by a more appropriate set.[8] Thus, social or economic history is intrinsically characterized by long periods of stability in paradigm, punctuated by relatively short periods of high instability: history as staircase, rather than ramp. This model fits the changing world of the corporation very well.

Look at the differences between the factors bearing on the design of an organization in the 1890s–1920s, the formative era for the traditional industrial corporation, and those emerging in the environment of the 1960s–1980s. The turn-of-the-century labor force was largely uneducated (in the formal sense), less skilled, often immigrant, with high turnover and high labor conflict.[9] The distinction between workers and managers was one not only of task but also often of language and social class. Production tasks were quite straightforward: moving objects, assembling mechanical devices, adjusting machinery, and using sheer physical energy. Contrast this with the emerging organization design factors.

TRADITIONAL ORGANIZATION DESIGN FACTORS (1890s–1920s)	EMERGING ORGANIZATION DESIGN FACTORS (1960s–1980s)
Uneducated, unskilled temporary workers	Educated, sophisticated career employees
Simple and physical tasks	Complex and intellectual tasks
Mechanical technology	Electronic and biological technologies
Mechanistic views, direct cause and effect	Organic views, multiple causes and effects

| Stable markets and supplies | Fluid markets and supplies |
| Sharp distinction between workers and managers | Overlap between workers and managers |

Clearly, we cannot use the organization of the 1890s to solve the problems of the 1980s.

Ironically, many of the new organization design factors were themselves created by companies barely out of their infancy in 1960: Digital Equipment and Control Data were three-year-old toddlers, Hewlett-Packard was a still-gawky adolescent, and Apple Computers was not even a gleam in its founders' eyes. Today their competition, and others', has moved giant IBM into personal computers, robotics, and the glimmerings of a more entrepreneurial stance.

There has also been enormous change in the context shaping organizational realities. In less than twenty years, a number of powerful social movements have changed the very ways in which we think about our organizations. The civil rights and women's movements, the environmental and consumer movements not only have given new people a stronger voice in our institutions, bringing new interest groups to political bargaining tables, but have also brought in their wake the "heavy hand" of government regulation—heavy in terms of litigation and paperwork requirements if not the substance of compliance. Here are some of the striking things that have happened to us as a working society in the last twenty years: [10]

- The proportion of married women who work doubled, now including more than half of all wives.
- The proportion of families with at least two wage earners passed the 50-percent mark.
- The median amount of schooling of the whole labor force moved past a year of college, and for employed blacks it went from tenth grade to some college. Clerical workers (by 1970) replaced operatives as the single largest occupational category, and professional jobs continued to grow rapidly.
- Unions began to bargain for reduced work weeks or flextime, and the number of employees on flexible work hours or staggered hours grew to around the 10-percent mark.
- Corporate collective bargaining agreements began to contain provisions for joint labor-management committees for special production problems; in 1973 General Motors and the United Auto Workers consummated the first major agreement on "quality of work life" cooperation.
- Monitoring of job safety through OSHA and state action grew, and federal and state legislation barring employment discrimination was created.

- Employee health benefits grew in level and kind—from cash
  alone to services, from medical only to dental, and with ever-
  greater employer contribution.

And in national surveys, more men complained about inconvenient
schedules, more women complained about sex discrimination, more
union members complained about union leadership, and more employees, in general, questioned the fairness of their companies' policies and expressed concerns about advancement.[11] In this context,
how organizations treat their people has to change.

It is harder to document changes in American corporate "culture" than to examine statistical indicators of labor trends, but I can
try to provide an approximation. One rough estimate of what preoc-
cupies the business community can come from looking at changes
in the topics emphasized in the business press. David Summers and
I analyzed the *Business Periodicals Index* for 1959–61, 1964–66,
1969–71, 1974–76, and 1979–80, measuring the amount of space
devoted to the most prominent topics and counting the number of
citations of articles on all topics concerned with the culture, environment, and human systems of business (e.g., tasks and roles of top
managers, compensation/reward systems, economic environment,
firings and resignations, laws, leadership, organizational change, relative attention to categories of workers, technology, treatment of
people, types of structure, and union–management issues). In total,
this study covered 7,297 pages of listings and 246 separate topics.

Appropriately, most space in the business press is devoted to
reporting business news, and by far the largest categories are industry groups such as insurance, advertising, or autos. (For example, in
1959–60 advertising had 23 *pages* of citations, compared with about
11 *inches* for employment management—which was one of the biggest of my "cultural" topics.) Not surprisingly, a large number of
topics remained relatively stable during the twenty-year period. And
a few fads flew by during these two decades. Peaking around 1970
and then virtually disappearing were these topics, many of which
have a 1960s "counterculture" flavor: drug problems in industry;
group relations and sensitivity training; labor supply; government
ownership; attention to engineers as an occupational category; general discussions about minority employment; debates about automation and its social aspects (but note that *office* automation
expanded in 1980 as a topic); and social aspects of business.

But there were also a number of illuminating changes in *Business Periodicals'* listings, changes which highlight the differences
between the corporate environment of 1960 and that of 1980. The
topics that declined or disappeared over the two decades spoke of a
traditional style of management, dominating and monitoring employees in an adversary relationship: e.g., management rights, work

measurement, work sampling, and collective bargaining. The topics that grew in prominence showed corporations struggling with a changing environment with increasing external pressures; more employee rights; an uncertain economy focusing more attention on selection, training, and motivation; and a "new" management style involving teamwork and participation. Overall, there has been more self-conscious attention in the business press in recent years to the quality of management and to management actions as a factor in corporate success, and human resource management has moved from backstage to center stage.

Although the themes that increased in visibility through these two decades included several broad business concerns—for example, regulatory affairs, general economic conditions, planning, and office automation—the majority of the issues growing in importance concerned human resources. They included compensation and incentives, dismissal, outplacement and resignation, employee counseling and appraisal, job analysis and satisfaction, employee rights, and a host of issues concerning labor, its costs, its productivity, and its turnover. There was also considerable attention to executive training and management development, affirmative-action issues generally, and women and minority groups in particular.

A few issues changed names, in a linguistic reorientation of corporate culture: "employment management" became "personnel management" and then "human resources." There was visible and increasing concern with "participative management," a phrase that grew out of "employee participation and management" and then became, strikingly rapidly, almost a slogan.

Some of these concerns had always existed in some small way; they simply became much more significant during this period. Others, however, appeared for the first time and thereafter grew in importance. Some can be associated fairly closely with specific dates. Business-school graduates, "M.B.A.s," first appeared as a significant topic in *Business Periodicals'* listings about 1965. In about 1969 and 1970, such topics as decision making, obsolescence of personnel, executive ability, management research and organizational change, government regulations, sex discrimination and equal employment appeared for the first time. These years also saw the first mention of work councils, alternative work schedules such as staggered hours and flextime, and matrix and group or project management—all reflective of new ways of organizing work.

Five years later, between 1974 and 1976, "affirmative action" (as against equal employment opportunity) became visible along with "equal pay for equal work" and "family life"—something not previously evident in business literature. At the same time, job enrichment became significant. Management by objectives and management information systems also appeared then, as did executive-search consultants, teamwork, group decision making, and middle managers. Finally, at the close of the decade, in 1979 and 1980, business and employee communication appeared, along with employee motivation, quality of work life (QWL) and industrial productivity.

These dates tell us nothing, of course, about the origins and first introduction of these concepts—many of them were "invented" much earlier—but they *do* tell us when the concepts became firmly enough embedded in corporate culture to receive sufficient attention in the business press to in turn earn their own listing in an index. This lag between invention and attention makes me even more confident in concluding that the world of the American corporation has changed dramatically since 1960 and that we are indeed in a transforming era. Our practices may not yet have caught up with our ideas about what those systems should be; but there is no doubt that our ideas have changed, toward a greater concern for people and a new range of organization designs.

Examination of the rhetoric of corporate leaders also confirms these shifts of cultural emphasis. To see what top business people were saying over these last two decades, David Summers and I did an informal content analysis of fifty-two speeches recorded in *Vital Speeches* by executives such as Thomas Watson of IBM and Walter Wriston of Citicorp. We read speeches made in 1960, 1965, 1970, 1975, and 1980, and we noted the major themes as well as the changes.

During 1960–80, business leaders were continuously concerned with just what we would expect: the existence of the free-enterprise system, the benefits of technology, and what they called excessive federal government regulation of domestic business and foreign trade. They defended profit making, complained about problems with unions (demanding too much), argued that "big is beautiful" in terms of corporate size, and worried about foreign competition, a concern growing stronger throughout the 1970s. They also asked for U.S. "national goals," but argued that national economic planning by government is bad for business. They agreed that business needs ethical standards, felt that internal organizational communication and planning could be improved, and expressed concern over the bad image of business.

But some changes also crept in. Whereas a 1965 theme was the desirability of organizational loyalty—managers who are "married" to the organization—by 1975, leaders were calling for new hours of work and rewards systems and acknowledging their responsibilities to employees. By 1980, motivating employees was a major concern. The bold statement that, in effect, "Workers can go somewhere else if not satisfied with the organization" (1960) gave way to an interest in more effective management of human resources (1970). Look at these shifts in views: from employees in unions considered hungry for power (1960); to calls to develop job-retraining programs, special bargaining committees, and more direct communication with employees (1965) and better human-resource programs (1970); to arguments for hiring, training, and promoting more women and minorities and to developing better reward systems (1975); and finally, interest in employee motivation (1980).

An air of humility and awareness of responsibilities began to replace *laissez-faire* arrogance in the rhetoric of business leaders. This shows up in other ways, too: from "fighting Communism" (1960) to "coexistence with Communism" (1970); from organizations' *causing* changes in their environment (1965) to organizations' needing to *sense* changes in their environment (1970). By 1980, corporate leaders were publicly acknowledging the defects in their own systems: too much occupational specialization, poor employee communication and incentives, and not enough concern for long-run health. Julius Heldman, a vice-president of Shell, talked about business's unavoidable role in public policy; Kenneth Dayton, chairman of the executive committee of Dayton Hudson, told Houston business people about the "5% club"—Twin Cities businesses contributing 5 percent of their pretax profits to charity. Reginald Jones, the highly respected ex-CEO of General Electric, commented on behalf of the Business Roundtable that "Public policy and social issues are no longer adjuncts to business planning and management. They are in the mainstream of it. The concern must be pervasive in companies today, from boardroom to factory floor." [12] And Roger Smith, the chairman-elect of General Motors, commenting on the "remarkable diffusion of economic power" in the last decade, expounded to the National Foreign Trade Commission in New York on GM's "long-term interest in helping developing countries become more competitive and prosperous" as General Motors is increasingly one world-wide company.

TRANSFORMING MANAGEMENT: WHAT IT IS, WHAT IT WAS, AND WHAT IT MUST BE TO SURVIVE

The changing corporate environment, and the emergence of new models, is also reflected in the critical management tasks inside organizations, the context in which people do their work.

The infallibility of management, the certainty of management tasks, and the predictability of management careers have declined;

but the potential of the rest of the work force for contributing to the solution of organizational problems has increased. As uncertainties and interdependencies rise, the past is an increasingly less appropriate guide to the future. For example, the sharp distinction between "management" and "workers" possible in the old organizational era is no longer as clear in the new one.

In the new environment, people at all levels of the organization are affected by the power or the control or the interests others have in their area. And so the unquestioned authority of managers in the corporation of the past has been replaced by the need for negotiations and relationships outside the immediate managerial domain, by the need for managers to *persuade* rather than *order,* and by the need to acknowledge the expertise of those below.[13] In short, regardless of organizational level, managers must take other people, outside as well as inside their areas, into account in order to do their work, and they must learn, in this new environment, how both to acquire and to share power.

One of the main results of this profound shift is an equally profound need for managers at all levels to shift their traditional emphases and to occupy new organizational roles involving very different tasks from those with which they were originally involved. It is not simply that organizations need to use their conventional capacity to attack different problems, rather, the very management structure of the organization and the roles of managers at all levels need to change because these new pressures systematically strain each of those levels.

External Pressures on Corporate Leaders:
Environmental Responsiveness or Strategic Blindness?

Top executives of large corporations may be less "powerful" today than ever before.[14] They are certainly *privileged,* and they can make or influence decisions with life-changing consequences for employees—in that sense they are very powerful—but in terms of the models that best explain their choices, they are now operating in open systems facing multiple constraints—retaining their jobs as long as they reflect the interests they are in place to serve, and as long as they can manage a wide variety of demands "external" to the organization.

At the top, limitations from the environment can inhibit action. The environment is increasingly activated, meaning that more stakeholder groups and interest groups identify themselves, feeling that they have a stake in their organizations' operations. They want representation, and they are willing to withhold their resources (material or symbolic) until they get it. Rather than simply assuming, as in the old models, that supply is certain and inevitable—that American corporations can get anything they want, whenever they want, from whomever they want—organizations now must develop strategies of bargaining for resources and of influencing the environment.

As one indicator of how much this recognition is now a mainstream part of elite corporate thought, the Business Roundtable, leading lobby for America's largest corporations, issued a new statement on corporate responsibility in 1981, which declared: "More than ever, managers of corporations are expected to serve the public interest as well as private profit." Four "constituencies" were identified—customers, employees, communities and society at large, and shareholders—and the needs of each delineated—e.g., for employees, financial security, personal privacy, freedom of expression, and concern for the quality of life, as well as fair pay. (Some progressive companies add a fifth constituency: vendors.) A leading economist pointed out, with disapproval, the shift of focus that this statement reveals:

> The Roundtable concerns itself with the expectations of constituencies of the corporation. This implies that the large corporation is a political entity subject to the votes of interest groups, rather than an economic organization subject to the market test for efficient use of resources. . . . Giving space to every group trying to politicize the corporation so as to make it the source of a gift or grant . . . must be the result of corporate executives not managing their companies, but rather becoming politicians.[15]

And "politicians" they are. Management of critical boundary-spanning issues is the task of the top: developing strategies, tactics, and structural mechanisms for functioning and triumphing in a turbulent and highly politicized environment. It is less and less possible to confine external interest to markets and competitors. The political tasks of top executives and the amount of time spent on them have grown enormously. Charles Burck commented in *Fortune* in 1975 that the political environment of the modern corporation was rapidly changing; an entirely new network of activist organizations and regulatory agencies confronts the top executive, leading to an increased public relations emphasis. A recent *Fortune* article pointed out that

> Few executives on the way up can fail to note that companies are putting a premium on people who are adept at handling corporate relations with the public—and the government. . . . Like ambitious politicians, ambitious executives today are campaigning for higher office so to speak, by projecting a "vote-getting" image.[16]

Top corporate leaders are apparently spending a higher and higher proportion of their time outside their organizations, developing relationships and alliances. Boundary tasks and institution building were always important, of course, but they now seem to dominate. Reginald Jones, the well-regarded chairman of General Electric, reported to *Business Week* that he devoted only about half his time to managing his company's operations and the rest to "externalities," making speeches on such political and macroeconomic issues as tax reform, capital formation, and inflation.[17] When Thomas Murphy and Elliot Estes ran General Motors as chairman and president, "Mr. Inside" (Estes) gave "only" about thirty speeches in 1979, while "Mr. Outside" (Murphy) gave more than ninety. At a recent lunch meeting with thirty CEOs from companies, universities, and hospitals in Pittsburgh, I learned that most of them had just had breakfast together to meet a political candidate, and a high proportion of them had also had lunch together the day before to talk about community action. The meeting convenor joked, "I think I'll tell all of you where I'm having dinner tonight, so you can join me." In short, these CEOs were spending virtually no time inside their organizations; they were spending time allying themselves and bargaining outside.

The CEO, of course, has always been largely focused on boundary-spanning issues. But these now seem to be an increasing preoccupation of other top corporate executives, especially with the proliferation of staff and staff organization in every sector. Some have commented that this has grown far beyond any reasonable rationale for the importance of planning. Even a *Wall Street Journal* columnist, a vice-president of McKinsey and Company, which specializes in strategic planning, wrote that perhaps planning has become a "fetish." "Fetish" in this context means a task providing an illusion of control in an environment that is clearly out of control: at least one can forecast, gather statistics, write reports, and hold meetings to provide the illusion of activity and control.

A turning away over the last decades from a concern with the technical side of an organization's functioning (the actual work process and product) to a concern with the environment is reinforced by the changing backgrounds of chief executives of major U.S. corporations. This has several aspects: education is one. Heidrick and Struggles, a leading executive-recruiting firm, noted in a recent survey of 971 top executives that only 16 percent of those over 50 years old had an M.B.A., rising to 31 percent of those between 40 and 49, and to an astonishing 48.7 percent of those under 40.[18] Another aspect is the shift to functional experience. It has been years since production and R&D dominated any but the newest high-tech industries (but those, after all, are largely still run by the founding generation). In the 1960s major corporations tended, by and large, to become market-driven and thus also market-run; but in the 1970s, executives with financial and legal backgrounds became preeminent.[19]

The impact of this kind of change is profound, so much so that an increasing number of analysts hold it in large part responsible for declining productivity in U.S. industry.[20] But it is not hard to see other implications: a stepwise move away from any interest in products (marketing, after all, is only one step away from production and closely linked to it) or even the industry in which the organization is located, and toward the freedom to make or break alliances with any other profit-making corporation to secure financial advantage: in short, mergers and acquisitions as the preferred investment strategy. We used to think, naively perhaps, that a university produced knowledge, not real estate holdings; that a steel company produced steel, not money; and that an airline moved people, not assets. But a shift away from product or market orientation at the top can also shift how the company conceives of itself.

The view of a corporation as merely a bundle of movable assets (a "portfolio") turns attention away from long-term productivity and innovation, both of which require internal investment. Faced with increased uncertainties and with "political" tasks in a politicized environment, chief executives sometimes find it easier to imagine shedding what is *not* working and acquiring what *is* working elsewhere than to undertake the longer, more tedious, more difficult, and less glamorous task of reorienting—changing—their own core company.

Moreover, decisions based on portfolio analysis tend to stress the benefits of broad diversification—distributing the eggs among as many baskets as possible. Especially in turbulent environments, such as the present one, this is argued to be the best way to reduce the risk of major loss. But it is *also* very much the *least* likely way to generate substantial benefits in any new area, and it is a complete reversal of the traditional entrepreneurial function, away from investment and innovation and toward conservation and fiduciary responsibility. Moreover, it probably doesn't even reduce the risk. As the recent set of events in the "merger wars" initiated by William Agee of Bendix Corporation demonstrated, Bendix' attempt to acquire Martin Marietta resulted in the swallowing up of Bendix by Allied Corporation, an increase in Marietta's debt, enormous fees to lawyers and investment bankers, and *no* net gain in the development of productive resources. At the very time when our leading banks are looking for opportunities for innovative programs and services, many of our formerly leading industrials (for example, in steel) are behaving more like banks.

The "New Politics" of Middle Management

For middle managers, some significant new pressures are also visible. Together, these require a very considerable reorientation of such managers, their development and their skills. For middle managers and professionals alike, changing times and new environments create a new set of pressures that need new insights, new skills, new orientations, and new roles.

One of the driving forces is direct career pressure, stemming from trends in the labor market, on both the demand side and the supply side. On the demand side, declining productivity and foreign competition have helped increase career insecurity and reduce the meaningfulness of formal tenure (e.g., university and civil service) or informal corporate tenure. On the supply side, the dramatic increase in jobholding women has increased the competition for lower-management jobs, as has the almost equally dramatic increase in years of schooling, which tends to be associated with increased ambition as well as a larger pool of competitors.[21] More young people have entered the work force recently than ever before in history, and they are not going to rise as fast as their seniors, just because there are so many of them. And now the whole population is aging, creating a bulge of middle-aged employees wanting better jobs and in the near future fewer younger people for entry jobs. Demographic shifts shape how all of us do our work. Whereas previously there was a limited pool of the "traditional" employees who sought opportunity, the pool is now growing: younger people, the baby-boom bulge, who want opportunity; women and minorities pushing for a fair share of the better jobs and for upward mobility; and the larger educated population that no longer seems content to accept limited jobs, except as a temporary expedient when unemployment is high.

Career anxiety affects organizational functioning. In the organizational arena, it is interdepartmental power rather than merely individual power that is at stake. While peers in the same work unit may be direct competitors for better jobs, they are also collaborators in the larger struggle to improve the entire unit's bargaining position in the organization. Resource scarcities increase internal bargaining for resources, which affect daily quality of life as well as ability to produce accomplishments that net career advantages. Turbulent environments keep shifting the focus of relevance and make whole functions or departments relatively essential or inessential, depending on their control over critical issues, as those issues themselves shift. Thus, the critical issues being managed at the top—from changing market conditions to regulatory pressures—shift the ways functions and units line up with respect to each other. And that affects both the opportunity structure (what career paths are likely to be significant, what fields will be included in dominant coalitions) and the power structure (who has access to resources, information, and support, as manifested in discretion, visibility, and relevance in job activities).[22]

The traditional struggle for power in the middle ranks of the organization concerned individuals *vis-à-vis* each other and sets of tasks. For managers, it used to take the form of career competition: succeeding in winning over their peers in the competition to take on ever-more-important sets of responsibilities. For professionals, it used to take the form of struggles over job control: succeeding in gaining desired degrees of autonomy and control over the conditions and standards for their work. But now the terms of the power struggle have grown to include departments as well as individuals.[23] Just as adversity can bind together the members of a collectivity as they struggle for joint survival, while prosperity may drive them to compete, and a "lean" environment causes organizations to form resource-sharing networks, so can the new environment for middle-echelon employees drive them to jointly seek to elevate and protect their own unit of the organization, while continuing to compete for advantage within it. It has long been well known that subunits suboptimize, and that differentiation creates dramatically different outlooks as well as conflicts the organization must manage.[24]

Thus, added to individual career issues is a struggle for survival between departments or functions: whether the individual gets "ahead" becomes a function of whether his or her department stays in existence in a time of scarce resources. In the middle, allies in the same field strive to prove the importance of their field to the survival of an organizations that has limited resources.

But one also acquires power by struggling to control those new issues which preoccupy the top of the organization, the new issues that crosscut old territories, such as issues of regulation, of political control, and of resource certainty, including human resources. Who should "own" those issues? Where should they go in the system? Management of an active, turbulent, changing environment poses a continuing series of new issues for organizations, issues that must be located in the structure. The middle-level power struggle is in part over "ownership" of new ideas, which by definition cannot be fitted into existing functional boxes and which throws the meaning of functional distinctions up in the air to be renegotiated.

For example, who should handle new issues in government relations—existing staff or a new department? At what level should productivity-improvement programs be designed and managed, and who should be included in their management? Where should the

EEO function be put? If the latter decision seems obvious, remember that all the legal mandate entails is that an EEO officer be identified and consider this finding: In thirty-two corporations, I found six different departments housing the EEO office: legal, administration, personnel, labor and industrial relations, operations, and division management. In many cases, the EEO office had moved several times. In one not untypical case, the formal EEO officer was in the industrial-relations function, but the primary "champion" of the issue reported to the vice-president of personnel, who had a working charter from the top to do something about EEO; there was a great deal of political maneuvering around who "owned" EEO, and how credit or blame would be distributed. The "blame" part should not be ignored, either. While some new issues seem highly desirable and therefore are candidates for power plays, others may seem risky, and those who are assigned to them may engage in a series of self-protective political maneuverings.

With the new issues, in short, have also come new staff departments to handle them, ranging from the more "defensive" positions, such as those handling regulation, to the more "offensive" roles, such as planning, forecasting, or market research. Staff or stafflike roles are proliferating in the corporation. Internal consulting positions have also grown, as demonstrated by the growth of "organization development" activities and the increasing membership in professional associations concerned with these matters. Furthermore, some companies (such as the ones examined in Chapter 5) routinely appoint "problem solvers" with stafflike roles in line departments. All of these jobs, considered middle management in status and privileges even though they do not fit the traditional definition of a manager, reinforce the burden on middle managers to operate by persuasion and bargaining rather than by formal authority. These staff professionals may have a small group of subordinates reporting to them, but the bulk of their impact comes from the work they do with and through line managers, perhaps attaching themselves temporarily to the line organization or simply influencing the actions of the line.

Staff–line conflicts are classic and well known. My point here is that a larger proportion of middle managers may find themselves, during some part of their careers, needing to act in staff capacities. Their mastery of political skills is essential.

Finally, new and more appropriate organizational structures also tend to make power issues more salient at the middle and put pressure on middle managers to adopt new styles. In these turbulent times, and particularly for fast-growing high-technology industries (as well as for those which are following "fashion" by adopting new structures whether or not they are optimal), new forms of organization structure need to be designed to maximize responsiveness. But adding responsiveness may mean minimizing traditional line authority, thus increasing conflict and ensuring that power struggles dominate much of the life of the middle of the organization. It is no accident, of course, that the new organization structures were invented in, or largely carried by, companies in post–World War II industries, such as aerospace and electronics.

In matrix organizations, for example, which grew out of aerospace firms, employees or managers may combine two or more dimensions in their jobs: a functional specialty (such as sales) and a responsibility to a particular product line or market area. This combination is reflected in reporting to two or more bosses, e.g., one for the function, and one or more for the product areas. Thus, whereas in the classic unitary chain of command authority could be directly and relatively easily exercised, in the matrix influence down the line must substitute for authority to gain compliance, since neither boss has complete control over the employee. Traditional authority virtually disappears; managers must instead persuade, influence, or convince. The subordinate is expected to be the resolver of conflict, integrating the demands of these two dimensions of the organization. Conflict is thus built into the matrix. Depending on the design of the particular job and the coalitions that form, the balance of power may be held by the matrixed manager, who plays bosses off against one another, or by one or another of his bosses.

Other new forms of organization also make influence—or informal power struggles—more prominent than traditional line authority. "Parallel organizations," such as the one I will describe in Chapter 7, add a series of temporary, rotating task forces managed by a steering committee to the conventional line organization, and "ad-hocracies" may encompass similar fluid nonhierarchical structures, including project teams and other professional or quasi-professional self-managed work teams.[25] In these situations, people are brought together from many levels in new groupings that are highly participatory. Leadership may be independent of level; participation may be based on skills independent of formal position or formal authority. Such designs also undercut traditional authority because, as in the matrix, people cannot fall back on functional authority, on the traditional line, on the chain of command, or on the reasons for compliance that come from the rules of the organization. Instead, they have to bargain for influence and status; they struggle for power enough to have some impact on ever-more-confusing systems.

Thus, for middle managers subject to any of these "new" forms of organization, authority and career success are not granted automatically. Old sources of security are disappearing, and middle managers too find their positions shaped by the trends of the last two decades. But this is a "problem" for those trying to operate segmentally, acting as though they alone controlled the resources they need.

Reducing the Authority of First-Line Supervisors

First-line managers and supervisors—those who supervise direct production and service workers—have always had a difficult job: exhorting workers to live up to standards and demands thrust upon them by higher levels. The new corporate era, which has brought new work systems, only increases the pressures on them.

One of the reasons for new work systems at the bottom of the wage and supervision hierarchy is the changing labor market. Some have argued that more and more people today are "knowledge workers" who cannot be closely supervised and controlled, because the organization counts on their knowledge and internal commitment to get the work done. Through developments in microcomputers, 1973 can be singled out as a watershed in which even shop-floor factory workers may have started to become knowledge workers because of new technologies. And one key to managing knowledge workers is to let them alone to use their knowledge.

This is related to the educational changes in the work force. In the last twenty years, there has been a great increase in the number of working Americans who are college educated.[26] While forty years ago only 5 percent of the whole work force had graduated from college, today 25 percent of all people who work are college graduates, and that is rising quickly. Whatever else people learn in college, they learn attitudes about dignity, entitlement, and using their skills. One cannot manage this educated work force in the same way that seemed acceptable for a low-skilled, largely immigrant labor pool. Thus, education creates another pressure for autonomy, flexibility, and freedom—even at the lowest levels of organizations, since education is growing in blue-collar jobs as well. Now 20 percent of all crafts workers are also college graduates, leading one giant American manufacturing corporation to engage in "blueblooding," their informal name for a process to retrain all of its production supervisors to deal with this more educated population—changing a supervisory style known as "knocking heads together" to a new style stressing freedom and flexibility for the newly educated workers on the shop floor.

Overall, through increasing sophistication on a number of fronts, forms of authority in many companies have moved away from the "direct controls" involved in close supervision (issuing orders and monitoring behavior) to what is sometimes called "bureaucratic controls," or indirect authority.[27] Instead of direct order giving, the organization sets a context making it inevitable that people do the right thing, via the setting of targets and standards and long periods of training. Then employees are left freer to do their work as they see fit. Managers design the organization to make sure that performance is as predictable as possible, and then they leave people free to make a large number of more immediate decisions on their own, measuring results out the other end through performance appraisals and other devices.

A part of this shift to indirect controls is the growth of explicit "internal labor markets," as economists call them, or career-development systems. The motivation for performance under these systems is not the hope of immediate punishment avoidance or reward, but the long-term expectation of a "career." Such career systems are moving downward, further augmenting the atmosphere of choice at the bottom—and limiting the power of supervisors. Indeed, "bureaucratic controls" often serve to protect the interests of the worker against the manager. New performance-appraisal systems, with ratings that both supervisor and employee must sign, and with third-party review (perhaps by a personnel staffer), protect the worker against an arbitrary exercise of hidden authority by the boss. And leave the boss who does not understand participative management out in left field.

At the same time, greater participation in workplace decisions is creeping into the American system, and it seems to be gaining momentum. In forms such as quality circles, there is increasing evidence that it works to raise productivity, a clear incentive for executives. Fad and fashion also play a role in extending more participative work systems. Organizations can be just as "fashion-conscious" as individual consumers, and when a few leading ones start adopting reforms, often the rest quickly follow because they want to be "modern" too. (Indeed, concerns about "image" are a driving force for executives in the new, more political and more public corporate environment.)

Furthermore, trends in unions contribute to reinforcing these changes. The growing edge of organized labor in America has been among white-collar workers, especially in the public sector, who think of themselves as knowledgeable, are better educated, and also want freedom from close supervision (to be let alone to manage their own work in the way they know best). Whereas all unions emphasize traditional bread-and-butter demands of pay and benefits, white-collar unions have generally been the leaders in bringing quality-of-work-life demands to the bargaining table; white-collar unions have pushed for such options as flextime, and the public-sector unions have contributed the lion's share of joint labor-management committees.[28] In short, many important interest groups may be converging in their acceptance of more participation at the shop-floor level.

Other forces also affect those managing the "bottom" of the organization. There are a growing number of innovations and reforms permitting time flexibility and schedule control for lower-echelon employees. Within some predetermined limits, workers under these programs can come and go as they please. In addition, there are growing numbers of third-party rights advocates ready to protect the rights of individuals in the workplace. There has been a dramatic increase since 1970 in the number of decisions in the federal and state courts supporting new employee rights, such as rights to privacy, rights to due process in termination, rights to conscientious objection to some employer demands, and rights to a variety of new freedoms and new aspects of flexibility at work.[29] (Malcolm Forbes, Jr., the son of the founder of *Forbes* magazine, recently told a group of executives, "If you don't support these new rights for your employees, you're only going to make a lot of work for lawyers, politicians, and management consultants.")

All these trends dramatically change the meaning of supervision and limit the authority of the first tier of management. This is happening at a time when the career prospects for first-line supervisors are also declining; more and more companies are hiring directly into the levels above, limiting the chances for supervisors to move up. It is no wonder that many firms find it difficult to get workers to take supervisory positions. Here too, these things create problems in the context of traditional assumptions. There can be "solutions" only as part of more systematic and extensive shifts toward new assumptions.

THE NECESSARY SHIFT
FROM SEGMENTALIST TO INTEGRATIVE ASSUMPTIONS

Only a few decades ago, before the transforming era began, ideas about the American corporation were dominated by four common assumptions of a segmentalist model, ideas embedded in the law, in management practice, and in organization theory.[30]

Old Assumption 1. Organizations and their subunits can operate as closed systems, controlling whatever is needed for their operation. They can be understood on their own terms, according to their internal dynamics, without much reference to their environment, their location in a larger social structure, or their links to other organizations or individuals.

Old Assumption 2. Social entities, whether collective or individual, have relatively free choice, limited only by their own abilities. But since there is also consensus about the means as well as the ends of these entities, there is clarity and singularity of purpose. Thus, organizations can have a clear goal; for the corporation, this is profit maximization.

Old Assumption 3. The individual, taken alone, is the critical unit as well as the ultimate actor. Problems in social life therefore stem from three individual characteristics: *failures of will*, or inadequate motivation; *incompetence*, or differences in talent; and *greed*, or the single-minded pursuit of self-interest. There is therefore little need to look beyond these individual characteristics, abilities, or motives to understand why the coordinated social activities we call institutional patterns do not always produce the desired social goods.

Old Assumption 4. Differentiation of organizations and their units is not only possible but necessary. Specialization is desirable, for both individuals and organizations; neither should be asked to go beyond their primary purposes. (Thus, in Milton Friedman's terms, corporations should pursue only profits and forget about social responsibilities.) The ideal organization is divided into functional specialties clearly bounded from one another, and managers develop by moving up within a functional area. As a corollary, it is not necessary for specialized individuals or organizations to know much about the actions of others in different areas. Coordination is itself a specialty, and the coordinators (whether markets, managers, or integrating disciplines) will ensure that activities fit together in a coherent and beneficial way.

Of course, many of these assumptions have been under revision or attack for many years. The notion of the corporation as an individual actor writ large once informed much legal thought, but there has been increasing acknowledgement that such social organizations are too complex to make the analogy to an individual appropriate. Beginning in the 1960s the academic study of organizations, as well as managerial practice, moved away from closed-system assumptions, especially as it became increasingly clear that organizations are highly dependent upon and sometimes shaped by turbulent and uncertain environments. Tracing social problems back to individual characteristics has similarly been challenged, and neither "blame the victim" nor "blame the leader" arguments have been nearly as prominent in American social thought over the last few decades as previously. Consensus about the proper conduct of social actors and the proper ends of institutions, if it ever existed, has been undermined by events. It is no longer possible to talk about *the* American family or *the* American community, for example, as though there were only one type rather than a diverse and pluralistic group. In today's view, organizational goals are not "natural" and "given," but defined by an organization's "dominant coalition" as the result of a bargaining process that favors some interests over others.

Moreover, segmentalist models themselves arise under certain predictable social circumstances: they are a response to particular situations. These situations include economic expansion, where opportunity and power seem limitless, and where it thus appears that only individual limitations prevent success. They include circumstances in which one's own social group is dominant over forces in the environment, and is able to control its activities by predicting and therefore mastering all the elements needed to operate. Furthermore, in such times, opposing forces or groups are unorganized, unactivated, or quiescent. The environment is stable rather than turbulent, permitting the illusion that differences in the effectiveness of individuals or organizations are based largely on the quality of their own decisions. (This is an illusion because under these circumstances it is difficult to see the conditions in the environment that make such success possible; they are so predictable and so taken for granted that they simply become part of the background.) And consensus appears natural because clear challenging groups have not arisen.

Most of these conditions no longer apply to American society. Despite periodic longings for the establishment of simple and bounded "perfect communities" which can wall themselves off from the outside and operate consensually but mechanically, we must reconcile ourselves to a world that is contradictory and puzzling rather than orderly and controlled. No single social group or set of organizations dominates, and America no longer controls those who supply the resources it needs to carry out its activities. Even the best leader may not be able to control an organization, or accomplish all its objectives, in a turbulent environment in which the organization's success may depend less on its *own* decisions than on decisions made elsewhere, by others, according to different criteria.

We may not *like* the external pressures on corporations, from foreign competition to government regulations to activist groups to changing employee attitudes. We may not find them necessary or appropriate. But we cannot fail to notice that they exist.

And so these old assumptions need to be replaced by a set of new, more integrative assumptions, stressing the relationships between organizations and their environments, as well as the interdependence of an organization's parts:

New Assumption 1. Organizations and their parts are in fact open systems, necessarily depending on others to supply much of what is needed for their operations. Their behavior can best be understood in terms of their relationships to their context, their connections—or nonconnections—with other organizations or other units.

New Assumption 2. The choices of social entities, whether collective or individual, are constrained by the decisions of others. Consensus about both means and ends is unlikely; there will be multiple views reflecting the many others trying to shape organizational purposes. Thus, singular and clear goals are impossible; goals are themselves the result of bargaining processes.

New Assumption 3. The individual may still be the ultimate—or really, the only—actor, but the actions often stem from the context in which the individual operates rather than from factors purely internal to the individual. Individual actions occur in response to the expectations of others with whom they are involved. Leadership therefore consists increasingly of the design of settings which provide tools for and stimulate constructive, productive individual actions.

New Assumption 4. Differentiation of activities and their assignment to specialists is important, but coordination is perhaps even more critical a problem, and thus it is important to avoid overspecialization and to find ways to connect specialists and help them to communicate. Furthermore, beyond whatever specialized roles organizations or the units in them play, they also have a responsibility for the consequences of their actions beyond their own borders. They need to learn about and stay informed about what is happening elsewhere, and they need to honor their social responsibilities to act for the larger good. These tasks call for managers with general perspectives and with experience in more than one function.

In short, our transforming era requires not only that we change our practices in response but also that we change the way we *think* about what we do.

RESPONDING TO TRANSFORMING TIMES: CHANGE AS THREAT VERSUS CHANGE AS OPPORTUNITY

Historians wisely try to refrain from the writing of history until well after the events to be explored have passed and a decent perspective can be gained. It is always extremely risky to talk about the historical significance of the present, one's own times. But I believe from all the evidence that American industrial organizations stand today at a critical watershed. Their response to the new environment, for better or for worse, is likely to determine the path of our economic system over the next several decades.

If, faced with these changes and transformations, American organizations use their strength, accept the challenge, take the risks of which they were once proud masters, and extend the capacity of their organizations to innovate, we may yet emerge strengthened and prepared to capitalize on the future.

If, on the other hand, our organizations continue to operate as if, in Marshall McLuhan's memorable phrase, they were driving into the future while looking out of the rearview mirror, we will probably see, at best, a stagnation of American capacity, and at worst, a continuing decline in our competitive abilities.

The more optimistic path is clearly within our grasp; some companies have tried—and are continuing to try—to deal with these issues. But innovation cannot flourish where segmentalism prevails. Under segmentalism, change is a threat. It is perhaps because there has been so much segmentalism in large American corporations that so many of the managers I talk with seem to feel dislocated, disoriented by the changes that have taken place in the American economy in recent years. What has slipped away for many managers and executives is not just a sense of supremacy ("America as #2") but a sense of control. That is what they find so unsettling, so frightening, so frustrating, so intolerable. They feel at the mercy of change or the threat of change in a world marked by turbulence, uncertainty, and instability, because their comfort, let alone their success, is dependent on many decisions of many players they can barely, if at all, influence. Where segmentalism has prevailed, security comes in the form of control, and loss of control is the supreme threat.

Thus, the sight of some corporations seeking government and union concessions in order to avoid cutbacks or closings is disturbing to those who are control-oriented. How far the mighty American corporation has fallen, the headlines seem to imply, when management symbolically wraps itself in rags and goes begging. Management turns its pockets inside out to show that they are empty, pleading poverty in order to gain wage concessions from the union, agreeing to profit sharing in return. A bargain is struck: financial assistance in exchange for a share of control.

Then the management experts tell executives to give up still more control—or so it seems to many of them. One message of the "how to manage like the Japanese" books is that companies should provide employment security, in effect losing control over layoffs and terminations as a smoothing strategy, while increasing consensual or participatory decision making—in effect relinquishing exclusive managerial control over decisions, and taking more time to boot. Interpreters of labor-force trends—myself included—report that employees want more rights, a greater voice in decisions.

And so it goes. Those in the corporate drivers' seats must sometimes feel that they are being asked to share the steering wheel while the vehicle is skidding on icy roads.

I do not mean to exaggerate the disturbing quality of contemporary lives in organizations. For many people and companies, daily life goes on as usual, there are numerous sources of security, and external pressures seem minimal or distant. Some of them find security as individuals in the rhythms of family or personal life; others feel they have relatively predictable careers.

But while individuals can try to wall themselves off from the effects of change in the private sphere, corporations do not have that luxury. Interdependence—and hence dependence—is even clearer in the world of organizations than it is in the world of individuals. The long arm of economic slowdowns, for example, reaches far down the raw-materials-to-market chain. Problems in the American auto industry not only affect the Big Three and smaller Fourth, not only threaten to board up the entire state of Michigan, but affect legions of other partsmakers, suppliers, and dealers. When Xerox and Polaroid, two of the reputedly most progressive companies in the United States, known for their *de facto* lifetime employment policies, begin laying people off, then the reverberations are felt more broadly: another security barrier knocked down by the winds of change.

This is the downside of change: feelings of loss of control and helplessness in the face of decline, change as "enemy." It implies loss when people are unprepared for it, when they have nothing in reserve, when their current capital fund of assets and skills is rendered obsolete, when no resources are available to help them make the transition to a new state—and they cannot even envision what the new state might be. Change brings pain when it comes as a jolt, when it is seemingly abrupt and shocking. The threat of change arouses anxiety when it is still just a threat and not an actuality, while too many possibilities are still open, and before people can experience themselves in the new state.

But not all change is negative, even though it may create uncertainty. Not all sharing of power implies loss; it can also lead to bigger gains. Not all turbulence is a mere distraction from business; it may lead to useful new inventions. There is, in other words, an upside to change. Change can be exhilarating, refreshing—a chance to meet challenges, a chance to clean house. It means excitement when it is considered normal, when people expect it routinely, like a daily visit from the mail carrier—known—bringing a set of new messages—unknown. Change brings opportunities when people have been planning for it, are ready for it, and have just the thing in mind to do when the new state comes into being. And it hardly needs pointing out that change also provides a chance for entrepreneurs to offer "change-management" products and services—turning other people's confusion into profitable businesses.

In short, change can be either friend or foe, depending on the resources available to cope with it and master it by innovating. It is disturbing when it is done *to* us, exhilarating when it is done *by* us.[31] It is considered positive when we are active contributors to bringing about something that we desire, or at least to making something valuable out of what is inevitable—lemonade from the economy's lemons.

Staying ahead of change means anticipating the new actions that external events will eventually require and taking them early, before others, before being forced, while there is still time to exercise choice about how and when and what—and time to influence, shape or redirect the external events themselves. But this does not mean turning into wild-eyed futurists or believing science fiction. In a practical sense, it means "leading the pack" without getting too "far out." I am reminded of a Woody Allen short story about an advanced civilization. Usually when we think about advanced civilizations, he recounted, we have in mind one that is thousands or millions of years ahead of us. But what worried Allen in the story was a civilization that was just *fifteen minutes* ahead: Its members would always be first in line at the movies, and they would never be late for an appointment. In short, a little lead time might be all the competitive advantage one needs!

I argue that tools already exist to "save the American corporation" or to "meet the Japanese challenge."[32] Along with the disruptions of the last twenty years, a proliferating number of social and organizational inventions have been developed, with demonstrated impact on productivity and motivation. Furthermore, the experience of growing numbers of companies with participative employee problem solving has shown that employees themselves more often than not know what needs to be done to improve operations.

Thus, *the problem before us is not to invent more tools, but to use the ones we have.* In many cases, as I am about to show, segmentalist companies are not even taking advantage of their own successful innovations, letting them disappear or confining them to narrow uses.

Living with change need not imply insecurity but, rather, developing new forms of security. In the traditional corporation, security was based on control. It was based on knowing where everyone and everything belonged, on having categories into which to place jobs (tidy boxes on organization charts) or people ("woman's place") or events (guidance by precedent).

In an innovating organization, in contrast, security will come not from domination but from flexibility. It will come not from having everything under control but from quick reaction time, being able to cut across categories to get the best combinations of people for the job. For their people, security will come not from staying in the same field or department or area but from identification with the whole company, with its unity of effort. The new security will be based on pride in individuals and their talents—reawakening or reinforcing the spirit of enterprise in all employees at all organizational levels.

The corporations that will succeed and flourish in the times ahead will be those that have mastered the art of change: creating a climate encouraging the introduction of new procedures and new possibilities, encouraging anticipation of and response to external pressures, encouraging and listening to new ideas from inside the organization.

The individuals who will succeed and flourish will also be masters of change: adept at reorienting their own and others' activities in untried directions to bring about higher levels of achievement. They will be able to acquire and use power to produce innovation.

PART II

WHY WE'RE IN TROUBLE: THE QUIET SUFFOCATION OF THE ENTREPRENEURIAL SPIRIT IN SEGMENTALIST COMPANIES

CHAPTER

Innovating Against the Grain:
Ten Rules for Stifling Innovation

Here's the way to avoid taking action on a problem. First, reject the problem and its solutions intellectually, saying that they won't work. Second, attempt to force the problem into existing processes in a mechanical way. Third, slice it up into pieces and allow people's natural tendencies for territorial assertion to take over.

—Senior executive,
"Southern Insurance"

Of course we have an open-door policy. It means that if you don't like the way we do things around here, management will be glad to show you to the door.

—Common employee joke

RARELY DO BOSSES in tradition-bound organizations actually have to say "No" directly to a subordinate's idea. A few well-placed frowns or eyebrow raises, some pregnant pauses, a reiteration of the *real* assignment, and citation of accumulated years of company wisdom can be enough to make it clear to people that new ideas are not welcome. And if that does not discourage them, then a conspiracy of silence from the rest of the organization might: fellow employees too busy with their own work to care, service departments too slow in returning phone calls to provide service, or cost controls making it impossible to squeeze out a few spare hours or extra dollars for experimentation.

And so, in such organizations, most people never bother to pursue ideas for improvements. One leading company found to its surprise that nearly 50 percent of its employees admitted on a survey that they were contributing below their capacity. These lost opportunities can never be quantified, nor can their potential benefits—should they have been acted on—be assessed, but occasionally later events make clear to an organization what it has been missing. A textile company had lived with a high frequency of yarn breakage for years, considering it a cost of doing business. Then in 1982 a new plant manager interested in improving employee communication and involvement discovered a foreign-born worker with an ultimately successful idea for modifying the machine to reduce breakage—and was shocked to learn that the man had wondered about the machine modification for *thirty-two years*. "Why didn't you say something before?" the manager asked. The reply: "My supervisor wasn't interested, and I had no one else to tell it to."

There is a cliché about the "failure of success" that is often recounted by people explaining why some major companies are so uninnovative. Success breeds complacency, the cliché warns, and so organizations with a formula that works well are doomed to replicate it, handing over their operations to people who control things so that there are no deviations from the formula.[1]

Folk sayings in management, like those in other realms, always have a grain of truth, but sometimes no more than a grain. If we buy the "failure of success" argument, then we would expect "Chipco," GE Medical Systems, Polaroid, or numerous other innovators to soon fall into the same trap. But to understand just how different the innovating organization is, and how complicated the problems are for less innovative firms, it is instructive to look at what happens when people try to innovate in opposition to their organization's traditions, encountering closed doors, stalled engines, roadblocks, obstacle courses, and only a few narrow paths or back alleys through which innovation can occur.

Not all less entrepreneurial companies show the same profile. Across the range of American companies, there are at least two major types that produce little innovation. The first are *innovation avoiders*. They are generally capital-intensive, like oil or insurance, and see little economic leverage in their internal systems or operations. A few grand strokes of strategy occasionally are all these organizations think they need to survive and prosper. There is another type that are simply *unschooled in how to innovate*—naive rather than unconcerned, happier to learn. They are likely to be labor-intensive, like many consumer-goods manufacturers, or essentially service businesses like telephones, and they may even have progressive policies for the treatment of their people. But they grew out of earlier eras and have become encrusted with innovation-defeating traditions over time.

While both kinds of companies are likely to set up similar barriers to innovation, the novices hold more promise for changing in the future. But in both types, the tragedy is that there are many people ready to contribute more, without sufficient access to the means to do so.

Two of the ten core companies in my systematic research are particularly good representatives of each of these types, because I found the patterns in them typical of a large number of other less innovative organizations. Both are prominent enough that I will disguise a few details and introduce them by pseudonyms: "Southern Insurance," the change resister; and "Meridian Telephone," the change *naif*.

In addition to coming to know the companies over a period of more than a year, I included them in my direct comparison of managerial accomplishments in six companies. Overall, they received the two lowest scores on the "innovation index"; moreover, the 20 accomplishments examined in detail at Southern and the 39 at Meridian were rated by my research team as generally reflecting much lower levels of initiative and as having much lower payoff for the companies than the accomplishments at Chipco, Polaroid, General Electric Medical Systems, and Honeywell—all much more innovation-supportive companies. And practically all the innovations that occurred involved internal reorganizations—not new products or new technologies.

So I was interested in two questions about innovation-suppressing organizations like these: First, what is it about the structure and culture—the roadblocks erected—in contrast to a Chipco or a General Electric, that keeps managers and employees from contributing more to productive change? And second, how does any innovation contributed by employees and managers creep through when the corporate environment discourages it?

Any of us who have ever felt stepped on by our company when trying to contribute a new idea or get support for an innovative project will recognize the experiences of the people who tried to innovate, under difficult circumstances, at Southern Insurance and Meridian Telephone.

"Southern Insurance": The Change Resister

Entering the main floor of Southern Insurance headquarters is almost like walking into a museum. Southern's history is long and dignified, and the portraits on the walls of Southern executives stretch back into the nineteenth century. There is an understated richness and marble-pillared elegance about the public places at Southern that make it clear that these are places for display rather than work, and voices tend to be kept low in honor of the surroundings. Away from the entrance, the offices at Southern are quite modern and bustling, but they are reached only through the display areas. Employees gaze upon Southern's history daily when they go to work.

Southern's organization is a kind of museum too; it is largely relics of the past that have been celebrated in Southern's structure and culture, and that makes change difficult. Southern was founded before the insurance boom supporting late-nineteenth-century business expansion, and it has held a respectable position in a conservative industry, slowly making room for new business. When I first encountered it, it had a core insurance operation, several other financial-service groups, and some much newer insurance-related human-service subsidiaries.

Recognizing the demands of changing times, Southern had recently gone through a "shake-up" several years ago, an attempt to pry Southern loose from its traditional ways of operating and to bring in modern, market-oriented management thinking. There was an almost complete turnover of top management, with five new executives brought in from outside, including the CEO. One of the new CEO's early moves was a major reorganization of district offices that consolidated a number of loose operations. Investment strategies, not internal operations, were seen as responsible for Southern's financial returns, however, and so top management focused less on the organization than on where the money went. While the new executive team spoke of the need for more innovation, only two new product opportunities came up in my study (an insurance package targeted to a new market and a new money-market fund), and both were imitations of products already common among Southern's competitors. Southern was not an industry leader in either financial performance or services.

The organizational changes shook Southern up in more ways than one. By the time I came to know Southern, the atmosphere seemed fragmented; employees were hard-pressed to provide a clear image of the company, its style, or its direction. "Idiosyncratic" was the word I felt best described Southern's culture. The environment for managers and their treatment depended very much on the particular boss of the moment; one-on-one relationships and private deals prevailed (just as in the selling of insurance itself). Very few people had any sense of where they fitted in the "whole." It was easy for people to retreat into the security of defined tasks in isolated departments—tasks with which the armies of paper pushers and numbers analyzers at any insurance company are quite familiar.

Some of Southern's nonfinancial subsidiaries were greatly different: newer, in growing markets, with a younger, professional group of enthusiastic managers. Indeed, four out of Southern's nine innovative accomplishments in my study were carried out by managers from a noninsurance subsidiary; without the subsidiary's contributions, Southern's scores on my "innovation index" would drop from 45 to 33—even lower than Meridian's. But these four accomplishments—all new structures or methods—involved entrepreneurial managers responsible for autonomous field activities who introduced effective (and profitable) new ideas, from modernized information systems to the creation of new services headed by supervisors with profit responsibility. It was difficult for many of these people to live comfortably in the detail-conscious systems designed by their parent company.

"Meridian Telephone": The Change Naif

Meridian Telephone had none of the museumlike quality of Southern Insurance; the small touches of humanity on hallway displays were celebrations of the achievements of individuals and their service to the community rather than ancestor portraits of executive patriarchs. Meridian was one of those progressive/paternalistic combinations that have often come a long way to bridge the distance between "corporation" and "person" through good benefit programs; strong social responsibility, including leadership in affirmative action; top executives who are aware of the need to be visible and available to their people; and a network of employee clubs and social activities that have developed through the years. At Meridian, there were large numbers of long-service employees who had grown up with the company, often joining it out of high school—including a high proportion of top management. Even in its big-city locations, Meridian retained a small-town atmosphere that made me feel at home at all the meetings I attended. This was not a high-pressure, city-slick company.

But the kind of business Meridian was in meant that its actual operations were part of a vast, finely specific bureaucratic maze—almost like a telephone network itself—and all of these fine cuts and intricate arrangements were reflected in a segmented organization. The question facing Meridian was whether this kind of organization was suitable to a new environment. Rock-solid in its quasi-monopolistic status for many years, Meridian was now, because of regulatory changes, coping with significant new competition for the first time in decades. Its entrepreneurial CEO—something of a "maverick" around Meridian's world, but a home-grown one who had always performed—was all in favor of the changes and eager to be among the first to demonstrate the value of new organization structures or methods. I met with him just after a "matrix organization" had been installed as a link between market areas and services, and just as employee task forces were helping the executive group define the issues needing attention in the future.

Then in a major speech, one of the executives firmly supported the need to make management more flexible throughout the company:

> Meridian must sell our services now; we're in a competitive mode . . . we're going from an administrative mode to an entrepreneurial mode. Employees are being asked to use their brains, be flexible, be leaders, take risk and worry about consequences later. There are less routinized approval processes now.

It takes a long time to change a company as complex and procedures-conscious as Meridian, however, and the Meridian I saw was still highly centralized. Managers and their subordinates a few levels away from the top were still rules-and-procedures-bound; they had little discretionary use of resources and few incentives to speed adoption of needed new organizational methods and products. Often, the "culture change" the top was trying to make disturbed and frightened people more than it awakened their entrepreneurial spirit, because they were not sure what they were being asked to do —itself an indicator that top-down dependency prevailed—and they could not see that they had the tools to do it, anyway. Meridian produced the least innovation of any company I examined closely: only 14 of 39 managerial accomplishments at Meridian had an innovative flavor; the rest were just part of the basic job.

The one thing Meridian people seemed sure of was that consideration for people was valued, as it had always been in Meridian's progressive tradition. And so, effective managers at Meridian confined their accomplishments, by and large, to the small-scale, close-to-home arenas they did control: treatment of people. The kinds of things Meridian managers spoke of most proudly rarely affected products, markets, policies, productivity, or any other innovative change, unlike the companies high on the innovation index. Practically all of Meridian's achievements were in new structures (mostly creative ways to reorganize in the face of cutbacks) or new methods (new budget-control systems or computerization). These kinds of innovations have a "sinking ship" flavor, not one of a company on the move in its marketplace.

Furthermore, Meridian had a highly disproportionate number of purely individual-level accomplishments (almost two-thirds of all such accomplishments found in the six companies came from Meridian). More than a fifth of Meridian's effective managers saw their highest achievements as including "developed people to become self-motivating"; "made a lateral move and mastered a new job"; "placed people facing layoffs in retraining program"; "improved communication within my department"; "handled a complaining subordinate"; "improved my relationship with my boss." By saying that this is *all* they did, I do not mean to denigrate accomplishments aimed at individuals and their situations. But in the innovating companies I compared with Meridian, these benefits for individuals come *along with* the design of new products or systems increasing the organization's future capacity. At the kinds of places that Meridian represents, in contrast, the arena for innovative action—if it exists at all—is tightly circumscribed away from company strategic needs and back into small moves within tightly bounded segments.

Whether companies actively avoid innovation, like Southern, or are not yet ready for it, like Meridian, they share the universal problems of change in the face of roadblocks. *Why* innovation is so difficult is only part of the story, however. After we look at what corporate entrepreneurs face in these settings, we will also applaud the achievements of those who manage to innovate anyway—the heroic figures in less innovating companies.

ROADBLOCKS TO MANAGERIAL ENTERPRISE

The sad thing about many older and tradition-bound American companies like Meridian and Southern is that they may recognize the need for innovation as a matter of corporate survival but not know how to go about getting it. Accustomed to setting up controls to avoid risk, they may have forgotten how to permit experimentation. Used to setting policy at the top, they cannot easily free the levels below to contribute new ideas. And yet these companies are increasingly seeking, they say, a new style of entrepreneurial manager who can guide each of the company's operations into new, more competitive postures.

But as long as segmented structures and segmentalist attitudes make the very idea of innovation run against the culture grain, there is a tension between the *desire* for innovation and the continual *blocking* of it by the organization itself.

Thus, the primary set of roadblocks to innovation result from segmentation: a structure finely divided into departments and levels, each with a tall fence around it and communication in and out restricted—indeed, carefully guarded. Information is a secret rather than a circulating commodity. Hierarchy rather than team mechanisms is the glue holding the segments together, and so vertical relationship chains dominate interaction. Each segment speaks only to the one above and the one below, in constrained rather than open exchanges. The one above provides the work plan, the one below the output. Preexisting routines set the terms for action and interaction, and measurement systems are used to guard against deviations.

The Elevator Mentality: Dominance of Restrictive Vertical Relationships

It probably will not come as a surprise that less innovating companies are dominated by tall hierarchies, and that honoring the chain of command is a value. In some industries, particularly older ones, this is so much the norm that few question it, and when they do, it literally makes headlines. Insurance is one of those industries, and John Hancock's decision to flatten its hierarchy and remove "excess" levels was featured prominently in *The Boston Globe*.[2] The fifth-largest insurance company in 1981, and somewhat larger than the company I am calling Southern, Hancock became concerned about its narrowly specialized managers, high financial costs, and lagging sales—and not incidentally, its long management chain. In one division of two thousand people there were twelve levels of management between the chairman and clerks, creating an awesome organizational distance in a small unit. (Evidently Hancock executives were beginning to think so too.)

Meridian's industry had also been around a long time, and management layers had built up. At Meridian there were fifteen levels of managers, with each level further divided according to standardized pay differentials—e.g., four such "bands" of first-line supervisors. Duties were finely specified and changed slowly. For example, four years ago a manager at the corporate office wrote a plan to formally describe the duties of a sales manager. The plan set out the frequency, attendees, purpose, and background information to be gathered for each of several kinds of meetings—appraisals, field visits, meeting with low producers, brainstorming sessions, etc. As a result, Meridian top management was viewed by some as guarding its authority to make both large and small decisions. One manager's comment was typical:

Upper management is not really willing to give up their power. . . . Nickel-dime decisions shouldn't be made at the top. Many of high management don't have faith in lower management. This is pathetic—to waste lower management talent because of the egos of top management.

At Southern, in particular, the controlling nature of the hierarchy stood out. In a number of cases, middle managers felt that the best thing they could do for their own operations was to block interference from superiors or other functions that would hinder the activity of subordinates. A manager who was going to accomplish anything significant had to concentrate attention on removing the possible hindrances. There were frequent comments about Southern's being too "top-heavy with managers" or engaging in too much "paper communication." Top managers were more often than not seen as controlling or as imposing rather than creating enabling conditions for achievements. Managers felt that even the current emphasis in the company on enterprise, creativity, and risk was imposed on them from above in the same way that other activities had been commanded by the top in the past. "They used to jam control down my throat, and now they jam creativity," said one.

The experience of the manager I'll call "Bob Smith" is instructive. A gung-ho but meticulously professional district manager for the Southern subsidiary in a health-related business, he reorganized the internal structure of his district to improve its market orientation and created a new management information system to generate new accounts and increase volume—an impressive achievement. One would think that Southern's executives would have applauded. But Smith's initiative violated the protocol of the hierarchy, and he was punished rather than rewarded:

> The company did not allow a lot of freedom to make changes like this. . . . The region was always critical and resented me, threatening to fire me. It was an obvious conflict. . . . Informally, I told my manager that I would get better. But I always ended up going over his head to talk with his superior.
>
> They were not just against me personally, but there was a philosophy that the higher-ups know everything and there is no room for negotiation or new ideas.
>
> I overcame these roadblocks by hiding things. I got caught on some things that weren't going well; and on some of my projects that were going well, I would go public with them.

When the returns from Smith's innovations were in, he was willing and eager to share rewards with his subordinates. He made sure the supervisors under him were credited and got them raises—or, as Southern preferred to label them, "salary adjustments." Then Smith went even further. Because the clerical workers (all female) in his area received low salaries and were not entitled to bonuses,

Smith and his staff took up a collection and raised $2,000, which they distributed as their own "bonuses" for the clerical people in his district.

One can see Smith's bitterness rising to the surface as he tells the rest of the story, mentioning his own generosity first to contrast it with the wrist slapping that was his own "gift" from above. Imagine his anger when he found that his own performance review was negative, with an overall comment from his boss in that cold, spare language of such documents: "He fails to attain expectations." "Fails" indeed! Here was a high-performing district with sparkling new professional management systems. Smith was not willing to let that one go. So he got in to see Southern's president and challenged the appraisal statement. The president was more supportive and liked Smith's achievements. Smith walked out of that meeting with a raise and bonus and then felt that the recognition was commensurate with the work he had done. But the performance evaluation itself was let stand, the records of a bureaucracy taking on a life of their own that perhaps makes them sacred even to the company president—and serving as a reminder to Smith that *protocol, after all, must be honored, and does not take second place even to results.*

It is typical in American corporations that leaders at the very highest levels are somewhat more supportive of risk-taking innovators, even if they bend rules, than are their innovators' immediate superiors a level or two below. Corporate entrepreneurs more often complain that their own boss is a roadblock than that they fail to receive high-level support—if they can ever get in to see the bigger bosses. (The 59 accomplishments at Southern and Meridian received high-level support more often than that of the immediate boss —50.8 percent of the time at the top, compared with only 23.7 percent for bosses). Particularly in the more entrepreneurial companies, those who rise high often embody the values of enterprise and achievement that make them support risk takers, whereas those left behind in the middle ranks may operate by playing it safe, even unconsciously squelching their subordinates' attempts to innovate, so the pattern of support's rising with level holds. In noninnovating companies like Southern and its counterparts, this phenomenon is exacerbated by the exigencies of power and powerlessness in very tall hierarchies. Those at the top with power can afford to be generous, can ask for exceptions without being greeted by raised eyebrows, and can easily protect those with desirable innovations from the rules of the system. But the long chain of command often means that bosses of middle managers and below feel powerless and thus need to cling to any shred of control; that they feel measured by rule obedience and are therefore prone to hold others to that standard.[3]

Then there is the "good cop/bad cop" routine. Top executives can afford to espouse participation and responsiveness to people—and even mean it, even practice it—as long as there are authoritarian managers below who will make sure that nothing gets out of hand. This accounts, perhaps, for why there was such a discrepancy at Meridian between a risk-taking, innovative, iconoclastic, and people-conscious mode of operating at the top and a tight, controlling, numbers-first style that lower-middle managers felt they saw in their immediate bosses.

Departments as Fortresses: Poor Lateral Communication

Sheer height of hierarchy contributed to a second common characteristic of less innovating companies: the virtual absence of lateral cooperation or communication and support across areas. Peers in other parts of the company can seem irrelevant when the structure is segmented and the culture segmentalist. People at middle levels are more likely to guard their turf, mend their own fences, and concentrate on pleasing their own boss than to assist others outside their box on the organization chart. Most often this is passive: noncontact, noncommunication. Occasionally, it takes the form of active conflict. And it shows up in managerial accomplishments.

Because vertical relationships are so important, power exists only through the hierarchy, and middle managers seldom cross organizational lines to secure money or support. That is why when it happens it seems to people so noteworthy. One Meridian manager's use of a team from different organizations to conceive and develop a plan was novel; that crossover was his real accomplishment. He called the members' cooperation "Utopian." Another manager explained, "No one had crossed organizational lines to work on projects [because] there was no incentive or initiative to communicate with each other." A manager with long experience in the company was starting to see more rapport across lines than in the past, but he was frustrated with the problem of teaching subordinates the "teamwork" mandated by the top when there was little precedent or reward for it. Still, there were friendship networks that helped those lucky enough to be in on them to get extra information or help.

Friendships aside, the isolated instances of formal teamwork apparent at Southern occurred mostly in one subsidiary, and among people in tightly clustered areas, such as the same office or close in function. These were downward teams of subordinates created by rather new, young managers who decided that they could operate more effectively if they made a team out of their subordinates. There were a few other rare instances, but all concerned "new style" managers brought in from outside. One such newcomer manager built a task force which turned out to be a key factor in the creation of an important new department. Some staff personnel brought about limited collaboration laterally by contacting line managers about their preferences, but this was very brief and informal.

Lack of cooperation was coupled, in some instances, with active noncooperation. Information was hoarded rather than shared, even on request. At Meridian, the people in corporate complained that it was hard to get information from the field, while field officers said the same thing in reverse. In one case, data already on the computer had not been processed to permit analysis. The manager had to contact the staff people in the field, one level above, who asked, "Who needs the information?" He answered, "You're the only person who can provide it." Many "games" were then played, he said, and nothing was resolved. The manager and an assistant went back and got the others to agree, though they were still balky.

It was not hard for conflicts to escalate; as one unit withheld information from another, the second would retaliate with its own hoarding. One manager described this cycle:

> I wanted to get a marketing plan from corporate. I knew they had it, but they have continued to treat managers like me at an elementary level, and kept the information away from me and my counterparts. We began to take control of the information and discover it ourselves, and withhold it from corporate.

The "information pathologies" of overly hierarchical, overly specialized organizations are well known, but it is always disturbing to see them acted out. That information, one of the theoretically most abundant commodities of all, should be hidden, with requesting departments viewed as "enemies," does not seem compatible with an emphasis on results. Yet companies let this happen. Harold Wilensky has identified at least four kinds of groups that are likely to restrict the flow of information in a bureaucracy. The "time servers" (those whom some companies might call "retired on the job") neither get much information in the first place nor have much motive for acquiring it. "Defensive cliques" restrict information to prevent change, because any change threatens their position. "Mutual-aid-and-comfort groups" have settled into a comfortable routine, and so they resent their more ambitious colleagues. Finally, "coalitions of the ambitious" would rather keep information in their own hands, to monopolize power.[4] It is so much easier for these groups to arise in a segmentalist organization than in innovating companies like a Chipco or a Hewlett-Packard where the system supports open communication.

Indeed, the "opposition" that American managers describe receiving as they try to develop innovative projects in nonreceptive systems nearly always includes turf issues: ownership of issues, jealousy of a manager's visibility to higher levels. One outspoken Meridian manager described vividly his problems with another area supplying his function with materials:

> You have to be a good hair parter or bouncer to persuade them. You have to crank tails to get their attention. I don't pull out a gun and shoot, but I've thought about it.
>   I sometimes need to explain, "Look, we have this date scheduled. I can't perform unless you get off your dead ass."

But then, the response of peers in other areas to a potential innovation is likely to be mixed if there is a feeling that power is scarce or that disrupting changes are imminent. In one well-known company where new products are essential to survival in the marketplace, it is still common for the announcement that a new one is on the way to be met with resistance to the developers. Marketing gets concerned about lower prices, sales frets about commissions, and manufacturing worries about the costs of changeover and a drop in its profits; so all feet drag.

In some cases, it is not just organizational but also personal territory that feels "under attack." At Southern there were signs that individualism, competition, and conflict were encouraged. Executives egged people on to attack one another's positions in hopes of discovering the best answer or decision; the meetings I attended had a tense and protective air. Indeed, the time-consuming battles produced by this combative style made some managers feel they were constantly "in a war." Ironically, I saw more cooperation and less "combat" orientation in Honeywell's defense operations, populated by retired military officers, than among the three-piece-suited, Wall Street–produced Southern Insurance executives. No wonder hostility at Southern was often overt; corporate staff, for example, were sometimes referred to as "the jackals" or simply "the jerks."

With this emphasis, it is not surprising that informal types of recognition and rewards, such as praise, pats on the back, or being singled out for attention, were so rare within Southern. It was often stated flatly: the highest officers do not give pats on the back. This feeling of lack of positive support and recognition seemed to be true all the way down to the middle-manager level of the hierarchy. When rewards like this are scarce, they tend to be given begrudgingly by superiors, and peers are also likely to resent one another's successes. There was none of the generosity of spirit I will show is associated with a culture of pride like Chipco's or GE Medical System's. Instead, Bob Smith, while a high-achieving Southern district manager, was criticized and resented by people in his region. Meridian managers too reported pervasive feelings of concern every time a group or a person was singled out for special attention.

I am shuddering myself at what a negative picture this paints, and so I should hasten to say that these roadblocks and tensions between peers are not the whole story in *any* company. People carve out pockets of cooperation and warm interpersonal relationships, and those who managed to innovate at Southern and Meridian were no exception. Peers *could* cooperate, but often only if a personal relationship had preceded, or stood apart from, the organizational one. Some people did receive encouragement from bosses and found other departments ready to assist. My point is not that good communication is nonexistent or deviant but that it is not systematically encouraged as much as it could be, or as much as innovators need it to be, since innovation requires more power tools—more information, more resources, more support—than the norm. Relying on idiosyncratic relationships that individual good communicators establish is not enough.

Limited Tools

A third, related, problem for middle managers in the less entrepreneurial companies was that tools or assistance for activities were difficult to obtain—even for the routine tasks that had been assigned to managers. They were ready to take action, but did not feel supported for it.

Problems at Southern, for example, included a lack of personnel, so that managers had to handle routine office functions by themselves; lack of training, including both managerial skills and skills in a specific new area such as marketing; and too much paper communication but not enough face-to-face communication. Even long-service managers who considered themselves organizational loyalists mentioned problems with getting enough "power"—especially resources—to act: difficulties, due to high cost, in transferring people into their units; losing resources as other people were transferred to newly formed regions; uncertainty of organization structure (role definitions and who "owned" certain tasks); and difficulty with communication.

The case of a Meridian manager is instructive; she eventually succeeded at a small-scale innovation, but not before she had run up against the problems of hierarchy, turf, and scarcity of tools.

"Martha Evans" was a manager of training in a personnel function when "Bill Louis," a line manager, told her he needed help in retraining some "surplus" people for other jobs. Confidently making

a risky claim, Evans told Louis she could train Louis' people faster
and more cheaply than Louis expected, with the help of a small
computer. Louis paid for the equipment and manuals needed for the
training out of his expense budget, with no fuss. But Evans had to
get the computer purchase approved, and since a computer was
deemed a capital expense, this "took more ink than you would believe."

Louis needed the training program to start soon; the consolidation that would make many people surplus could not wait. Thinking
the approval for the purchase would probably go through, Evans
bought the computer with a personal check for $850—knowing that
if the purchase was not okayed, she had just acquired a home computer. The training program got under way while the days of the
authorization process stretched on.

The opposition of the data-processing people threw a wrench
into the slow-moving process of authorizing money for the computer.
The data-processing people thought Evans should use *their* big
mainframe computers for the retraining. Salesmanship was involved. Evans explained, "It was a combination of knowing people's
personalities and what they're afraid of, and anticipating their questions, knowing their hot button." Luckily, Evans and Louis had the
support of Louis' boss—a computer nut who owned a home computer. After a few phone calls, Louis' boss persuaded Evans' new
boss to back the computer purchase. The head of data processing
was finally persuaded by the small size of the computer to be
bought, a long memo that Evans wrote to him, and hearing from
Louis' boss that "Even if you don't like it, we're gonna buy it!"

The purchase also required the approval of the vice-president
of personnel because it had not been budgeted. On a visit to the
VP's office late in the support-gathering process, Evans accidentally
found that her letter to the VP backing the purchase had been mailed
without a key memo explaining the good reasons for the purchase.
She quickly sneaked a copy of the memo into the envelope. The VP
finally approved the purchase, thinking the amount needed to buy
the computer was small. (Here is the organizational distance issue
again: finally get to the right person, the person at the top, and
approval is easy.)

Evans was doubly relieved: she was reimbursed for the purchase, and the training program was succeeding even beyond her
optimistic estimate. The employees' union supported the retraining.
The trainees willingly invested their own time in learning the new
skills, learning them more quickly than Evans had estimated.

Evans was the first in the company to arrange such a training
program. A report she wrote describing the accomplishment eventually reached the vice-president and was returned with favorable

comment. An article in the company paper and much inquiry followed. Evans has since given a demonstration of the computer-based training to the vice-president.

Martha Evans succeeded despite roadblocks in the system. But if she had to go through all that for a relatively minor accomplishment, what was the fate of innovation at Meridian?

For both Meridian and Southern, then, roadblocks to managerial enterprise involved a number of barriers to the circulation of power at their level: the dominance of vertical relationships, lack of horizontal cooperation or communication, and lack of tools to help in defining and carrying out innovative activities. This is a familiar litany of bureaucratic woes. Yet, as we shall see later, power is allowed to circulate much more freely in highly innovating organizations.

But change *was* an issue at both companies, and it *was* occurring. And some effective middle managers succeeded in developing or contributing to innovations. So the next question is *how* less entrepreneurial companies go about the business of change when they do permit innovations to creep in, against the grain.

CHANGE PATTERNS IN LESS INNOVATING COMPANIES

There are four occasions that permit a few "corporate entrepreneurs" in less innovating companies to contribute to innovation and change, but only one of them resembles the free atmosphere for experimentation characteristic of high-innovation environments, and all of them severely limit the benefits to be derived.

The company may *mandate change from the top*, allowing a few at levels below to shape tactics. It may set up *formal tests of ideas initiated at the top*. It may bring in *outsiders* with freedom from the constraints facing the rest of the employees. Or it may be unable to maintain tight control at all times, thus allowing a few *"holes in the system"* constituting unplanned change opportunities. In the first two cases, the decision about the shape of an innovation will already have been made, limiting the chance for a potential entrepreneur to introduce new ideas; in the last two, only a few people can find themselves in the right place at the right time, and the impact on everyone else can be negative.

Top-Down Dictate

The typical pattern at less entrepreneurial companies is to try to force change by top-down dictate. In essence, top executives might decide to reorganize or to change policies, and then order people in the middle to pick up the pieces and make the new system work, with little warning and little assistance. This creates an atmosphere of uncertainty and insecurity which immobilizes managers rather than stimulating them to innovate.

At Southern, change was seen as occurring abruptly, often without warning, and managers commented that they wished the organization were better about explaining upcoming changes. There was impatience to do better faster, and favorable comments were made about a competitor with a high growth rate achieved quickly.

The perception of unexpected change at all levels of the organization made reporting relationships and procedures unstable, thus focusing the energies of some on simply coping with change and keeping up with new developments, and inhibiting others who felt they had to know where things stood before taking action. These were typical comments:

> We used to be run by the book and now I don't even know where the book is.

> If you don't like the organizational chart, just wait until next week.

> Yesterday's procedures are outdated today.

In listening to the Southern managers talk, and adding my experiences in other companies where abrupt policy changes occur as a matter of course, I was reminded of a classic psychological principle out of learning theory: Random reinforcement—an uncertain, apparently senseless cycle of rewards not clearly linked to behavior —provokes more anxiety than does negative reinforcement. To me, employees were responding to their lack of knowledge of where things stood, or why, with the same degree of tension anyone would under random reinforcement. There were anxious questions about what would change next, and whose staff would be the next to be displaced.

Instability in practically everything lowers security and trust. And security—especially knowing that one's job will still be here tomorrow, in a reasonably recognizable setting—is associated with higher flexibility and lower resistance to change.[5] But insecurity with respect to some basic continuities in the system can produce ritualistic conformity and fear of change—change as threat, rather than change as opportunity.

Low trust was expressed at Southern in the form of negative comments about higher levels of management, and it also came back to haunt middle managers. Concerns about the likelihood of more change made it harder for many of the middle managers to get the support of their subordinates or other key people because the others were afraid to invest in something that might turn out to be temporary. One manager said that colleagues were afraid to give assistance and invest energy, for fear that the manager would leave or be removed. Thus, even those likely to take risks and to generate an enterprising project found that the uncertainty in the air made it harder than it would have been otherwise to get the resources they needed.

The constant threat of more change without warning and without participation made people focus on the short run, not the long run. This understandable short-term orientation also limited the kinds of incentives the managers could offer to others to join them in their accomplishments: instead of building long-term collaborative relationships with customers promising future rewards, managers had to provide immediate "payments" (such as raises) for services rendered. (After all, what was the point of promising something that would pay off only in the long term if subordinates were not clear that the manager would last?)

Consequently, the reorganization to decentralize parts of Southern, for example, provided a set of *constraints* on managerial action more commonly than it provided a set of opportunities. For about a third of those interviewed, the reorganization *inhibited* positive accomplishments because it created the necessity to ensure that there were no negative impacts from it. For some managers, both traditional and new-style, there was a great deal of uncertainty around their functions, and they had to spend time calming their subordinates, particularly through face-to-face communication.

A large number of the "significant accomplishments" identified by managers interviewed at Southern involved nothing more than bringing things back to normal in light of the top-driven changes: retaining agents, making sure new procedures were understood, calming fears, and so forth. The uncertainty in the system made nearly all kinds of inventive or innovative accomplishments seem very risky to the managers carrying them out, but there was probably more *avoidance* of risk in this company than in others I have seen. (Maybe the industry and its need to offer "safety" contributes to this.) Traditional managers were particularly likely to report that they had to spend their time keeping things manageable and preventing losses rather than improving performance.

Meridian was less prone to the command style of change than Southern, because of an enlightened cadre of top officers who were very much interested in participatory management techniques. So Meridian was more likely to involve middle managers in making some of the *tactical* decisions around change once a general strategy was decided at the top. But the way this would occur is related to a second kind of change strategy at less entrepreneurial companies: to set up carefully controlled formal mechanisms for planning innovation.

Formal, Restricted Vehicles for Change: "Trials" and Tribulations

Formal mechanisms make "innovation" itself a kind of specialty, done by a specified group of individuals, and neither a routine expectation nor a routine event. These included committees or task forces, special assignments with clear parameters, and "trials," or tryouts, of a top-dictated change. Three-fourths of the accomplishments I saw at Meridian involving new structures or methods rode in on such vehicles, ordered from above.

Yes, innovation can occur in this way, but it is unlikely to be *very* innovative, if the distinction makes sense, because the new products or processes are unlikely to move very far beyond what the system already knows. It can be just as segmentalist and limiting to set up an R&D department if that becomes an isolated unit, specializing in "change" while everyone else specializes in maintaining the routine. Confining "change" to one part of the organization, or to a special role, may restrict its value—and create implementation or diffusion problems.[6] The problem is not with the existence of a formal vehicle itself, of course, but with its tight control and segmentation.

Sometimes the larger part of the manager's accomplishment under a tightly controlled change assignment may be to get through it without *losing* anything. "Ronald Peterson" was one of those who knew that a special "change" assignment might involve being the scapegoat instead of the hero. He had been around Meridian long enough to greet the project with fear and trembling rather than pleasure at a challenge. He anticipated more negative than positive recognition at the end, and he was right; Peterson felt lucky that the "flak" subsided to a modest level soon after he finished. Here is what happened:

Peterson got a call one day in November from his boss saying that he would like him to do a "resizing" study at the request of a vice-president. "Resizing" was one of those corporate euphemisms for cutting back and laying off; American companies seem reluctant to admit they ever fire anyone, preferring instead to "terminate" or "dehire" or "unfund" a job—even hiding a divestiture sometimes behind a term like "fractional acquisition." But Peterson knew what he was being asked to do, and so he did not want to get involved at first. But since the initial request came from a high level, he felt he

had few choices, and pushing his anxieties down, he decided to take
it on. He had the option of creating a committee—which in retrospect might have spread the responsibility and, with it, the flak—
but because of the tight time frame, he decided to handle it himself.
He was given a very scant file of material on the project and some
strict parameters that held him in check: there was no option to
recommend an increase in staff; he should observe three other units
in which reductions had been carried out; and he must be finished
by the end of December. So much for innovation.

As Peterson had guessed, the three weeks of observation elsewhere were a waste of time—but try to tell *that* to the boss in advance, Peterson thought. Furthermore, the numerical formulas that
so appealed to Peterson's bosses had no relevance to the jobs he was
analyzing. So he invented his own method, and then went around to
interview job occupants. He said:

> It spread like wildfire that I was doing this. I was afraid that
> people wouldn't talk to me. But until the study was published,
> people were friendly and cooperative rather than nervous and
> afraid, maybe because the vice-president has so much power—
> positive *and* negative. And I took computer printouts with me.
> Only one manager gave me trouble. That manager needed "re-
> sizing" most.

There was also some politicking. Peterson's boss barraged him
with paper justifying his own staff size, and this was awkward because Peterson had to continue working with him. Peterson had to
maintain his objectivity, though, because his boss was really trying
to justify keeping the most questionable staff.

Meanwhile, Peterson was also being squeezed on the deadline.
The vice-president changed the due date of his report midstream, to
December 15. (Peterson managed to argue for one more week to
finish.) This change was announced to Peterson just at the point at
which he decided he had wasted his time following instructions to
observe others. He became quite worried, and for a few days was
not sure *how* to proceed.

Finally, Peterson handed in his report, eight days before the
original due date. He expected that his recommendations for cutting
staff were going to be turned over to the committee on resizing and
then watered down. But he was in for another surprise. One of the
ground rules given to Peterson was that he was not allowed to reveal
any results until the vice-president saw them; so here was Peterson,
a more junior member of the department, unable to tell his boss or
the department head what he was recommending about their personnel situation. The vice-president did not allow the results to become public until several weeks after Peterson finished, when they

were announced at a series of meetings. To Peterson's surprise, he had to stand up in front of his peers and his boss's peers to defend his study. "These meetings were very emotional; I was raked over the coals."

The recognition he got, he said, was quite negative. After all, he was skipping his own peers, studying his boss and levels above, and reporting directly to a vice-president, in secret—what other kind of reception would this activity get? "There's still some flak," Peterson commented several months later, "but at least not much."

A "trial" was a generally more positive example of a formal change vehicle at Meridian than a tactical assignment to implement a change strategy. Although top-down in initiation and control, it was more likely to unleash the creative, enterprising talents of middle managers. A trial was the adoption of a new idea, generally a work method, on a small scale to determine its feasibility for use throughout the company. A trial would be held only if top management okayed it, and it was often initiated at the top and then assigned to middle managers to carry out within carefully predetermined guidelines.

Running a trial of a pet idea was an infrequent, exceptionally visible opportunity for a manager or professional to show top management his or her skill—if the person succeeded. Empowered only temporarily, the manager was often laying his or her reputation on the line to make the idea work. Occasionally, a mid-level person could help initiate a trial, with sometimes unfortunate consequences. One manager described another's diligent but finally pathetic attempt to prove the utility of a pet idea then undergoing a trial. The manager first used persuasion and cajolery to get others on board. When the numbers showed the idea a failure, he "corrected" them. Later the failure was discovered by others, and his reputation and that of others up to the VP level were stained. Another manager considered his major accomplishment stopping the very trial a colleague had bragged about starting and implementing.

These illustrations tell us that the irony of using the word "trial" for a pilot program may not have been lost on Meridian managers. It was "trial by fire" indeed for many of those chosen for projects, and "trial by jury" for those caught in adversarial crosscurrents.

One of the most successful trials at Meridian involved a mixture of top-down change forcing and middle-manager collaborative/participative action. "John Burke" and "Peggy Holton" worked on this trial, and both considered their parts of it a significant accomplishment. It was initiated after a yearlong series of committee deliberations, set up by top officers, recommending a reorganization to pull apart two kinds of technicians who set up and repaired Meridian's service equipment. John Burke, on the corporate staff, got involved initially in preparing presentations for the committee because his boss chaired it, and Burke was eventually assigned to create a trial to make this change work. It had been tried in another part of Meridian once before, with negative results.

The first issue for Burke was to find a place for the trial. That was not easy. When he first approached the management team in the area he had picked, "they were unfavorable. They felt besieged, and they didn't want to change." It helped only a little that he had talked with Peggy Holton and another second-line manager in this group before the meeting, and they had expressed some support for the idea. The meeting was a flop, and Burke should have known it would be. But the trial really got started when the area department head was given no option by his boss. Burke also felt he had no choice but to pull rank because of an extremely tight deadline:

> There was a stalemate. *My* assignment was on the line. I *had* to escalate it by getting my own boss to call an officer, who had phone calls and one meeting with the department head . . . He accepted the decision, and we went forward. He was very helpful after that, coordinating activities and turning over activities to the two managers I had already interested in this idea . . .
>
> We overcame criticism because there was a prevailing attitude that we would be going ahead with the trial, and that they were there to make it work.

The trial involved a radical change in the way Meridian did business, involving procedures about loading and dispatching service vehicles and a reorganization of work teams. Resources were not a problem; Burke's boss simply went to the appropriate officials and got the money. This was a high-level initiative, and budgeting decisions for new vehicles and equipment, or for staff time, were routine.

Peggy Holton was one of the second-line managers involved in the trial. Holton was interested, with a few reservations, when Burke approached her, because some of the existing procedures struck her as wasteful, and she had long felt performance could be improved. But, she recalled:

> My boss wasn't interested and wanted to avoid it. *His* boss was mildly interested but wasn't that anxious to try it. But John Burke advised me to get involved. We could see the handwriting on the wall that it was gonna be done.

Holton's supervisors and crews were going to use new equipment, packaged in new ways, sent out in new vehicles. Burke's efforts made getting most of it no problem, from Holton's perspective, but there was one kind of equipment that the corporate office was reluctant to have used. Holton said that they "had to lean on corporate real hard," and their persistence in hounding one of Holton's high-level acquaintances finally worked. The acquaintance "dropped the ball six or seven times by making something else a priority. I got mad at him from time to time and finally threatened to escalate it."

Holton was involved in defining and carrying out the specifics of the new system, while Burke put in about a day a week as a kind of "coach." He met with dispatchers, with repair crews, with two groups of first-line supervisors, with managers from neighboring areas. He worked with the union. He got dispatchers together with managers to make decisions about equipment loading. In short:

> I got their input, and it became their project. Their attitude was excellent. . . . There was no blueprint to change; we put it together in evenings and hotel rooms, writing on scraps of paper. . . . It consumed us. . . . We were writing procedures as we went. . . .
>
> It took some positive strokes. We got the local executive together with the workers after three weeks for lunch. The boss told them, "You're pacesetters" and made them feel good. He went out to the loading docks to look around. Then we had a reporter from the company newspaper interview a couple of the workers, and their pictures appeared in the paper.

A fair amount of explaining and persuasion was also necessary to prevail on neighboring areas to cooperate, especially as the trial was indicating how their fields would need to bend under the new system. The best selling point was always that the new system would ease everyone's time pressure. But Holton was still having trouble convincing her boss; "He has a paranoia about his feelings; if something wasn't his idea, it wasn't a good idea." The boss expected measures of productivity—a must in numbers-conscious Meridian—to rise immediately, and they did not, except in one area. That was a touchy relationship anyway. The boss was constantly threatening to take Holton off the promotables list for one or another minor bending of a rule (such as failing to discipline an employee for a minor and understandable violation), but his threats were never credible as far as Holton was concerned because he talked that way to all of his reports.

Supervisors were fairly supportive of the change, but there were occasional breakdowns, Holton remembered:

> It's like the Army and Navy. When they were in a war against Japan, they fought together, but in a bar they'd always fight each other. The same thing happened with the two sets of work crews. Supervisors would be defensive of their group. I'd try to

stay on top of it by meeting every two weeks with the supervisors. Somebody might throw a stone at the boss at those meetings. Sometimes that's healthy, and sometimes not. . . .

But I'd tell them, "Hey, *here's* the objective. We should work as a group. Tell me what you've done to solve this problem." And I'd give them the Army–Navy example. . . .

I got *lots* of new ideas from the supervisors—so many I can't count them.

Burke and Holton both spoke very highly of each other; they had never worked together before. Holton said Burke "gets things done. He was powerful in his last job, too." Furthermore:

You know who's competent, who gets things done, by word of mouth. . . . Most corporate staff aren't respected by line people, but you spend ten minutes talking with John Burke, and you know he's honest, competent.

The feeling was mutual. Burke said Holton was:

Excellent to work with: progressive, imaginative, willing to take risks. She and the other second-line manager participated a hundred percent, though it was a large burden.

Burke also spoke enthusiastically about the project:

It was really fun, exciting, the most fun for me. There were many unknowns, a deadline, we were blazing a trail. We felt *we* were responsible for the decisions, so we'd make it. . . . There was almost instant feedback.

The company newspaper printed a big article about the trial, and most of the other areas implemented the change. Burke "practically went on a road trip to sell the idea," describing it in directors' meetings, with first-line supervisors, to workers, and to the executive committee in the boardroom. "It wasn't a difficult idea to sell." Burke agreed. "The company *expects* you to succeed when they hand you something like this, but we were more successful than they expected."

Thus, a trial was formal and top-driven, but it could provide an opportunity for enterprising managers to design major accomplishments and even to sneak in smaller changes as well. But they had to brave the displeasure of peers and bosses to do so.

Outsiders: Reinforcing a "Culture of Inferiority"

Southern used a third change strategy more than Meridian: to employ outsiders to bring in innovativeness. Southern's recent policy was to "import talent" rather than "promote from within." Five out of six highest-level executives had been in the company less than four years, and 75 percent of the senior officer group was similarly new-but-experienced-elsewhere. Almost the entire personnel function consisted of staff hired away from such companies as General Electric and IBM.

The awareness that managerial talent was being purchased rather than groomed reinforced a widespread perception on the part of Southern managers that the company was not a leader in its industry. There were frequent negative comparisons of Southern with larger insurance companies and a desire by some managers to emulate companies with a high growth rate and high return on equity. The obvious implication of the follower image and the import strategy was not lost on Southern staff: the company did not think much of its own managers. (To be fair, Southern was in the process of introducing extensive new management-development programs— but these too had "outsiders" at the helm.) It is hard to be creative or give more than the letter of the job under these circumstances. What loyalty was expressed by older Southern managers was more often directed toward individual executives than to the company.

Furthermore, expertise was frequently imported in the form of consultants. Not only were consultants heavily used, but they tended often to be handling rather routine tasks. In one case the consultant helped a former government official, himself relatively new to Southern, to figure out something as trivial as how to speed up the photocopying of manuals.

Appearing to expect all innovation to come from outsiders conveys a clear message to the people already in the organization: long service in the company unfits people for change; new talent must come from outside. A "culture of mediocrity" results, one in which people feel that anyone inside must be less able than innumerable others from outside— in striking contrast to the mutual respect which encourages cooperation for innovation at companies like Chipco and GE Medical Systems.

Indeed, there seems to be an assumption in many less entrepreneurial American companies that change can come about only by either a new person, with experience elsewhere, or a new mechanism that is also "foreign" to the system. Perhaps in an attempt to produce more product innovations in the 1950s and 1960s, many companies kept bringing in a series of new research directors; turnover in this position was noted as unusually high, especially in the firms that might have been ambivalent about innovation anyway, as technology expert Donald Schon reported. Furthermore, in industries with low rates of product innovation, such as textiles and machine tools, there was a familiar pattern. Segmentation prevailed

inside (e.g., separation of production and marketing functions, fragmented work areas controlled by different units, top-down management, private-club atmosphere, little external input), with
innovations coming essentially through "invasion." Established
firms in other industries would get into the business, and because
they were less bound by tradition, they often became the source of
product or process breakthroughs.[7]

Such experiences only reinforce the notion that it takes outsiders to make change.

But even outsiders brought in to make changes may be restricted to the "honeymoon" period of the first few years when they
are still sufficiently new, before they are swept into the depressive,
initiative-suppressing culture of mediocrity. Newcomers received
change mandates on being hired—to build, create, improve something—and they could take advantage of their credibility as outsiders and the tolerance for their unfamiliarity with standard
procedures extended to newcomers. The managers with entrepreneurial accomplishments at Southern, for example, tended to be
both younger than the rest of those in my research pool and much
newer to Southern: thirty-eight years old, on the average, and median Southern service of four years for "entrepreneurs," compared
with fifty-one years old and twenty-six years of service for the others.
As a group, they were also judged by one of my researchers as three
times as risk taking and twice as proactive as the traditional managers (92 percent took risks, and 67 percent were proactive initiators,
versus 38 percent of the others in both categories). In short, I found
that these unusual managers' behavior profile was more like that of
all managers at Chipco or GE Medical Systems or Polaroid (but with
the critical difference that at the innovating companies "old-timers"
could be "entrepreneurs" too).

"Jonathan Jones," Southern's most obvious middle-level entrepreneur, was an outsider—one of the group brought in to modernize
Southern—and a veteran of years of successes in a related business.
Even his office looked different; he had pulled some antiques out of
a Southern storeroom by being brash and insisting. There was an
aura of elegance and "class" around his department because even
his subordinates had offices "above their station."

Hired with a change mandate—"Get us into modern marketing"
—Jones developed a new market-research function for Southern,
building a small department into a significant one and developing a
variety of innovative methods that were uncommon in the insurance
business. In the course of introducing new functions and operating
methods, he had also developed a number of people: an $8,000-a-
year clerk quadrupled her salary in two years; a secretary was

trained to be a marketing assistant; and several people he groomed moved away to head their own departments.

Jones began with several empowering conditions: a new function, thus involving a great deal of discretion; undertaking this function in a time of change, so that few established routines or procedures stood in his way; a great deal of relevance to the future business needs of the organization, and thus organizational power and support from the top; and because of all this, a great deal of visibility in the organization. Thus, Jones was one of just a few managers for whom the recent changes at Southern represented an opportunity rather than a limitation. The shake-up and reorganization provided both a vacuum into which he could step and a lack of clear rules or procedures that would prevent him from doing exactly what he wanted. Of course, he could afford to break rules because of his power base and the "honeymoon" period for a new manager.

Jones invented and developed his activities as a result of extensive discussions. He first traveled to various field offices, getting to know the Southern culture while doing other work for Southern. He used his status as an outsider to ask blunt, almost naive questions. He asked people to tell him what they needed to do their jobs better. Then he wrote up several hundred conversations and circulated them to other managers for added ideas. His plans for the new department developed out of the feedback to his memos. Not incidentally, he now had visibility and a support base at Southern.

Jones sought support upward and then mobilized peers and subordinates. After defining the nature of the new department, he prepared a proposal. He asked his boss to let him have a budget to develop his scheme with a six-month timetable for showing results. He received the blessing of his boss and then his boss's boss and finally reviewed the program with the president, who took it to *his* superior in the hierarchy. This approval was based on a detailed account of what he planned to do—and of course, was made much easier by the credibility of his previous experience and the fact that he had already gathered important data on needs from the field and could show the numbers of managers who would back these innovations. He then did something that was unusual for Southern: he used a variety of methods to get other people involved as part of the team. He hired outside consultants, formed task forces, and shared resources with subordinates.

Jones tried to make sure that people felt adequately rewarded for their participation. At meetings, he distributed pieces of a presentation to allow others to share in the limelight. He made special efforts to make the work of other people look good, often giving them credit, so that they would rush the work that he needed or give him critical support for his ideas at critical times. Persistence was important here. Some deadlines were missed, and some projects failed, but he kept going and generally "things got ironed out on their own." There was no opposition to or criticism of what he was doing, because of the careful problem definition and coalition building, but he had to make sure that there was no slippage in schedules from inertia.

When the new department was in place, he publicly shared rewards with his subordinates in order to make his whole department look good. "All of my people have offices beyond their statuses. They deserve it. It's good also for people to know that the boss can get things done."

What was unique about Jones's accomplishments was not only the organizationwide scope of the new function's services but also the degree of high-level support and coalition building across the organization that was involved. Jones could do this precisely because he was not Southern born and bred. His previous experience gave him the personal conviction that he was right and credibility in the eyes of Southern executives, who seemed to value outside expertise more than Southern experience. His status as a newcomer made it easier to move across the organization to ask questions without losing face or appearing ignorant, as a Southern veteran might feel. And he could use new methods without guilt at violating a cultural norm.

But the very factors that made Jones succeed only made longer-term Southern employees more depressed about *their* chances to contribute to change.

"Holes in the System": Unplanned Change Opportunities

Finally, a certain amount of unguided innovation can ride in on the coattails of the more structured forms of change. New systems can create enough confusion, in the most tightly controlled companies as well as looser ones, that a smart manager may find a vacuum, an opening wedge, or a loophole. Routine "holes in the system" also occur—new-budget presentations, leadership transitions—that create opportunities. At Data General, for example, a vice-president commented to me that he thinks that innovative activity is attuned to the budget cycle: more innovations while budgets are still in flux.

The innovators at Southern were often the beneficiaries of higher-level ignorance. They were allowed to go ahead with a significant project because their bosses did not want to be bothered; a frequent comment from the boss was "Do what has to be done."

Sometimes this freedom was granted by default and seemed to be one of the few positive aspects of the reorganization; for example, one manager who was in the midst of a region being created said, "No one knew what was needed, so I got what I asked for." (Confusion had its benefits.) In other cases, "benign neglect" made it possible for some managers to make significant improvements in operations by doing things that would have been forbidden had their bosses or superiors known about them.

At Meridian, a few of the managers with innovative accomplishments grabbed resources during reorganization or turnover of bosses. Of thirteen accomplishments at Meridian that produced new processes, methods, or systems, a change in unit structure, reporting relationships, or bosses occurred in all but two. (These accomplishments were judged by my research staff to be quite risky, and only one was rewarded.) There were also special opportunities caused by changeovers to new systems, such as new budgetary processes. One manager explained with a broad smile, "A three-tiered money-authorization process for computerization allowed the project to be an excuse to put money in your budget."

A new boss could also open a hole in the system—through the boss's ignorance, need to concentrate on getting up to speed, or more responsive style. At Southern, turnover of bosses provided, in a few cases, a superior more receptive to a suggestion for change.

ACCOUNTING FOR INNOVATORS IN UNRECEPTIVE SYSTEMS

In organizations dominated by segmentation and segmentalism, the power to innovate is restricted to a limited number of people under a limited number of circumstances. There are few incentives to move beyond the boundaries of the job unless an assignment from the top demands it, and power tools are restricted. And this is not surprising. There are some who have argued that the organization's keepers act as if their job were to prevent change.[8]

But then, why would anyone at middle levels *risk* taking initiative and behaving entrepreneurially under the kinds of circumstances I have described?

Remember that generalizations do not describe *everyone's* situation. Remember also that some people were favored with the opportunities to get involved in the design of a new product or system, and in those cases, the controlled hierarchy worked to run interference for them while they zoomed ahead to the goal. After all, it can often be easier and faster to implement nearly anything in a centralized, authoritative system, if it does not require much change from others.

A few special people—what companies like to call "superstars" or "water walkers"—in even the most rigid organization can dare to do the extraordinary because of the power of their track record, the sponsorship of key executives, the sheer force of their competence, or the awareness that they are being groomed to step into top slots. Or they may be explicitly designated as rescuers or turnaround experts who come in to fight fires and save lives; in a crisis, rules are suspended, blank checks are handed over, and management makes sure there is cooperation. (Noting this, of course, some managers may avoid putting out or preventing fires until there is a crisis enabling them to grab attention and power.)

Still, if we leave aside the people who were handed a change mandate—who are "entrepreneurs" only in the loose sense in which those with inherited funds can risk starting businesses—it is interesting to look at the situations in which some people will initiate innovation on their own, even in a company where this action goes against the grain. I found three kinds of contributing circumstances in Southern and Meridian that seemed typical.

First, companies may encourage enterprise and initiative in their *rhetoric,* if not in their operations. Meridian, in particular, was espousing a new management philosophy of participation and collaboration in the interest of a more competitive market posture. There were noncynical middle managers who believed this and liked it. Even though executives can often support innovation verbally while resisting it by setting impossibly high standards, or oscillate between support and resistance,[9] any support at all may be enough for a few people. At Meridian a few corporate entrepreneurs felt encouraged by top management's stance to go around a more traditional, preservationist boss to seek support for an initiative from a higher-level manager, and they sometimes got this support. There were public displays of participative processes at Meridian—e.g., a set of large task forces of volunteers contributing ideas for the shape of Meridian's future—which signaled a new era to younger managers schooled in the changing styles of organization that have emerged in the last two decades. Meridian was on the way to developing a more participative organization culture. A visit to Meridian today would undoubtedly uncover a more innovative climate than the one I saw in 1981.

Second, there are often a number of *rewards* for people who "innovate against the grain" that are abundant and not controlled by the management hierarchy above: praise and gratitude from below, or the intrinsic satisfaction of a challenge mastered. Meridian was more forthcoming in rewards from above than Southern was, and so my researchers and I were likely to hear about these, but we were also struck by the amount of enterprising behavior not driven by formal rewards or acknowledgment. Some rare managers shared their gains with their staffs, and they took pride in the spirit that their achievements created. After all, some people play "pinball" in less entrepreneurial companies too, looking for the chance to win a try at a bigger game.[10]

Third, and perhaps most important, many enterprising managers are part of *"dissident" subcultures* in which initiative and risk are a badge of membership. Peer groups are powerful entities, and they can arise, as we well know, even when the structure of the organization does not formally encourage lateral cooperation and teamwork. Relationships with key actors that transcended the bounds of the job were not only a tool that innovating managers used for their accomplishments; they were also an occasional stimulus to enterprise. The personal friendships that people developed over the years sometimes gave them the confidence to suggest something daring; for a few of the older, more traditional managers at Southern and Meridian who managed a more entrepreneurial accomplishment, it was the fact that they had "grown up" with key executives and could count on personal support which allowed them to initiate an innovation. (Of course, individual friendships can be a conservative as well as a "revolutionary" force.)

The largest amount of innovation that was either self-initiated or not directly ordered by the top occurred where there were pockets of a strong peer culture, in a peripheral rather than a core area. Far-distant divisions or areas less important to the core technology or less responsible for financial results may have less power, but they may also gain a freedom to experiment while the attention of management focuses on the central functions. Among the innovating younger managers at a Southern subsidiary in a noninsurance business, there was a greater sense than anywhere else in the two companies of a positive and explicit peer network and culture that people could count on. Managers were more aware of one another's activities and of events in their part of the organization than were any other group of managers in the firm. Signaling their dissident subculture, they tended to describe themselves as "the rebels" or "the hot ones."

Finally, a small number of managerial entrepreneurs in unreceptive environments are what Ken Farbstein called "Lone Rangers," organizational loyalists acting on their *values* to remedy what they see as less-than-optimum situations for a company and a job they care about. A person who is this kind of "bureaucratic insurgent" can be an activist reformer who remains loyal to the organization and its mission while working gradually but persistently to "convert the heathen. . . . He takes advantage of loopholes, skirts the edge of regulations, evades formal orders, and is less than fully compliant when he cannot ignore them." [11]

The Lone Ranger image is a good one; half outlaw as well as hero, the innovator may be ready to break rules to reach a greater goal. He or she may engage in illicit budget transfers, using funds for a purpose other than the official one; hold off-site meetings to raise the morale of the troops even though the company has forbidden it; create his or her own reward systems, as Jonathan Jones did; spend money before it is allocated, like Martha Evans; or even get a product into production before receiving official approval, as I saw an entrepreneurial engineering manager do at a leading manufacturing firm. Companies often derive great gains from those willing to bootleg funds or disobey orders. The IBM 360 system, for example, a $5-billion new product, one of the most important in IBM's history, benefited from Lone Ranger actions. A research lab in San Jose, California, had been told explicitly to stop working on a low-power machine but continued anyway, despite top management's repeated orders. Management then transferred some of the researchers out of the lab to another one—whereupon it turned out that the machine they had been working on solved a technical problem. Ultimately their machine was incorporated into the line and went on to sell better than any other in the series. [12]

It is encouraging to know that a few individuals with an entrepreneurial spirit may send out sprouts, even in otherwise uncongenial climates. Still, as encouraged as I am by sprouts, I would rather see gardens.

"Rules for Stifling Innovation"

Rhetoric of change aside, the people in the middle at noninnovating companies generally do not see top executives acting as though they really wanted enterprise and innovation. There are notable exceptions; newcomers or those involved in a special R&D effort like a "trial" can be very positive about the changes and change potential in their companies. But for most, the message behind the words of the top is that those below the top should stay out of the change game unless given a specific assignment to figure out how to implement a decision top management has already made.

In many instances, the behavior of the top can be understood as though it were acting according to a set of "rules for stifling initiative." Imagine something like this hanging on an executive's wall in a segmentalist company, right next to the corporate philosophy:

1. Regard any new idea from below with suspicion—because it's new, and because it's from below.

2. Insist that people who need your approval to act first go through several other levels of management to get their signatures.

3. Ask departments or individuals to challenge and criticize each other's proposals. (That saves you the job of deciding; you just pick the survivor.)

4. Express your criticisms freely, and withhold your praise. (That keeps people on their toes.) Let them know they can be fired at any time.

5. Treat identification of problems as signs of failure, to discourage people from letting you know when something in their area isn't working.

6. Control everything carefully. Make sure people count anything that can be counted, frequently.

7. Make decisions to reorganize or change policies in secret, and spring them on people unexpectedly. (That also keeps people on their toes.)

8. Make sure that requests for information are fully justified, and make sure that it is not given out to managers freely. (You don't want data to fall into the wrong hands.)

9. Assign to lower-level managers, in the name of delegation and participation, responsibility for figuring out how to cut back, lay off, move people around, or otherwise implement threatening decisions you have made. And get them to do it quickly.

10. And above all, never forget that you, the higher-ups, already know everything important about this business.

These "rules" reflect pure segmentalism in action—a culture and an attitude that make it unattractive and difficult for people in the organization to take initiative to solve problems and develop innovative solutions. Segmentalism may not be the *only* problem of these companies with respect to change; there are also human failures, poor management, or absence of a real drive/need for change. But in general, segmentalism is the handicap. Segmentalist companies may not suffer from a lack of potential innovators so much as from failure to make the power available to those embryonic entrepreneurs that they can use to innovate.

And, as we are about to see, when innovations do occur, segmentalist organizations may not even be able to take advantage of them.

CHAPTER

4

The Withering of the Grass Roots:
The Fate of Employee Innovations
in an Indifferent Environment

> Working with employees on task forces . . . is
> like watching a steam locomotive with a big
> load of cars begin to chug-a-chug. Eventually
> it gets such a head of steam that its momen-
> tum is difficult to slow down. But it is irritat-
> ing when someone subtly refrains from
> throwing the switch that would keep the mo-
> mentum going. In other words, to not follow
> through in a timely fashion in implementing
> management commitments.
>
> —Douglas Wallace,
>   Vice-President of Social Policy,
>   Northwestern National Bank

> The worst sin towards our fellow creatures is
> not to hate them, but to be indifferent to
> them; that's the essence of inhumanity.
>
> —George Bernard Shaw,
>   *The Devil's Disciple*

IT IS ONE of the more anomalous features of American corporate life
that even when a few hardy internal entrepreneurs succeed in producing innovation, officials in the company may not know what to
do with it—or even know *about* it. Companies that are set in their
ways often ignore small advances that can mushroom into larger
ones. In one striking example, an American steel company sold a
fledgling invention to Japan, and then had to buy it back later at a

higher cost, even paying the Japanese an extra fee for training in the related work processes.[1]

Here is a trade-off the potential innovator faces: make it clear that an innovation may have large impact on the organization—and thus mobilize all the defensive resources of other segments with a stake in the status quo; or else take advantage of the freedom to innovate at the periphery, keeping the innovation quiet—and risk being sold off, remaining invisible, or fading into irrelevance.

In segmented organizations, keeping it quiet is often the preferred strategy: each area simply seeing what it can do on its own, in true segmentalist fashion. That is what the Marketing Services Department at the company I'll call "Petrocorp" did in its attempts to improve productivity via employee involvement. The goals and strategies of this project were similar to those I later show working successfully at "Chipco's" Chestnut Ridge plant in Chapter 7. But the atmosphere at Petrocorp was more like that of "Southern Insurance."

Petrocorp is a company that may desperately require innovation in order to turn itself around, let alone prosper. A multibillion-dollar company at the time I studied it, Petrocorp managed a range of businesses concerned with the preparation of raw materials for industrial and consumer use. Over the most recent ten-year period, a time of occasional double-digit inflation, Petrocorp's net profit margin remained almost flat at about 5 to 6 percent of sales, rising only to 9 percent twice, once early in the decade and once owing to a changeover in accounting systems. Such a low profit margin in a capital-intensive industry requiring higher margins in order to maintain investment was a cause of grave concern to the banking and investment community viewing Petrocorp—especially since the company's earnings were temporarily swelled by the sale of assets, and operating margins were significantly lower or essentially wiped out by inflation. While the increased costs of its raw materials accounted for some of Petrocorp's problems, its major competitors faced exactly the same situation but still operated at much higher profit rates, in some cases twice Petrocorp's performance. Some sectors of Petrocorp were doing much better than others because of the introduction of some highly innovative new products; but the return in its core business, on which it depended for over a third of its sales, was way down.

Characterized at one point in a major business publication as a "behemoth" known for its "pratfalls and sluggishness," Petrocorp finally brought in a new chief executive toward the end of the decade who initiated an upper-echelon overhaul to "cut back bureaucracy and encourage criticism from below."

But the recognition of a need for change did not guarantee that anyone would know how to go about it. Only the Marketing Services Department at Petrocorp launched an attempt to increase its productivity through innovations guided by employee teams.

When I first saw the Marketing Services Department, it consisted of 125 professional, clerical, technical, and managerial employees, handling marketing support, public relations, advertising, technical literature, internal employee communications, and meetings and conferences.

The Marketing Services Department reflected Petrocorp's segmentalist style. It was a top-down organization in which the boss was the only link between work areas, and information flowed up and down separate hierarchies. At the top of the MSD was "George Bunce," my pseudonym for the director (reporting to "Sam Marcus," a marketing vice-president), who was affable and charming but thought to be controlling, conservative, and somewhat inflexible by the managers below him. There were three associate directors, "Ron Landers," "Paul Nieman," and "Charles Smith," each of whom was in charge of a small fiefdom staffed with a set of functional managers and professionals, fanning out into four more layers of personnel. The middle managers and professionals included a large number who had reached career plateaus and were generally considered to be at their ultimate level. The lowest-echelon employees were nonexempt secretarial and clerical personnel, almost exclusively female, and a few male technicians; the professionals and managers were largely male. Typically, the women were relatively junior and lacked managerial responsibilities, except for two who managed the nonexempt population.

A cultural gap as well as a status gap divided top from bottom, and the boundaries of specialties separated each section. People were generally uninformed about anything outside their small unit. There was no history of routine meetings of either the directors or the managers, and certainly the entire department had never been involved in a joint meeting. In part because professional territories seemed to divide so neatly, no one had really seen the need for anything but separate baronies, and that seemed to fit both George Bunce's and the company's style. This was symbolized by the physical layout: the department occupied two nonadjacent floors in one of Petrocorp's downtown office complexes, and Bunce's office was on still another floor.

So, except for a certain creative, nonconformist flair on the part of some of the employees involved in design and advertising (the building's only beard resided in the department), and a long-standing avocational interest in people-related matters on the part of two of the associate directors, there was nothing about the Petrocorp Marketing Services Department that made it possible to predict it would be the location for what the corporation's senior organization-development professional called "the most exciting thing we've seen around here in years."

When I visited it the first time, the Petrocorp Marketing Services Department looked like untold numbers of other white-collar units in big, matured corporations. Getting off the elevator at the first department floor, a visitor would see nothing to distinguish the people or the facility from those of its counterparts around the country. The atmosphere seemed formal, rather cool, and slightly passive —even to members of the department itself.

Everything appeared calm on the surface, but there were a few signs of problems facing the MSD. Despite a large number of external awards for professional excellence, the MSD's image inside the company was poor. There were frequent complaints from internal users of its services about slow response time and poor quality. This was troubling to George Bunce for both professional and political reasons. Politically, he knew that Sam Marcus was a productivity fiend, and with an increasing corporate emphasis on cost cutting, Bunce's department might be an obvious target. This was also a matter of great concern to Paul Nieman. As the newest associate director, recently brought in from one of the favored parts of the marketing function to help increase internal services, he saw his own career future looking grimmer unless the MSD's image improved. Furthermore, Nieman, an active, people-centered wheeler-dealer, found it hard to work with the more control-oriented Bunce.

A little over a year later, the department's performance had turned around, largely as a result of a series of efforts to get all levels of employees engaged in improving communication, problem solving, and work methods. The MSD was turning out more work than ever before in its history with a 20-percent reduction in staff—and with higher perceived quality. For the first time, the MSD was getting recognized for excellent work by its internal clients, and there were numerous positive comments from many Petrocorp executives who had not been its supporters in the past. Indeed, three vice-president/general managers of product lines had elected to increase their budgets for communication projects even at the cost of cutting funds elsewhere. Use of internal publications was way up, and an MSD survey of readers indicated a 50-percent jump in satisfaction measures from 1977: from an average of 35 percent indicating the highest satisfaction level, to an average of 85 percent. Professional pride in the department's efforts on the part of its own staff was also high. Both tangible and intangible signs of productivity and quality improvement signified great changes in the MSD. A once-sleepy activity was now bustling with life and energy.

THE BEGINNINGS OF CHANGE: DEPARTURES FROM TRADITION

Marketing Services Department members begin the official history of its productivity-improvement project with the explicit strategic decision by department executives that resulted in a daylong whole-department meeting. But like most organizational changes, the project really had its origins further back, in "prehistory."

The initial impetus for the changes in the Petrocorp MSD did not come from the department's organizational problems or its desire for more innovation and productivity. In the beginning, George Bunce and his three associate directors would probably not even have been able to articulate their concerns and wishes for the department. Instead, the first issue was a personal one: Bunce and Paul Nieman did not get along. So Nieman used his charm and corporate credibility to persuade Bunce and his fellow associate directors to let a consultant Nieman knew come in to help them run better meetings. But Nieman's hidden agenda for the consultant was: Fix Bunce.

That such a slender and personalized thread should lead to major organizational performance improvements is not at all surprising or atypical. The tendency to translate problems in the larger world into discontent with the people close to us is characteristic of all relationships that get into trouble, familial as well as organizational. Thus, the first-named problem is often a person, and the boss is certainly a likely candidate to be the embodiment of all that people feel is wrong with their work or environment. After all, the boss not only bears the burdens of hierarchy; he or she also represents the organization's culture by virtue of his or her success in it; the boss's emphasis and the organization's style are likely to match.

So what George Bunce personified for Paul Nieman was deeper problems in Petrocorp's segmentalist culture and structure. Bunce was a top-down manager who kept the parts of the department separate; the advertising, creative, and internal-operations sections were run as separate baronies by the three associate directors, with little need or desire for contact. Bunce also bowed to the segmentation surrounding the department in how he operated outside it; he worked only through his boss, Sam Marcus, and he was very cautious and conservative, keeping information to himself and avoiding anything that might incur the disapproval of higher management. After all, a good success strategy in segmentalist environments is to stay quiet about problems, make your peace with isolation from information, keep your boss happy, and don't do anything that might call attention to yourself for anything except output.

Fortunately, Bunce liked Nieman's consultant, and he had some agendas of his own: to prove the department's importance to his boss, Sam Marcus, an officer reputed to have the ear of the chairman. After all, the Petrocorp MSD was the sales-support arm of a sales-and-marketing-oriented company; it produced technical material for a wide range of complex products, and it also handled internal employee communications. Bunce could see a power base in the making. So he authorized the consultant to meet with him and the three associate directors, and also to interview a range of employees.

And thus, a small departure from tradition laid the groundwork for what became a major change.

It was immediately clear to everyone—though Nieman still held on to his complaints about Bunce—that there were problems in the organization itself that were being played out among the directors, problems that looked remarkably like the results of segmentalism. Three baronies led to a war of the barons. Minimal communication across department areas prevented cooperation between the creative people and those managing projects for business areas who had to rely on what the creative people did. Bunce's controlling style was in part a function of the total lack of involvement of others in planning or decision making beyond their own narrow areas. So most employees, managers and clerical staff alike, simply abdicated responsibility and retreated to the minimum acceptable performance level, in true bureaucratic fashion. Meanwhile, the nonexempt staff were alienated and took out their discontent in slowdowns and subtle sabotage, two of the ways people who are powerless can fight back.

The meetings with the consultant began to give the directors a new sense of their options. The experience of some joint meetings that tackled departmentwide issues made the four men see that teamwork was satisfying and helped them clear up a few long-standing problems—this small innovation in problem-solving laying a foundation in experience for major changes later. They were now a "team."

The establishment of a working team at the top made all the other department changes possible, and it made the directors feel comfortable with creating more team-oriented vehicles for innovation. They were now ready for a more radical departure from tradition: a daylong departmentwide problem-solving meeting for all 125 employees, nonexempt secretaries as well as exempt staff. This would be the first time all of them had come together—indeed, the first time many of them would have thought about the fact that they belonged to the same department. Furthermore, planning for the meeting would be done participatively, by a task force representing a diagonal slice through the MSD, including the secretarial coordinator and a senior manager, reporting to the directors' group.

The directors varied in their commitment to this step. Nieman was cheering enthusiastically at finally breaking ranks; Landers was on board because it sounded as if it might inject some positive excitement into his dull routine; Charles Smith said little but was privately concerned that chaos was imminent; and Bunce himself conveyed that he was showing his stature as a leader by making a gesture of "noblesse oblige" to the troops—but he was not going to publicize the meeting to Marcus.

"Mass" meetings were not the norm at Petrocorp; after all, segmentation means that only smaller units ever have to get together. The choices of a meeting place for 125 people plus facilitators was between the cafeteria, which cut out a good half of the day, and an auditorium with fixed seating, which meant that people would be glued to their chairs. The auditorium was chosen, but a series of nearby conference rooms were secured for small-group meetings.

The all-employee meeting proceeded in smooth fashion. The directors made general presentations of department plans, and task-force members acted as moderators and recorders for the cross-sectional working groups. The issues identified were not surprising to anyone and rang true: noncooperation across sections, a lack of overall information, little knowledge of job opportunities, lack of training, unclear duties but passivity about taking responsibility because of a fear of encroaching on someone else's territory, and a feeling that talents were underutilized. It also came out that several valued professionals, including two young women, were planning to leave because they thought there was no career future in the MSD. A symbol of the department's problems was hit on: George Bunce's office was two floors away from everyone else's—a seemingly vast organizational distance.

Finally, vocalizing one's thoughts, and getting confirmation from others, is a relief, and it releases energy otherwise used to segment off that knowledge from other areas of experience. People in the MSD were ready to define possible actions, and the suggestions for productive change poured out fluently. The meeting ended on a note of optimism, with a general perception that management was committed to improving both the quality of work life and the productivity of the department. Many new relationships were established, and there was a general recognition that "we *are* all part of the same department."

One tangible outcome of the meeting was the appointment of a Communications Council to manage communication within the MSD. The directors chose six employees and a manager to serve as the chair, and gave the council a mandate to tackle some of the problems that had surfaced at the departmentwide meeting. I think there was a collective sigh of relief, especially on Bunce's part, that the effort to change the MSD could be passed on to the council, and the directors could turn their attention to something else.

But "abdication" was not feasible, because the council got off to a bumpy start. Its chair, annoyed at being "volunteered" for this assignment by the directors, missed the second meeting. A professional, who had been absent from the first meeting, "took over" the second, and the lower-level employees felt railroaded. Then the ball was dropped over who was supposed to call a third meeting, and no one did it. To department employees, the Communications Council was invisible; a quick survey of a dozen managers three months later showed that they had forgotten it had even been recommended at the department meeting. Furthermore, one of them commented that he saw "no visible import from the department meeting; George Bunce should have immediately moved his office down to our floor." The associate directors were no help in resolving these problems; they had no experience with integrative activities, change management, or even simply taking charge; Ron Landers faulted himself for this: "I tend to go passive when things are not working well."

The ball was back in the directors' group's court. The time had now come for strategic decisions to be made that would mobilize the department's resources toward solving its problems and improving its performance.

"Project Opportunity":
Forming a Steering Committee for a "Parallel Organization"

The directors took up the challenge of changing the department in a more coherent way. They went offsite to a resort to work on a plan for what they called "Project Opportunity": a project to improve productivity, innovation, and quality of work life by better matching department work and department people and promoting teamwork. This was now six months after the departmentwide meeting.

By this time, the directors were working together smoothly, and it was not hard to define objectives and a project plan. They recognized that an important next step was to translate their own feelings of participation into change downward. "We currently have a two-caste system," a director commented, "this group and the rest of the department." They decided that they would bring in the next level of management to help guide Project Opportunity, as a top-management Steering Committee. The Steering Committee would guide a "parallel organization," parallel to the regular MSD hierarchy, managing a set of flexible, temporary groups involved in innovation and problem solving—like the one at Chipco (in Chapter 7) or at Honeywell.

The overall plan for the program involved supporting the Communications Council by giving it guidance, providing education seminars for all key managers, and launching three sets of activities.

The first involved *job/task analysis*, an attempt to increase efficiency. Information would be generated on all department tasks to provide the basis for better distribution of work, while helping both managers and professionals to become more thoughtful and reflective about their own activities. What was essential or nonessential? What could be delegated? What should be done as a team?

The second element was called *planned opportunities*, and this was the innovation piece. Special projects would be launched based on suggestions from employees; people could work on them on a volunteer basis. This would give the department a chance to work on new ideas, while recognizing and rewarding the skills of employees who asked to participate, regardless of their formal job category.

The third activity, to come later, was a *survey of employee needs and resources*. Who could do more? Who wanted to do more? How did these skills match the jobs that managers and professionals could delegate?

Project Opportunity was officially launched two months after the directors' planning meeting, with two workshops for 19 key managers: the 4 directors and 15 managers below them. George Bunce was now ready to "go public" with the plan, and so he happily invited Sam Marcus to the first workshop, where the plan was reviewed. Marcus liked what he saw—there were logic and formality and technical content in the plan, as well as a compelling rationale —and expressed his willingness to endorse Project Opportunity.

The work began. The key managers met, talked about the department, practiced communication skills, speculated about the ins and outs of teamwork, and used the job/task analysis forms to reorganize the work of their own sections.

The Communications Council also met, with greater guidance about agendas from the directors and an outside facilitator available to help, and it finally identified tasks other than just talking. In fact, one of the goals of Project Opportunity—to identify new talent and stimulate innovative approaches—began to be met. "Laura Mintz," a rather quiet young woman, a junior professional, who had not said much in the first few meetings, said that she would like to head a task force which would create a central resource library to improve information access and transfer in the department—the first official Project Opportunity task force, and a chance for a woman buried in an invisible staff job to show her skills as a leader. The council also created display areas for department work, a department newsletter, and a "bio book" with photos and descriptions of department members. Ron Landers, the associate director over creative activities and an avid amateur photographer, was brought in by the council to take the pictures.

Now there were tangible signs that things were changing. A third of the department was actively involved in learning, reassigning work, or participating in new projects. Out in the formerly bland corridors by the elevators were the Communications Council's displays, and several times a day people would be clustered by them to find out what was happening and to admire someone's achievement. Meetings of a whole area to discuss plans, priorities, and operational improvements were becoming common. At the key managers' meetings, and at the directors' group meetings, new accomplishments and new performance levels were reported.

THE ENVIRONMENT INTRUDES

But while the MSD was pulling things together, the rest of the corporation threatened to pull them apart.

A period of financial austerity was beginning at Petrocorp around the time that Project Opportunity was officially launched. Initially, the MSD directors were determined to ignore the negative side and use this climate as a chance to demonstrate that their innovations in participative management would produce results in a leaner environment.

The first challenge for the MSD was to keep morale, quality, and output up while losing staff. The order to reduce the MSD size was handed down to George Bunce by Sam Marcus. Marcus wanted the department cut back to 61 professional and managerial (exempt) staff. Seventy-seven had been the entry budget for the year; the head count now stood at 69. The 50 nonexempts would similarly need paring. Marcus put this to Bunce in a jocular statement with a hard edge underneath: "If you guys went out of the golf-contest business, you wouldn't need so many people. . . . If I knew more about your department, I'd probably cut you back further."

Bunce told his associate directors, as they sat around a conference table one gloomy afternoon, "It's a floating crap game out there. Marcus has his pet projects to build up, and the vultures are hovering around." They talked about the company's typical boom-and-bust cycles, with speculation about the people the department would be losing. Ron Landers said angrily, "Just because someone cut a bad deal on widgets, they shouldn't go around jerking everyone's chain and cutting overhead."

Vague worries that there was more to come were confirmed a few months later at the next meeting Marcus held for his direct reports, including Bunce as head of the MSD. The tone was confrontational, as though the company's financial woes were a direct result of these managers' laziness and excesses: "We got the riot act read to us," Bunce reported. New Petrocorp policies were announced. There would be no "resort-type" meetings, travel expenses would be scrutinized carefully, the next Christmas party was off, there would be no moving expenses and no consultants, and executives would no longer be able to "entertain each other at lunch." Three sales meetings were cancelled. Even coffee wagons were eliminated because they "take secretaries' time." Some of these cuts seemed to the managers to be designed to punish—e.g., the plans were posted to revoke employees' telephone credit cards. No one expected that this would save much, but as a spokesperson for top management commented, it was a move to set the stage psychologically for further cuts, since "it hit everyone in the gut."

People would clearly be the next to go. Bunce also learned at Marcus' staff meeting that two of his key executives, including Paul Nieman, the associate director acting as "prime mover" for the MSD's innovations, were being scrutinized for possible transfers.

The announcement of the transfers followed quickly, coming just as rumors of layoffs began to hit the MSD, and interoffice mail began to be stuffed with notices of changes, cuts, and rules about what *not* to do anymore. Professionals and managers complained of receiving memos from officials above Bunce with "cost-conscious, capricious comments about five-hundred-dollar expenditures for core activities like sales brochures." Anxieties were even higher for the clerical staff. A lead secretary said, "We're all losing commitment fast. Jobs are being cancelled right and left, and nothing is being done to replace them. There is no sign of a clear direction. It's very demoralizing."

The rumors were confirmed by what MSD employees read in the newspaper; many felt this was a better source of information than company channels. The business press reported: "Plagued by excess capacity, Petrocorp cut capital spending for the next fiscal year by several hundred million dollars over their most recent projects. The president also predicted a net reduction in the work force." Then, a month later, Petrocorp carried out its first payroll cuts in ten years, mostly by inducing early retirement and "performance-related" firings. This sent shock waves through portions of Petrocorp accustomed to a Civil Service–like atmosphere. Though the president tried to reassure the remaining employees, doubt, insecurity, and pressure reigned. Anxieties were fueled by the secrecy surrounding company plans. Managers who had information about changes were expected to keep it confidential. Consequently, the average employee mistrusted official communications and preferred to believe the worst.

Almost every week there was a new surprise, often resulting in a scramble to reorganize MSD work to fit the change. Information on "consolidations" continued to reach the Marketing Services Department under the bloodless headline "Organization Announcement." One of them told the MSD that, as part of Petrocorp's continuing examination of the businesses in its portfolio, three operations teams were assigned to the business-development department for orderly divestiture, five operations teams were transferred, and two were combined to be added to a third department. The Marketing Services Department program managers with responsibilities in those areas would need to readjust their priorities.

The next Marketing Services Department directors' meeting had a general flavor of pressure and concern. Bunce reported on a meeting with a senior vice-president: the company had made a better profit than expected, but below plan. Petrocorp was suffering because of losses in domestic industries and a general "disaster" in international markets.

There was no expectation, Bunce said, of across-the-board cuts. Any cuts were to be made on a business basis, based on a reassessment of priorities. Reappraisals of individual performance were going on. But inside the Marketing Services Department, people were checking on their retirement status and getting information on their payoff if they decided to retire early. Marcus was developing a list of people to be "encouraged" to explore early retirement. The directors then discussed the ins and outs of demotions and probationary status and the legal issues involved in "encouraging" early retirement. Bunce reminded the group that he had been asked not to say anything about this to other employees.

Much of the tension at Petrocorp around this time was subtle, in the atmosphere rather than in tangible behavior—sidelong glances as people walked through the lobby, barely formulated questions about who would stay and who would leave. One of the directors found himself asking that question in his head after nearly every meeting, and he half-expected every internal phone call he made to get a secretary saying that the person he was trying to reach was no longer with Petrocorp.

THE PROJECT CONTINUES:
THE CRISIS AS GALVANIZING EVENT

Since, true to the nature of a segmentalist organization, no one
had any real information about upper-management thinking, it was
convenient for the department to take refuge in segmentalism too,
and retreat to a preoccupation with internal activities. Project Opportunity began to be seen as a way to be innovative in a lean environment, a way to keep morale up, and a way to demonstrate to the
corporation that productivity could have a human face. Furthermore,
the loss of Paul Nieman without a replacement provided a chance
for reorganization that would better integrate department activities
under two directors working as a team, and the project could help
accomplish that.

At first, then, the financial crisis galvanized the department to
make greater use of its new tradition-departing practices to solve the
problems. So, in another strategic decision, Project Opportunity was
given the new subtitle of "Productivity Improvement/Personal Satisfaction" to signal its connection to solving corporate problems, and
activities continued. Bunce remained a little detached and preoc-
cupied, but the associate directors threw themselves in with enthusiasm.

The pace of events began to speed up, and the project began to
produce a concrete result here and there. The key managers started
to feel more like a team, and communicated more readily outside of
their biweekly Steering Committee meetings. They worked out the
bugs in the job/task analysis, and began to use it to make decisions
about managing their time, cutting out nonessential activities, and
delegating to subordinates. Each section was beginning to reassign
tasks in light of the job analysis, improving efficiency and giving
more responsibility to clerks and junior professionals. Perhaps more
important, sections were holding their own team meetings; each key
manager was transferring teamwork downward. The first Communications Council task force for a central resource library overcame
its initial inability to organize itself and became a working group. It
issued its first progress report, with a great deal of pride apparent
between the lines.

People were being pulled together all over the MSD to work on
issues jointly. The nonexempt secretaries and clerks were in on this,
too. The consulting team met with them to bring their concerns to
the attention of management—and management listened, the first
time any of the women felt they had had a hearing. Secretaries had
been annoyed, they said, at the method employed in assignment to

bosses. They felt that they were not consulted and that they were not given adequate notice about the decisions. Many nonexempts were skeptical that anything would change in the department because, despite the enthusiasm following the department meeting, few changes had occurred for *them*, in contrast to the managers. Furthermore, work was unevenly distributed among nonexempts, and it did not seem fair. Several expressed frustration over the low ceiling on job grade levels. This served as a deterrent to taking on additional responsiblities, since there were no conventional rewards for doing so. And a general, unspecific fear of possible reprisals affected people's willingness to speak up, whether in public or with the nonexempt administrator.

One old-time manager, "Ralph Caldwell," listened to this report with mounting interest. He was close to retirement—"retired on the job," some said—and had not found much pleasure in his work recently. But here was a cause he could get behind and contribute to uniquely, because most managers simply weren't interested. So under Project Opportunity: PI/PS auspices, he formed a task force of nonexempts and managers to develop action plans for solving nonexempt problems. After a few weeks of leading this effort, it was as though the grayness had disappeared from his personality, and he had come alive in full color. "I don't believe Ralph is for real," a colleague commented. "He bounces in in the morning, whips through his mail, and calls his task force with another new idea." And the nonexempts liked the results: career-development plans and improved communication with bosses.

A Philosophy and a Set of Action Vehicles: "This Department Is a Nice Place to Be"

Everything was definitely clicking in the MSD. The Steering Committee of key managers reached another milestone. Like most groups "commanded" to get involved in change, it had begun with passive acceptance or outright skepticism, then moved into a time of more active involvement with an eye to getting the approval of the directors. But it was not until about five months into the formal project itself that the managers began to understand what was happening and internalize why it was important and commit themselves enthusiastically to the changes.

The managers and directors started to see that what they were doing in Project Opportunity had implications for the way they managed *all* of their activities. One of their meetings got into a discussion of "philosophy," and at the next one, Ron Landers brought in a statement he had drafted which he thought might represent the sentiments of the group. Now there was a "department philosophy," owned by all its managers: *We work better when we're doing what we like to do/are good at. Everyone benefits when individual goals can be aligned with organizational goals. Commitment and enthusiasm flow out of learning, challenge, and a chance to contribute. Empowerment (giving influence and access to resources) increases ability to get things done (both giver's and getter's). Most people have better ideas and could contribute more than organizations usually allow. Management style should be "supporting" rather than "controlling."*

Events that had been oddities at the beginning were now the norm: spontaneous meetings, attempts to communicate, teams instead of one-on-one relationships. Another whole-department meeting was announced, with the Communications Council serving as planning task force and moderator. The meeting introduced the new PI/PS subtitle and the management philosophy, results to date were conveyed to everyone, and the process of spreading "planned opportunities" to participate in task forces was introduced. Particularly noted was the praise of clients for improvements in the quality of the department's work and its responsiveness.

The Steering Committee sought employee suggestions for task forces, but they were also particularly interested in having task forces work on some problems that had long plagued them: How well spent were advertising dollars? How could communication goals for profit centers be better coordinated with market area needs? Could a technical career ladder be developed for professionals in the department, so that they could "grow" without becoming managers? Could a solid system of time recording be developed, based on the findings of the job/task analysis?

Volunteering for a task force was not to be a distraction from the employees' jobs, and it was to be a well-managed process. The directors set up clear ground rules for how "planned opportunities" would be handled. First, a member of the department develops an idea for a new project or task. That person gets permission from the directors' group to go ahead with the idea, and either becomes "project leader" or, with permission, identifies another to take that role. The project leader sends out a memo and/or holds a meeting describing the idea and soliciting volunteers to work on it. Respondents react to the idea, get more information from the project leader, and some may volunteer to participate. Volunteers then negotiate with their own manager and the project leader to make the appropriate arrangements (which may require negotiating with co-workers who might be expected to help absorb some of the respondent's original work responsibilities).

There were three principles to be upheld in creating planned opportunities: (1) that the program be voluntary, (2) that people not be penalized for volunteering (subject to their ensuring that their original work continued effectively), and (3) that ongoing tasks be performed with a minimum of disruption.

The next Steering Committee meeting took a whole day and was used to reward achievements and lay plans for keeping up the momentum, especially now that the task-force phase was under way. The corporate organization-development professional, a well-respected Petrocorp figure, was invited to participate in the meeting: as guest, as link to top management, as possible resource for the next stages. One of the central figures in the meeting was Laura Mintz, the young woman who had started the central resources library via a Communications Council task force; she spoke eloquently about what the project had done for her. On the basis of the superior job done by the task force, she was about to be promoted into a job no one would have considered her for a year earlier.

The comments made by people at the meeting spoke to the progress of the MSD—how much hard work had been done to overcome roadblocks and avoid discouragement, and how much change had already taken place:

Nonexempt Manager: This has been a real learning experience for me. There have been real ups and downs in my enthusiasm. How we're going to find the time has been bothering me. This meeting fixes that for me. I'm beginning to get up again.

Charles Smith, Associate Director: I'm dealing with frustration and impatience. Frustration that I don't know enough about the people in the department. Things might have been different if we had started this project earlier. Impatience that there's so much we have the skill to do and on a much broader scale. This thing could grow.

Manager: Collectively this group knows more about this business than any group that could possibly be assembled. That's my bias. We have the potential for being the best organization of its kind in the United States, if not the world.

Ron Landers, Director: I will never forget the expression of a year ago: we're running this department like three duchies. I've been working on this ever since. Making progress. I found out years ago that my management style was called "Country Club" —important for everyone to get along, even if nothing gets done. This project has made me realize we have a way to get a lot more done and be very happy doing it.

Manager: Thinking back, I've seen a lot of things happen. One can become jaded and indifferent. It's nice to be doing this. We have private notions that reflect one's own agony rather than reality.

Manager: This project has shown me that everything we do is "cooperative." The exempts cannot do everything themselves and take all the glory. Bringing the nonexempts on board has been the most important part of this project for me.

Manager: Everything we're trying to do in this project is immediately pulled down when a boss pulls rank. Don't destroy what we've worked on for months with a "flick of the tongue."

Ron Landers, turning to George Bunce: I want to tell you, George, how much I value your real support, willingness to try some things, and consistency—particularly in view of the cutbacks.

Bunce: There's got to be a reason, other than the paycheck, for coming to work. I enjoy the work I'm doing. I think the "personal satisfaction" is slightly more important than the "productivity improvement" in this project.

Corporate OD Manager: I really want to be involved in this project. This doesn't go on everywhere. I'm impressed by what I see here: "Raising hands" to volunteer; helping each other; constructive disagreement; lots of flexibility, energy, and enthusiasm. *You can feel it.* This department is a nice place to be.

About fifteen months after it began, the directors proudly pointed to the results of Project Opportunity's attempt to improve productivity and personal satisfaction: getting more work done, of the same or higher quality, with fewer people; designing a more efficient and better-integrated organization structure; making a better match between people's skills and people's tasks; running a number of special projects to solve departmentwide problems; identifying promotable resources and encouraging leadership among women in junior professional and lower-echelon clerical jobs; "turning on" some of the "deadwood" in the department, whom management had given up on as productive workers and yet who became "live wires" in the project, taking on significantly more; beginning to move the overall management style from "controlling" to "supporting" or "enabling"; opening communication between and among every level of the hierarchy.

The last important task for the Steering Committee, in light of such positive results, was to consider how to "institutionalize" the project activities, moving them from experiment to a way of life. They were aware that they had done more than "change attitudes" or "improve teamwork"; they had a number of structures and action vehicles that must be supported and continued.

THE END OF INNOVATION

Then external events got in the way again. Managing the large number of changes as Petrocorp's financial woes grew began to take large blocks of Bunce's time away from the MSD's own affairs, and because Bunce had not yet fully let the "team" take over for him, his absence was critical. Then the MSD's external sponsorship disappeared for the same reason.

Just two months after the glowing report about Project Opportunity was issued, Bunce was already making gloom-and-doom pronouncements, acting as if all the changes in the MSD were about to disappear. He told his colleagues at a directors' meeting, "We've had time to work on this project to increase performance and opportunities. It's helped us. That time is about over." Most of the meeting was spent discussing an impending relocation of headquarters, a "real problem" since there were still questions which professionals would stay and whether any of the lower-level staff would go with them. Bunce was now spending more than half of his time just on this. He could no longer commit time to the department's innovative efforts to raise productivity.

Next to go was the MSD's slender base of external support. Petrocorp began to shuffle its executive deck. A new vice-president of marketing, "Thomas Fitzhugh," was appointed, replacing Sam Marcus, the Marketing Services Department's previous boss. His background was in operations and product management, and he knew even less about marketing support than Marcus. "What does this mean for us?" department members wondered; the appointment was the single "hot" lunch topic for nearly two weeks. Furthermore, the one other link to top management also weakened. The corporate organization-development professional was no longer available. His time too was taken up with other matters: outplacement rather than organization improvement.

Continuing financial losses plagued the company, and executives redoubled their efforts to slash as much as possible, putting in cost cutters rather than organization builders. Two months after the changeover in marketing VPs, Petrocorp cut its losses on a marketing agreement with a small high-tech company and wrote off several million dollars in payments. This was the latest in a series of sell-offs. Eight small businesses and international interests had recently been sold, another ten to fifteen were targeted for sale, and slashes in capital budgets and employment continued. The business press commented: "The impression given analysts was of an unwieldy giant run amok, plunging into often mindless new ventures. Growth was the theory—but now, with declining earnings highlighting its situation, Petrocorp is mounting a major effort to transform itself into a manageable organization."

One of the ways Petrocorp was handling its personnel cuts was to ask for reappraisal of employees with "satisfactory" ratings on its performance-appraisal forms and to put all those with the lowest grades on "probation"—meaning that they still had a job, but could be cut at any time. This made it hard even for those with "secure" positions to get their work done. One manager wondered out loud, "What's it going to feel like to be here in the building with all those people—God knows how many—on probation but still in their offices? What will it do to us? How will it feel to be here? I don't even know *who* they are. When I walk through the halls now, or go to the cafeteria, I don't know whether I'm looking at people who are staying or who are about to leave. I don't know what to say to people. Going about our jobs is like rearranging the deck chairs on the *Titanic*."

Throughout this period so little information was available, and so many "decisions" were being reported and denied, that the entire organization took on some of the character of *Alice in Wonderland*. Decisions announced one day were countermanded the next. Authoritative statements flatly rejecting the possibility of option A were no sooner stated or promulgated than option A was announced by someone else. At the beginning of a conversation, executives talked about the possibility of a certain action which, by the end of the same conversation, they utterly denied.

One manager, who had taken a two-week vacation at the time people were being put on probation, was astonished at the change. Rumors were rife, jokes were bitter or sardonic, and the most visible sign of activity was the endless scheduling and rescheduling of people as they rushed from one meeting to another, meetings at which, because too little information was available, plans were made that were unlikely to be carried out because subsequent decisions would render them obsolete or inappropriate. And no prospect of getting the kind of information that *would* enable serious planning was in sight.

Thus, by this time, the managers pushing to continue to innovate and improve MSD performance could not even get internal support. People began to cover their career bases by competing rather than cooperating with others, by using slack time to hunt for jobs rather than attend task-force meetings. There was so much uncertainty and insecurity that no one saw any reason to put time into productivity gains or to invest in anything that would take more than two weeks to show a return.

The final curtain fell two and a half years after Project Opportunity was initiated. *The Petrocorp Marketing Services Department was broken up, and pieces of it were placed in three different departments located in the new suburban facility.*

THE POWER OF ENVIRONMENTS

The moral of the story is: It is hard to get people to look for ways to improve the productivity of plowing if the earthmoving trucks are hauling away the soil.

Hard, but not impossible. It is striking that the MSD was able to do as much as it did; that people were willing to rally around and give their all, to experiment, to volunteer, to commit. That gives the events at Petrocorp a tragic underlay echoing other ultimately tragic human dramas. Here were people trying hard, trying to believe in their own leaders, while the world was falling apart around them— and then it turned out that their own leaders were not so powerful after all. People in the MSD were pulling together while forces outside were ready to push them apart.

Of all the reasons for a system's decline and death—atrophy due to repetition of obsolete programs, the vulnerability of newness, loss of legitimacy, or declining ability of its environment to support it[2]— surely sheer indifference is among the most disturbing. In the case of the Petrocorp MSD, few of the usual problems plaguing work innovations applied; only its corporate environment was wrong. There were no issues of hostility or greed, resistance or poor management surrounding the Marketing Services Department's productivity and quality-improvement program. There was no lack of skill within the department and its management, and no failure to establish new structures and mechanisms to support carrying out of the program's lofty goals. Moreover, the project *worked.* But demonstrated success, consistent with corporate goals, was not enough to save it. The truth is: no one at higher levels "killed" the program because of what the department was doing; the fact is, in true segmentalist fashion, no one at higher levels even *cared* about what the department was doing. And that, along with other pressures from the environment, was at the heart of the matter.

The Marketing Services Department experience illuminates the problems of segmented organizational units: areas within departments within divisions within corporations within industries—or in this case, locales within organizations within environments.

The MSD was a dependent cultural island. It controlled only its own culture and not the flow of people or resources or task demands in and out of it. Even its structural arrangements—although there

WHY WE'RE IN TROUBLE

was some freedom there—were limited by what the corporation permitted in job grades and thus task assignments. But while the MSD was dependent, it was also isolated. Its transactions across its boundaries were minimal, confined to specific tasks. Few outside its borders cared that it existed, and for even its most supportive clients the important thing was the direct service provided by program areas and not the unified existence of a new way of organizational life. The MSD was controlled, but it could not exercise control in turn.

The problems the MSD's participation project encountered stemmed from this situation. It could not protect itself against the intrusions by the corporation (which was itself responding to Petrocorp's market problems). And this, in a sense, was appropriate. The Marketing Services Department should not have been able to erect protective boundaries around its activities. That would not be suitable for any arm of a modern corporation, just as it is increasingly inappropriate for corporations themselves to see themselves as "closed systems" or bounded entities impermeable by stakeholders or by the forces of external change. But this simply makes clear the paradoxes in trying to create vehicles for involvement in unresponsive corporations. The settings through which participation and involvement can be made meaningful need to be local units with clear identities. But in centralized organizations, these settings are vulnerable to intrusion and replacement by indifferent authorities. Furthermore, the rest of the corporation's stresses create distractions, and management under pressure cannot take all the steps it should. This is why, for example, a well-designed job-enrichment program in a major bank was a failure.[3] Stress and crises also caused a reversion to more authoritarian management and infighting at a Corning Glass plant trying to be more participative and at General Foods' Topeka showcase for workplace innovation.[4] *mgmt*.

Here is clear irony: *change requires stability*. To make employee participation a permanent part of its operations, any department needs to be able to hold other things still for a while. Time is one of the first requirements for significant long-term organizational changes. There has to be sufficient calendar time to make it work, as well as enough available participant time to engage in planning, communication, and reflection about the appropriateness of job and project activities. There is less automatic or habitual action in any change effort and more action that is the result of conscious reflection. People have to be able to keep at it for more than an organizational moment: to keep trying, learning, and accumulating and transferring experience. For one thing, a flexible skill base—knowing more than one job—takes more time to acquire.

Continuity of people is a related requirement, or else one is constantly beginning again, without progress (see Chapter 9). Change demands new learning, but turnover requires endless relearning—unless the new people come in already schooled in the "new way," which is unlikely until an organization has shifted its whole culture sufficiently. But regardless of what the new people know, changes that increase participation and involvement rest on relationships and communication. And new people require new beginnings, to convey information, to learn to work together.

Change also requires a measure of security and an understanding that the flow of resources will continue and that some areas will work without much attention, easily neglected as energy and attention are directed toward the change effort. When everything is highly uncertain—as we saw in the case of Southern Insurance—it is difficult to invest in or to believe in change, or even to stop worrying long enough to have the extra energy it requires.

The Marketing Services Department lacked all of these cushions of stability. Calendar time for Project Opportunity was itself truncated by cutting of consultant budgets and a demand by Sam Marcus that the department produce quick results. (In the twelfth month Bunce said, "Time is running out. . . .") Mounting pressures from outside the department made it harder for people to attend meetings or to put energy into the project; their time was taken over by external affairs that were even less related to their core jobs. Turnover of key players began almost as early as the project itself. First, Paul Nieman, one of the project's initiators, was pulled out, and then other participants began to leave. A few of them were replaced by newcomers who came into the project in the middle and never did learn what it was all about; the costs of covering old workshop ground again would have been too great. And the presence of new people drove a wedge in old, solid relationships. Finally, the department had little control over resources, rewards, and perquisites. As the corporation removed former privileges and benefits one by one, there was little the department could do to compensate. Anxieties and insecurities began to preoccupy people.

It is not a mere lack of support from the top of the organization that hurts efforts like the PI/PS project, as most analysts argue. It is something more organizationally profound, stemming from the consequences of oversegmentation. The most immediate links between the Marketing Services Department and the rest of Petrocorp were its clients, its boss, and the corporate organization-development professional, who had the ear of top management. All were very enthusiastic about the changes in the MSD, but each link was like those in a spy cell: none to know of the existence of the others or to communicate with anyone but his or her immediate contact.

The clients never saw the department as a whole unit; they

124

WHY WE'RE IN TROUBLE

were involved with fragments, and anyway, they were a diverse and amorphous group themselves. I would not expect a group like that to try to protect a service department, and besides, the cutbacks at Petrocorp were affecting them too. The boss, Sam Marcus, was very positive about Project Opportunity and even recommended that the consultant speak to the officers of Petrocorp, but his own priorities lay elsewhere, and he too moved on before the project could take complete hold and before the decision to disband the MSD. The corporate organization-development professional soon had to spend all his time dealing with the traumas of the cutback; he was not available, even though he wanted to be. And Petrocorp protocol, like that of segmentalist companies it resembles, did not allow members of the MSD to go higher in the corporation to find sponsors or to argue its case. Each area works through its boss. Period.

The kind of participative structure that Project Opportunity represented is simultaneously more adaptive to some kinds of change and more vulnerable to others. The MSD project made it more responsive to client needs, better able to move resources around to work on problems, and capable of carrying out several reorganizations faster and with fewer problems than ever before in its history. The project even insulated the department against some of the shocks of the cutbacks, because the MSD was able to rearrange its people and rematch them to tasks to meet the demand for output maintenance with reduced staff.

At the same time, the new structure was more vulnerable to disruption by other kinds of change. With every reduction in time or turnover of staff, the participative structure was less able to function. A virtue of a highly specialized authoritarian organization is that it is up-and-running faster; the learning curve is shorter. But while faster to respond eventually, participative structures are also slower to get started. Hence, they begin life fragile and vulnerable to an unreceptive environment.

But after sliding into sociological fatalism—the environment dominates, and that's that—we should consider whether there is anything that the MSD could have done differently that might have saved the department as an intact unit. Keep in mind, after all, that the company considered the *functions* of its Marketing Services Department important enough to preserve, but spread through several other areas, so the question was not whether Petrocorp needed its MSD—it did—but rather, how the corporation would treat its component parts. It was the MSD as an integrated task and social unit that was literally destroyed by its environment—and with it, an innovation-producing project.

George Bunce and, by extension, his managers were trapped into reliance on vertical channels, with single links between levels

—the kind of structure we saw erecting roadblocks to innovation at Southern Insurance and "Meridian Telephone." Bunce was the only formal link the department had to external support, and Sam Marcus, in turn, was Bunce's sole noncasual contact with officialdom at Petrocorp, except for the OD professional. Thus, while there was internal coalition building within the Marketing Services Department, there was no equivalent activity outside. The clients who were so positive about the "new MSD" were never mobilized, coalesced, linked together, asked or permitted to play a role as a support group for the MSD's future. Yet the clients were important figures, and they represented potentially important sources of power. Nor were peers in other departments informed about or invited to see the MSD's innovative project, so they too could not help the MSD.

Perhaps it was Bunce's ambivalence about Project Opportunity that showed. Change requires leadership, after all, a "prime mover" to push for implementation of strategic decisions, and when Paul Nieman was removed, the MSD lost its real "prime mover" for innovation. Ron Landers took on this role in a fashion, but he was, appropriately, more concerned with his own section and with internal MSD doings to focus on the department's relation to its environment. Furthermore, Bunce as sole link was also sole information source, and Landers had little information about wider events. So that left Bunce, part enthusiast about the gains brought by Project Opportunity, part Petrocorp product who would bow to the bosses rather than fight for an innovation.

Experience after experience with innovations that fizzle after a bright start, be they new participative work systems or new products, shows that external relations are a critical factor: the connections, or lack of them, between the area initially producing the innovation and its neighborhood and beyond. It appears that workplace innovations do best at the extremes: when they occur in units with total autonomy, so that the area is much less vulnerable to the surrounding environment; or when the innovating units are well integrated with the larger units in which they are nested. The problems come about when there is dependence but not integration. A well-known work innovation at General Foods' Topeka, Kansas, petfood plant that showed great productivity gains with the use of almost self-managed worker teams slid away from its excellent system in part because the plant became culturally isolated and deviant within the company. In one of those familiar cycles, plant people viewed corporate staff as hostile and retreated into defensiveness; bridges were burned rather than built. In a similar case at a bearing plant, the innovators felt confident of hierarchical support but still "failed to establish a pattern of mutual influence" with higher levels.[5]

But if the organization's segmentalist tendencies to pull things apart and keep them separated outweigh its integrative possibilities, then how much can any one set of people do? They can at least be sensitive to the need to manage the context in which they live, aware that the technical details of the innovation itself are often the easier part, to be left to the foot soldiers of change. Corporate entrepreneurs like Tom West, the engineering manager at Data General who "stars" in *The Soul of a New Machine*,[6] or Stephen Fuller, the personnel vice-president of General Motors described in Chapter 11, leave most of the details to specialists on their team, while focusing the largest part of their own efforts on external relations, to protect and fight for the changes they are making. If innovation benefits from a receptive environment, it also gains from innovators who manage the environment, who act on it rather than allowing themselves to be dominated by it. Out of this interaction the fate of an innovation is determined.

The loss to a company from the smothering of innovation is hard to assess very directly. We can only look at the process by which opportunities slip away. A segmentalist organization is less innovating than an integrative one at three points. First, it is less likely that potential innovators will overcome the roadblocks in their midst to innovate against the grain. But if a few do, operating in the same integrative, participative way as their counterparts in entrepreneurial organizations, it is less likely that the organization will preserve or take advantage of their innovations;[7] the sad fact about many stagnating organizations is that they often have more potential innovation hidden in peripheral corners than they ever know about or use. Finally, even if an innovation in work methods or production technique in one area lasts, a segmentalist organization has few of the links and networks that can help it transfer the experience elsewhere.[8] In short, there are three successive screens that block most innovation, though a few heroic people and heroic accomplishments may get through each.

It takes more than a single organizational moment and a single cultural island to bring about change toward a system that can stay ahead of change. But even so, for that one brief moment, and in that one place, there was an organizational Camelot; the grass roots came alive—energized beyond anything in Petrocorp memory. Will the individual participants in such an episode carry the value of participation and the will for experimentation with them into the next job? That's one possibility. Another is that they will be more skeptical, cynical, and cautious about whether their organization really does want to take advantage of the contributions of individuals to innovation.

PART III

PLACES WHERE INNOVATION FLOURISHES —AND WHY

Cultures of Pride,
Climates of Success:
Incentives for Enterprise
in High-Innovation Companies

> They didn't have to name the bigger game.
> Everyone who had been on the team for a
> while knew what it was called. It didn't in-
> volve stock options. Rasala and Alving and
> many of the team had long since decided that
> they would never see more than token re-
> wards of a material sort. The bigger game was
> "pinball." . . . "You win one game, you get to
> play another. You win with this machine, you
> get to build the next." Pinball was what
> counted.
>
> —Tracy Kidder,
> *The Soul of a New Machine*

> "It's like riding a bucking bronco."
>
> "It's like riding a roller coaster."
>
> "It's exhilarating."
>
> "It's fun."
>
> "There's never a dull moment."
>
> "Time flies here."
>
> —Comments from "Chipco" managers

THE GRAY FLANNEL SUIT has so long been considered the entrepreneur's straitjacket that we sometimes fail to recognize how much the

entrepreneurial spirit lives in corporate garb in some leading American companies—and how much those places do to keep it dressed for action.

If segmentalism characterized all American companies, I would indeed be pessimistic about the American economic and social future. But fortunately there is another, infinitely more positive, model of what corporate life can be like. There is a kind of company that encourages the entrepreneurial spirit in its people and contributes to America's leadership in innovation in fields like computers, medical electronics, and aerospace—or in exemplary workplace practices. It is this kind of company that can more readily produce the adaptive response which helps an organization stay ahead of change, shifting posture and resources as circumstances require. It is this kind of company that stimulates and empowers its people to innovate.

I know this from investigating the ways in which the successful companies with high scores on my managerial innovation index, and others like them, differ from places like "Southern Insurance," "Meridian Telephone," and "Petrocorp." At "Chipco," General Electric, Honeywell, and Polaroid, nearly two-thirds of the effective managers viewed were responsbile for an innovation—a productive new work method, organization structure, product or market opportunity, technological process, or policy. Hewlett-Packard and Wang Laboratories, where these findings were checked with a smaller group of managers as part of a larger examination of their cultures and structures, showed a high innovation profile too. But in the less innovating companies, by contrast, not only did a good majority of the best managers fail to do more than their basic jobs but also, when they did innovate, the effort was both less significant in impact and perceived as riskier. It was clear that the 115 innovations out of the 234 accomplishments I examined in depth benefited from the more supportive and empowering environments in the highly innovating companies, environments where the structure and culture is integrative rather than segmented and segmentalist.

A company, after all, provides a manager with more than a title, a telephone, and a desk; it also provides an arena for the exercise of productive power. The nature of that corporate arena helps determine how much the person will contribute to the development of innovation and the mastery of change.

A TALE OF TWO COMPANIES:
DIVERSE ARENAS FOR INNOVATION

I cannot find a single mold out of which innovative companies are formed, even though they tend to share an integrative posture and a set of characteristics I will describe in the chapters that follow. Instead, there are a number of routes to the development of an innovating organization, from the expected—a rather new high-tech company like Chipco that is still close to its entrepreneurial origins —to the less-expected—a large division of one of America's oldest and largest manufacturers, General Electric, the kind of corporate giant not usually associated with an entrepreneurial spirit.

"Chipco":
A Company with "Ten Thousand Entrepreneurs"

"Chipco" is my pseudonym for an entrepreneurial computer manufacturer that has achieved record-breaking success and considerable fame in the years since its founding several decades ago in the post–World War II electronics boom; even in recent years its sales and profit growth has hovered between 25 and 35 percent. It is headquartered near one of those high-technology neighborhoods that have become famous addresses in their own right, like Silicon Valley, California, and Route 128, Massachusetts, but its flavor is totally its own.

In a variety of realms, including its products, its setting, its culture, and its structure, Chipco always seems to do its competitors one better. For example, if it is common among electronics firms to keep physical surroundings plain, then Chipco outdoes the others with the bare wooden floors of the old buildings it likes to convert. The signs of liberalism are also there: bins for recycling paper in the hallways, pictures of employee jazz bands, energy-efficient fluorescent lights in offices, van-pool and car-pool announcements, and a few male secretaries. Just details, perhaps, but revealing: The new generation of high-tech companies often pride themselves not on the opulence they can afford but on their egalitarian systems in which even big bosses have just a cubicle with partitions in a renovated factory; not on the orderliness of their environments but on freedom, looseness, and creativity.

And the best employee-relations policies. Chipco has been cited for leadership in affirmative action; for responsiveness to minority, inner-city issues; and for a variety of excellent personnel practices, from "open-door procedures" to benefit programs—perhaps one reason why there is no union at Chipco. Indeed, progressiveness with respect to the treatment of people characterizes Chipco's industry; of the forty-seven American companies nominated as the leaders in employee practices by the personnel vice-presidents I asked, by far the highest proportion came from electronics and computer industries.

To people in the industry, Chipco is one of a few companies setting the tone; I would often hear a Chipco-originated phrase or expression popping up at another, newer company—perhaps carried there by the ex–Chipco-ite who had come in to help with start-up. Chipco is frequently contrasted with IBM because of the great difference in their styles. A Hewlett-Packard manager once tried to explain H-P to me by saying it was about a "4" on a 10-point scale on which IBM was "1" and Chipco was "10." And I remember being at a seminar with some Chipco people at a large elegant hotel that had an IBM meeting next door to us. At coffee break, the beards, blazers, corduroy pants, and polo shirts of the Chipco men could not have stood out more from the dark suits, white shirts, and conservative ties that poured out of the IBM room.

The fact that three-fourths of Chipco managers in my innovation study could describe an entrepreneurial project is a key to understanding high-flying firms like Chipco. When I came to know Chipco, it was a rapidly growing organization with abundant resources priding itself on its stream of successes. The founding fathers still run the company—living proof of the existence of an entrepreneurial spirit. And a third of the innovative accomplishments I examined at Chipco involved new products or market opportunities, including some involving phenomenal successes, such as doubling market share in a remote region in just over a year by inventing a new product application.

Indeed, the word "entrepreneur" was frequently used at Chipco to refer to the kind of person who can survive and succeed in Chipco's fast-changing environment; ideas were supposed to bubble up, with top management selecting solutions rather than issuing concrete directions. Therefore, managers at Chipco made a point of demonstrating their initiative and inventive capacities. To admit to simply improved performance in a clearly structured job would be countercultural at Chipco, since managers were supposed to be *inventing* their jobs for themselves.

Employees portrayed Chipco with a variety of vivid images: a family, a competing guild, a society on a secluded Pacific island, a group of people with an organization chart hung around it, a gypsy society, a university, a theocracy, twenty-five different companies, and a company with "ten thousand entrepreneurs." Organization charts drawn by Chipco informants often resembled plates of spaghetti more than a conventional set of boxes. Such imagery described many of the striking features of Chipco: its large number of enterprising employees, their interdependence in a complex matrix organization, the emphasis on knowledge and teamwork, continuing vibrant growth and change, and Chipco's sense of its own uniqueness as a market pioneer with a culture of creativity. Its youthful exuberance was aided by a work force with a mean age under thirty.

Middle managers at Chipco typically had two or even three bosses in a matrix organization, making almost every manager accountable in two directions—e.g., for a product and a function (manufacturing, engineering, etc.). This was designed to help managers avoid pursuing the goals of a subunit at the expense of overall company goals. So was the aphorism taught to Chipco people: "Do anything you want to as long as you are predictable, do not fail, and do the right thing." (Chipco people thought they were alone in this, but it was a common expression around the industry.) Both freedom and responsibility were assumed to be available at Chipco.

Some of this was due to Chipco's spectacular growth rates and its pattern of growing by stretching and reshuffling, which made all parts of the organization vulnerable to constant change. Players and pieces of the system were continually shifting; about half of Chipco's work force at any one time had been with the company less than two years, and about 30 percent changed jobs every year. Thus, an obvious arena for innovativeness at Chipco concerned the organization itself. Managers who developed new methods or new structures accounted for about half of all Chipco accomplishments, doing such things as implementing a "program office" of cross-functional teams concerned with a single product, or developing a "phase review" process in engineering, or establishing a cross-product-line pricing committee. Chipco managers themselves sometimes wondered how much of the entrepreneurial activity around Chipco was really beneficial to the company; frustration was occasionally expressed with duplication of effort and difficulty of transferring good ideas across organizational boundaries, slippage of schedules and change of ground rules as new players entered the action, and a certain amount of unfocused entrepreneurship—"Ready, fire, aim!" was the way one manager put it.

To manage such change as a normal way of life requires that people find their stability and security not in specific organizational arrangements but in the culture and direction of the organization. It requires that they feel integrated with the whole rather than identified with the particular territory of the moment, since that is changeable.

Thus, Chipco appeared conscious of itself as a culture, not just a technical system, and took steps to transmit its culture to newcomers in the managerial and professional ranks, through legends, stories, and special orientations at offsite meetings that were like boot camps. Just learning the job was not enough for success at Chipco; one had to learn the culture of the organization as well, and this could often be disorienting for the stream of new arrivals; "It was like trying to swallow an elephant," one recalled.

But while Chipco's flavor was its own, this general problem was shared with other firms in the industry, like Wang Laboratories or Hewlett-Packard. In these companies personnel or organization-development specialists often worried about how to give people a common focus within a shifting structure, especially when the company got too big to transmit its philosophy by osmosis. At Hewlett-Packard, for example, the legendary "H-P Way" was a philosophy involving a commitment to people that pervaded the company. When I asked a number of employees if there was such a thing as the "H-P Way" as articulated in company documents, the answer was an overwhelming chorus of yeses; teamwork, consensus, communication, and a sense of the person's value were the most frequently mentioned elements, expressed in the principle of "management by walking around"—being present, being accessible, dealing with people on the spot in "real time." The "H-P Way" was supported by careful selection procedures, like Chipco's (a candidate for a managerial or professional job at Chipco might be interviewed by as many as twenty people), and by offering virtually permanent employment with excellent benefits to keep turnover low and thus retain a stable work force. But still, H-P was finding it necessary to begin formal training in its management philosophy in order to ensure transmission and continuity.

Chipco was concerned about the same issue. A personnel manager worried that with its increasing size Chipco's "*esprit de corps* might be fraying at the edges." But doing much about this at Chipco was slightly countercultural, since a survival-of-the-fittest myth had elevated Chipco's figure-it-out-yourself-to-prove-you-belong-here system to a virtue. Less formal and less paternalistic than Hewlett-Packard, Chipco asked its middle-level people to be self-starters and to live with ambiguity.

Growth is helpful, of course, in encouraging people to see change as an opportunity rather than a threat; this is one reason that Hewlett-Packard's official statements find value in growth as not an end in itself but as essential to attracting and holding competent people. Chipco's success meant that people felt good to be part of such a dynamic environment where there would always be an interesting problem to solve. Jobs at middle and upper levels were continually expanding in scope and responsibility, and there seemed to be unlimited opportunities just around the corner;[1] a successful middle manager might spend the first six to eight months in a new job totally immersing himself or herself in order to stay on top of it, then look for an entrepreneurial project to carry out for about a year, and at eighteen months begin to think about moving on.

At least this pattern held for some people, particularly when Chipco was smaller; larger size also meant that there were a few left out in the game of musical chairs, some who did not know how to crack the informal network to find another job, and others who wandered a little aimlessly for a time after their jobs were "unfunded" or reorganized around them. (And, as we shall see in Chapter 7, life could be different at the bottom of Chipco from what it was in the middle.)

GE Medical Systems: Innovation Adopters and Nurturers

The setting for General Electric Medical Systems' headquarters is interchangeable with those of hundreds of high-technology firms all over the country: a low, modern building sprawled across a grassy hilltop near a superhighway outside a big city—in this case, Milwaukee. It also shares a track record of sharp, upward growth; a highly professional, scientifically oriented work force; and a consequent emphasis on outstanding employee-relations practices—a necessity to keep growing and keep that work force happy.

There's a fork in the high-tech road, though, that divides the "independents" (more narrowly focused companies in nothing but high-technology businesses, like Polaroid, Hewlett-Packard, Wang, and Chipco) from those that are subsidiaries of diversified companies (like Honeywell's Defense and Marine Operations Group and, in this case, GE Med Systems). Many people find it easy to account for a high degree of innovation in the first instance, and even write it off as a function of the industry, organizational youth, recent entrepreneurial origins, or market pressure for technical achievements rather than internal organizational practices. But the more instructive case is really the second, the Honeywell DMGs and the GE Meds, where innovation flourishes within a larger, much older, and more traditional business. There are certainly abundant examples of how conglomerates have taken over entrepreneurial firms and then bled them or destroyed them (ITT and Avis, for one), so it was encouraging to me to come to know GE Med Systems, where high-tech-style innovativeness gets along rather well with a giant corporate parent.

General Electric is widely regarded as a well-managed company turning in a consistently solid financial performance. In the 1970s GE's earnings grew at a compound rate of 14 percent and profits tripled on a 163-percent rise in sales; at the helm was a financial executive, Reginald Jones. In 1981 a new chairman was appointed, John Welch, young (45) and from a technical background that was said to signify a shift in strategy away from quick profits and back to investments in technology, a shift begun with an increase in reasearch-and-development spending under Jones and an emphasis on building more, smaller-scale innovative ventures instead of just a few large ones like jet engines.[2]

The Medical Systems division, focused on medical electronics, is one of the big contributors to both profits and innovation. Among its core products is the CT (computer tomography) scanner, frequently described as "the hottest medical diagnostic tool." Med Systems' early achievement was a breakthrough design that reduced time and exposure for a scanning. It then bought the scanner business of an English company, EMI, and now holds the leading market share worldwide.

Med Systems' leadership does not stem from technology alone, nor is technology its only realm for innovation. Indeed, it is perhaps marketing and production efficiency, general business expertise, and good management systems that contribute disproportionately to the organization's success. Among the forty-four managerial accomplishments included in my innovation study, 64 percent, almost as many as at Chipco, were innovative achievements in a variety of realms: methods, structures, and market opportunities, including new-product development. Managers reorganized an X-ray manufacturing plant, designed new quality-control systems, masterminded the introduction of a computer monitor for a patient's heart cycle, instituted a new financial reporting system, beat all timetables for a new-plant start-up, and opened up the Japanese market for CT scanners, scoring a "big win" for the company.

What struck me most in my first contacts with a few Med Systems people was a remarkable open-mindedness—willingness to listen, nondefensiveness, and ability to let go of an investment in their own ideas in order to pick up on a different idea that might produce results. The circumstance was one likely to provoke defensiveness in corporate managers: a discussion of the possible causes of high turnover and "failure rate" of women in the sales force (a problem shared, I might add, with practically every other company at that time, though not very many were appearing as concerned about it). Med Systems wanted us to see if the customer environment was a difficult one for women—hospitals, and particularly male radiologists. When we suggested that the first place to look for causes might be inside Med Systems rather than outside it, the employee-relations director listened carefully, decided that that made sense, and endorsed a study. (The study did uncover internal barriers, and Med Systems moved quickly to correct them, with a subsequent rise in women's success rate.)

Later I saw that this ability to switch gears and pick up on someone else's idea was true not just of the particular individuals I had met but seemed to be the attitude encouraged at Med Systems, in its product development as well as its treatment of people. As much as any organization can—given that ambition, competition, and envy are human characteristics—Med Systems seemed to avoid the worst problems of the two extremes that trap people into the "NIH" (not invented here) syndrome: the *bureaucratic trap*, in which powerlessness and the need to defend established territory lead people to resist other people's good ideas; and the *entrepreneurial trap*, in which the need to be the source, the originator, leads people to push their own ideas single-mindedly.

Instead, Med Systems' organizational image seemed to be more that of innovators who applied ideas, who brought others' inventions to use. Results rather than origins were considered the important thing; the realm for competition—and there *was* competition, interdepartmental rivalry, and political maneuvering—was not authorship of ideas.

Thus, Med Systems did not put a premium on invention or originality for its own sake. Its managers seemed to feel comfortable bringing in and applying someone else's idea or building on the foundation left by a predecessor. For example, after deciding to enter a new area, Med Systems might purchase a small company already manufacturing the product and build on that company's ideas, assembly procedures, and facilities. One manager made a habit of visiting trade shows for the purpose of picking up new product designs and applications. And some of the managerial innovations in my systematic study were what David Summers termed "amplified accomplishments," achievements based on adoption of a task suggested or tried by an earlier manager (who had been blocked, had lost interest, or had been transferred).

Perhaps this willingness to be the adopter rather than the inventor stemmed from a position of security. Med Systems people lived within a more stable overall environment than those at Chipco. There was a clear central focus from headquarters (and, ultimately, from GE itself); there were more formal policies and budget limitations than at Chipco; and even entrepreneurial achievements might begin as assignments coming down the line—favored projects consistent with top-management strategy. (Thirty-seven percent of all the GE innovative accomplishments were assigned in some fashion rather than totally self-initiated, compared with an average of 31 percent for the whole six-company sample.) Focus from the top was balanced by local and managerial autonomy, however. Med Systems had a complex matrix structure, linking a number of areas, that required teamwork and lateral communication while providing what enterprising managers saw as "decision vacuums" which they could fill. In short, there was both general direction at Med Systems and some room to pursue one's own path to get there.

Of the other highly innovating firms among my ten core companies, Honeywell was closest to GE in overall characteristics as well as in its innovation patterns. Wang Labs was very similar to Chipco, and Polaroid and Hewlett-Packard occupied a position between GE/Honeywell, at one end, and Chipco/Wang at the other.

Even though it is clearly important to understand each company on its own terms, as a distinctive environment, it is also important to see the patterns that set these high-innovation companies apart from those which smother it. The clues are in the integrative nature of the structure and culture in the more innovative firms.

LEARNING TO LOVE THE MATRIX:
A STORY ABOUT STRUCTURE

Perhaps the most striking thing about the work of many managers in innovating companies is how far it departs from conventional notions of what a *job* is. Received managerial wisdom from the past holds that a well-put-together job should exist in definable organizational space set off from all other jobs, with clarity about whose responsibility each task is; after all, "Too many cooks . . ." Tasks should be clearly specified; people should know exactly what they are being asked to do. And most important, lines of authority should be unambiguous. A subordinate knows who his boss is, and a boss knows exactly who her subordinates are and has the full authority to direct them to perform in the areas assigned to them. There is a chain of command specifying ever-broader levels of decision-making responsibility.

That tidy world of the neat and orderly job in a simple structure barely exists for the middle ranks of an innovating, change-embracing organization. Instead, managers and professionals function in a world that often contains vague assignments, overlapping territories, uncertain authority and resources, and the mandate to work through teams rather than to act unilaterally. But then, creativity does not derive from order but from the attempt to impose order where it does not exist, to make new connections.

This can be a shock to people accustomed to the old kind of

"job." The man I'll call "Ed Quiller," a district sales manager and a well-regarded veteran of thirty-five years of service at Med Systems, was one of them. In 1975, Med Systems reorganized, replacing a simple, old-fashioned structure with a complex one designed to stimulate more innovation and better handle change. And it *did*. But it took Quiller several years to adjust before he could take advantage of the power the new system offered.

Quiller remembers graphically the difficult transition to the matrix and the new skills required of managers. In the "old days" (pre-1975), Med Systems was a conventionally line-organized company. In sales, for example, the sales representatives worked for a district manager, who also had responsibility for a range of specialists in service, product, and business areas. The district manager reported to the regional manager, who in turn reported to a national sales manager, then the division marketing manager, and finally up to the general manager. Quiller was a district manager with a high degree of authority, leverage, and autonomy. But as Med Systems developed more products with increased technical sophistication, and as sales grew, this organizational structure was becoming overloaded and unresponsive. It was impossible for a line manager to be on top of all the technical detail. And so the purpose of the change to a matrix organization was to maintain a balance between a focus on management of product lines and a focus on professional expertise —in effect, to promote more collaboration.

There were two rounds of organizational overhaul. The first split sales and service. Quiller became a district *sales* manager, on a par with a district *service* manager, who used to be his direct report. The second round pulled the business and project specialist areas out, further cutting down Quiller's formal authority and clout. People who had been Quiller's subordinates were now in positions parallel to his.

For obvious reasons, Quiller did not at first see the advantages of the new matrix organization. Not only had it forced him, in one sense, to go begging for support from people he had previously ordered around, but it also seemed to him that everything took more time. And the more Quiller resisted the new system (and conveyed this, however unconsciously, to his people), the more his unit's morale and performance slipped. In 1978 he received his first negative performance appraisal in years. One could not help wondering whether Ed Quiller was being set up for failure, whether the company was making something too difficult out of what ought to be easy. Why leave things so fuzzy and complex, Ed thought, forcing people to "waste" their time in endless meetings, when a simple structure could show all employees their places, and they could get on with their jobs?

The answer lay in what the new organization helped people achieve at Med Systems, especially compared with their previous performance.

Quiller finally got on board a few years after the reorganization. He decided in 1978 that there must be *some* advantages to the new system, and that he was going to figure them out—"I'm an idea person"—and change. He did just that, acknowledging that "we were managing change: I guess that's what to call it." Not only did he catch on to the virtues of the new system, but he adapted his management style accordingly and abandoned his once dictatorial approach ("I surprised *myself* by being able to adapt to change in the system"). When fresh goals were being set for 1979, Quiller even took the opportunity to "rally the troops" around an ambitious goal —to exceed $30 million in sales in the district, and to look for new opportunities to bill for service.

Getting the sales reps fired up was straightforward—at least compared with working with the other parts of the matrix. Sales people still reported directly to Quiller, he thought them an "awfully good crew," and getting their support was really not at issue, because he was, after all, their only boss. (The matrix stopped at higher levels.) But Quiller engaged them in new ways, stressing teamwork and setting common goals, providing a great deal of personal attention, and communicating a new process of planning for sales growth by targeting certain product areas for attention at particular points in the sales process. First, he wrote a memo to all of the sales people in his area, copying the district managers for service and products. The letter had these pieces of what became Quiller's standard pep talk:

> We were successful before. . . . We need something to hold us together. . . . We want to get the advantages of the old and new systems. . . . Let's get behind the new system; it's the only way we can beat our competition.

He then held a series of sales meetings, inviting commercial and service staff too. They came in small numbers out of curiosity at first, and later turned out in force. At these meetings the strategy was laid out, and participants helped to refine it.

Quiller explained and reexplained the benefits of cooperation across the sales/service/product boundaries to people from each function. For example, if sales reps sold enough of certain smaller products in areas where new hospitals and clinics were being built, then enough dollars of equipment would be in those places to justify putting in a full-time service person. In turn, with a full-time service person there, customers would be more willing to purchase the more sophisticated and expensive equipment that the sales reps preferred to sell. "The key was making people realize they were all working together. What appear as conflicting goals in individual areas are really part of a larger goal that should be seen as common and desirable to achieve together." Getting the players from all three areas together at meetings was itself novel.

Quiller also worked up the line to get the people he needed— in one case, a specialist in an important piece of equipment to replace a person who had resigned a few months earlier. He wanted to bring in a technician and to make him a product sales specialist without previous experience as a general sales rep, something that violated organizational policy. He appealed to his boss, the regional sales manager, who in turn wrote a memo to the general manager. The general manager turned him down, but Quiller refused to accept this decision. He came up with a "negotiating strategy" to get the person he wanted. He proposed to install him as a temporary sales specialist for six months, *then* to put him into a general sales slot. Quiller felt convinced that the man would produce, and he was then planning to argue that on the basis of his stellar sales performance thus far, it would be grossly inefficient to put in a new person, and therefore he should be given the job permanently. "So it all worked out," Quiller said. "We have an outstanding management team above me. They don't hold grudges. The important thing is to show you can get results."

In dealing with his peers in the service and monitoring organizations, Quiller had to count on his skills as a negotiator, hoping to persuade them to go along with his strategy. "Dialogue and interpersonal relationships became key."

Quiller had many ideas for the activities of the commercial, service, and product organizations that could support his sales efforts, and he successfully sold most of them to his peers at the district-manager level in other areas. "I didn't have to ask for anything special" in the way of support because "support was already established." Some of these people had been Quiller's direct reports under the prematrix system; he had hired many of them; and in fact, "some people still call me 'Boss.'" The district business manager was a good example. Quiller had trained her; she was his secretary for sixteen years before getting this job. (Clearly the new organization was also creating new opportunities for talented women who had been "stuck" in the clerical ranks.)

Negotiating was not just a matter of arranging for the staffing of special projects that involved the three areas adjacent to sales; there were also hard dollar realities. Every district manager had his or her own budget, but sometimes expenditures made by one area were charged to another. If sales incurred an expense that service should pay, it came out of the service budget. This meant that service's budget could easily be eaten up, because sales had no reason to skimp on it. Therefore, policies had to be established to avoid confrontations, to diffuse financial conflicts before they grew to any magnitude. So most of Quiller's time was spent "dialoguing with the other managers to get what the sales people needed from them: enlisting their support, helping them realize common goals, and accounting for expense dollars."

Quiller was now running "the most successful district they ever had"; it had broken all previous sales records and had exceeded the goal of $30 million. He was recognized for it, although "once I got the ball rolling, I tried *not* to take credit—not to be on an ego trip—because we were all working together." The most gratifying kind of recognition Quiller seemed to get was the fact that his peers respected him and "carried the message to their reports." His reputation was back where it had been in the old days. Official recognition consisted of the largest percentage pay increase he had ever had and getting the highest possible rating in the appraisal system. "It's surprising," he said. "I wasn't really expecting any special recognition. But it meant a lot that the sales record was mentioned in at least two national company meetings."

What would have happened if Quiller had failed? "There was no risk at all," he now claims. "It would have been pretty embarrassing, though. We had publicized the goal so widely that we had really committed ourselves to it. But it was no risk because I knew we could do it."

"One person can't possibly manage alone in a complex, changing system," he concludes. "You need more specialists, a lot more resources. . . . *This new system saved the business.*"

It is not hard to see what had happened as a result of the new organization: it forced Quiller to adopt an integrative, participative approach. Segmentalism simply could not work. To get things done, managers like Quiller had to reach beyond the bounds of their jobs and, as a consequence, work collaboratively with others to produce a higher level of performance. This, indeed, was characteristic of the actions stimulated by the structure and culture of more innovative organizations.

INCENTIVES FOR REACHING

Innovating companies provide the freedom to act, which arouses the desire to act.

The way innovating companies are designed leaves ambiguities, overlaps, decision conflicts, or decision vacuums in some parts of the organization. People rail at this, curse it—and invent innovative ways to overcome it. People wait for the decision makers to do what they "ought to do," and then, in frustration perhaps, conclude, "Darn it, I'll have to do it myself."

My point is that some kinds of uncertainty create opportunities, even if they are perceived as uncomfortable, and even if individuals feel that they are seizing responsibility only because "someone has to," "the organization won't," or "I'll be the one to finally solve this." In this kind of environment, people who are obviously very much attached to their jobs and love the challenge in their work will be the first to tell you all the things wrong with their company; this is the typical pattern at Chipco, for example. Companies that stimulate innovation are not those which hide their problems or solve them in secret at the upper levels before people come in to do their jobs in the morning, but rather those which make the problems available to people at lower levels to tackle. The true organizational loyalists, in these companies, are not the people who pretend they do not see anything wrong, but rather those who shout the defects from every rooftop while trying to pull together a team to eliminate them.

Individual responsibility seems to be aroused in situations in which the organization confronts the person with a problem: getting a job done that's bigger than the person's formal authority. The choices are either to fold in the face of this challenge—about two-thirds of Chipco's very low turnover occurs within the first two years on the job—or to rise to it. (The second choice is most likely when the tools not possessed are at least close at hand, when there is "power circulation," as I will discuss shortly.) To look at this from another angle, people are empowered when they are given not the answers (i.e., a preprogrammed, routinized job designed for comfort and convenience) but the problems to solve.

Thus, incentives for initiative derive from situations in which *job charters are broad; assignments are ambiguous, nonroutine, and change-directed; job territories are intersecting,* so that others are both affected by action and required for it; and *local autonomy is strong* enough that actors can go ahead with large chunks of action without waiting for higher-level approval.

In effect, in the innovation-producing companies, a degree of "independence" from higher-level interference is vested in a local group who are mutually "dependent" on each other. At these companies either there tends to be a note of scorn for higher organizational levels (epithets such as "bureaucratic waste" would be used to dismiss control from the next nonlocal level of management); or the local executive group would actively compete with their own bosses at corporate or at a group level, wanting to beat them to the

punch with pioneering ideas and practices. And local autonomy was
used to experiment. As one innovating manager said, "If my bosses
knew what I was going to do, they would have fired me." "Rules"
existed in the more innovation-producing companies, but managers
felt free to challenge or ignore them. But in the less entrepreneurial
companies, like Southern and Meridian, the pattern is reversed: individual managers are "independent" of peers, possessing their own
clear territory, but are highly "dependent" on higher-level officials.

Broad Change Mandates

Job assignments (new ones or simply those understood as part
of the job) stimulated a high proportion (51 percent) of all of the
innovative accomplishments in my research. Managers did not necessarily have to think up projects by themselves in order to be acting
as organizational entrepreneurs; their enterprise came from accepting the responsibility and finding a way to indeed build something
new for the company.

What is important is not whether there *is* an assignment, but its
nature: broad in scope, involving change, and leaving the means
unspecified, up to the doer. Characteristically, the manager's formal
job description often bore only a vague or general relationship to the
kinds of innovative things the manager accomplished. The more jobs
are "formalized," with duties finely specified and "codified," the
less innovation is produced in the organization. An emphasis on the
"numbers" (a quantitative versus a qualitative thrust in jobs) and on
efficiency also depresses the amount of innovation. "Low formalization" is associated with more innovativeness.[3]

In the companies with high levels of managerial enterprise,
managers frequently had special pieces added to their responsibilities by higher managers "reaching down" to give them a special
assignment. The organization chart itself was not sacred.

For example, it all began for "Fred O'Connor" the day the top
brass at Med Systems decided to develop a new product that would
combine technology from a traditional machine and their faster-sell-
ing space-age equipment. O'Connor, a manufacturing manager for
the traditional product, wouldn't learn about this for several weeks
—not until the company publicly announced its intention to design
and build this machine in time for display at a national trade show
in a year.

O'Connor noted the announcement in his company mail but
didn't think much about it. As a middle manager with two bosses—
the manufacturing department head and the general manager for his
equipment line—O'Connor would generally hear something about

this from one or the other without going out of his way. Then the vice-president of manufacturing—two levels up one of O'Connor's reporting lines—called O'Connor in to tell him that *he* would be in charge of the planning study—estimating time, development costs, and the feasibility of the whole new-product project itself.

O'Connor took it in stride, he remembered later, though this task was just piled on top of a load of other responsibilities without even a budget to draw on. His stride had to be pretty fast, because he was being asked to run to make a lot of top-management promises come true. *They* only had to decide they wanted it; Fred had to make it happen. . . . Fred's matter-of-fact, businesslike tone as he reminisced couldn't quite hide a note of pleasure and high excitement, the thrill of danger accepted and mastered.

Thus was an innovative accomplishment launched through a top-down assignment. Eventually Fred O'Connor, with the help of his supporters, came up with his own plan which contradicted many of top management's initial assumptions—a plan he successfully sold—and he took over the design of the manufacturing process for the new product.

Broad assignments are generally characteristic of staff managers in problem-solving or bridging positions who have a general change mandate to "invent something" or "improve something." The more innovation-producing companies are marked by a large proportion of problem solvers in operating departments who float freely without a "home" in the hierarchy and thus must argue for a budget or find a constituency to please. For them the incentive to enterprise is clear: doing *anything* requires taking responsibility to launch a project.

One of the most dramatic, and extreme, examples of this occurred at Chipco. "Steve Talbot," as I'll call him, was adventurous, and it was a good thing, too, because his assignment seemed a little like being sent off to join the French Foreign Legion. In his midthirties, physically fit and verbally energetic, Talbot was given the job of "operations manager" for his company's Middle East region, a complex with marketing, sales, and service functions. Though the title was lofty and Talbot reported directly to the general manager, he had no budget, no staff, no job description, and only a vague mandate to "explore options to improve performance."

This was not a job for the fainthearted or security-minded—or for those who want easy visiblity to headquarters. But Talbot tackled it the way someone might scout a new area on a backpacking expedition: staking out a territory, then exploring to see what resources were located in the neighborhood. . . . Within eighteen months, Talbot had been instrumental in a rise of market share for the firm in this lucrative region from 35 percent to 65 percent, an accomplishment which the company marked by flying him in to talk to the executive committee and then promoting him to a corporate marketing directorship.

Managerial assignments, then, in the more entrepreneurial companies, sketch out an area and point people in a direction but leave it to them to define their means of transportation and traveling companions.

Intersecting Territories: The Virtues of Forced Collaboration

Initiative is also encouraged by a combination of relative independence from higher levels and relative interdependence among peers across functions,[4] a highly integrative rather than segmentalist focus.

It was lateral relations that counted at Chipco, and almost as much at GE Med Systems. Contrast this with the overwhelming vertical emphasis at Southern Insurance, Meridian Telephone, and Petrocorp, and the absence of effective lateral contact across functional areas. The complex, matrix-type structure of the innovating companies, with its network of intersecting lines on the organization chart, provided an incentive for people to move beyond the box of their own job to seek projects with wider relevance. In a matrix, action in any managerial area is likely to simultaneously affect someone else and to require his or her cooperation for action. And what's more, managers can get it; the proportion of accomplishments receiving cooperation from peers was much higher in the four innovating companies than in the two others (61.1 percent versus 45.7 percent).

Chipco's matrix grew organically; at GE it is a "hybrid matrix" that came rather abruptly out of a period of dramatic changes in the mid-1970s. Under the present structure, Med Systems has three primary areas: functional departments devoted to the core work of Med Systems (e.g., manufacturing, sales); program offices, responsible for coordinating each product line; and support or staff departments responsible for ancillary activities and support of program offices and line departments (e.g., finance, legal, employee relations). For example, at Med Systems an engineering manager would report to a superior in the engineering department and have a dotted-line report to the manager of the program office for the particular product line to which he was assigned. This kind of structure is common wherever there are large, complex projects or products and rapid technical advances[5]—e.g., at Honeywell's various aerospace-oriented divisions. At Polaroid a functional manager may "work for" up to four product-line managers as well as his or her boss.

A ubiquitous feature of the matrix structure at all of the companies is that managers from two or more functions would generally need to collaborate in reaching a decision or taking some action. More so than at any other level, middle managers were the ones involved on a day-to-day basis with the matrix structure. This was frequently characterized as a "dotted line" relationship to another department, which signified a working relationship but not always direct authority.

Decisions were made slowly in the matrix environment. Major decisions usually required a series of consultations; one manager concluded, "Never make a decision alone," implying that one didn't want to have the ultimate responsibility, or if it came down to it, the utlimate blame for having single-handedly chosen a particular solution to a problem. Concomitantly, there appeared to be a great deal of uncertainty over who possessed the proper authority and power to select among possible alternatives in a choice situation. As one manager put it, "The matrix left us begging for someone to make a decision." Frequently, managers in the matrix were unwilling to exercise their authority on their own. Thus, by requiring extensive cross-functional consultation, the matrix diffused authority among a group of managers. In many instances, this opportunity was used in a positive manner by particularly entrepreneurial managers who were willing to take risks and assume responsibility. Ed Quiller at GE was one of them—and his district's phenomenal sales record was a result of the extra collaboration the matrix organization forced on him and the extra opportunity it provided.

Likewise, the matrix structure introduced an element of considerable complexity into the companies which managers could learn to deal with only through experience. One person reported that it required "about three to five years for a manager to learn to work within the matrix." Combined with the high degree of technological complexity that is inherent in products like computers and CT scanners, there is a great potential for problems resulting from both knowledge and power diffusion, unless managers work actively to communicate across fields.

But in spite of the protracted nature of decision making in matrixed situations, managers can accomplish significant tasks in a relatively short period of time after the initial decision is made. More than one manager commented that after reaching consensus, or obtaining permission to proceed, the matrix makes people act in a "highly focused mode" once the team is on board.

It is important to neither glamorize nor criticize a matrix form of organization in seeking the roots of managerial enterprise. Indeed,

complex structures have come under attack in recent years, perhaps
as a reaction to the confusion inherent in a matrix organization
where people have not been given sufficient support or training to
work within it, where dual authority is extended inappropriately, or
where too many assignments are left unclear. Thus, Thomas Peters'
report of the McKinsey Study of "excellent companies" stresses simple rather than complex structures.[6]

But as the aphorism has it, we should mistrust simplicity even
in its pursuit! Under stable conditions, where the emphasis is on
incremental improvements rather than innovation, simple structures
might work best. *But to produce innovation, more complexity is
essential; more relationships, more sources of information, more
angles on the problem, more ways to pull in human and material
resources, more freedom to walk around and across the organization.** One does *not* need a formal matrix structure to do this: looser
boundaries, crosscutting access, flexible assignments, open communication, and use of multidisciplinary project teams would work. So
specifying multiple links between managers in a formal sense
(through showing more than one solid-line or dotted-line reporting
relationship on an organization chart) is merely a way of acknowledging the interdependencies that complex products and innovative
projects require.

In short, a matrix organization is not, by itself, a necessity; it is
one way to accomplish the organizational integrativeness that fosters
innovation. Furthermore, there is good evidence that a matrix organization is only a transitional step in a decentralizing process, helping move toward complete responsibility to a product line rather
than a split between product affiliation and functional reporting.[7]
And, as I shall show in the next chapter, the complexity of managerial relationships encompassed by the matrix, and the other options
for ensuring integration, ought not to extend too far below; the rest
of the organization should enjoy more stability, formality, and singularity of focus in its jobs. Above the matrixed middle manager is
the stability and simplicity of focused plans; in the ranks below, the
stability of simple department structures. The innovating organization may need to be at its loosest—and most complex—in the middle.

* The 115 innovative accomplishments I examined were much more likely than the
routine job or a "basic" accomplishment to need to pull together more "resources."
They tended to require large dollar investments—25.8 percent of the innovatives
versus only 12.6 percent of the basics; to require a great deal of technical information
—78.3 percent of the innovatives versus only 57.9 percent of the basics; to require
political information—58.3 percent of the innovatives versus 31.8 percent of the
basics; and to involve work outside the primary area—35.8 percent of the innovatives
versus 21.6 percent of the basics.

Culture of Pride, Climate of Success

The final incentives for enterprise stem from a company "climate of success"; they are less tangible and more difficult to measure.

First, there is emotional and value commitment between person and organization; people feel that they "belong" to a meaningful entity and can realize cherished values by their contributions. The images Chipco people use exemplify this: "family," "guild," "gypsy society," "society on a secluded Pacific island." There is a sense of uniqueness and joint-ness that is supported by a feeling of being a *member* as much as being an employee. Hence, there is usually more innovation in organizations with more job satisfaction and with less "stratification" (with fewer hierarchical distinctions that carry sharply differentiated rewards).[8]

The second related "intangible" incentive is the company's culture: whether the company's culture pushes "tradition" or "change." Innovators and innovative organizations generally come from the most modern, "up-to-date" areas rather than traditional ones with preservationist tendencies, and they are generally the higher-prestige "opinion leaders" that others seek to emulate. But opinion leaders are innovative only if their organizations' norms favor change; this is why the values of the leaders are so important.[9] Most people seek to be culturally appropriate, even the people leading the pack. There is thus more impetus to seek change when this is considered desirable by the company. Chipco images also convey this sense of the cultural appropriateness of innovation: "a company with ten thousand entrepreneurs."

Pride-in-company, coupled with knowing that innovation is mainstream rather than countercultural, provides an incentive for initiative. A feeling that people inside the company are competent leaders—that the company has been successful because of its people—supports this. Chipco people often point to its extraordinary growth record and frequently make the assumption that anyone succeeding at Chipco must be unusually competent. Polaroid knows that it is the technological leader in its field. Hewlett-Packard prides itself on its people-centered corporate philosophy, the H-P Way, as well as on its reputation for quality, important in its retention of customers. Honeywell's DMG employees report perceptions of very high technical competence throughout the division. And General Electric's medical-electronics sector maintains a high degree of pride in its management and its market success.

Such cultures of pride stand out in sharp distinction to the cultures of inferiority that lead less innovating companies to rely on outsiders for all the new ideas, rather than on their own people. Southern Insurance, for example, replaced most of its top management with new faces from other companies, staffed its developmental areas (marketing and personnel development) with similar imports, and used consultants for apparently trivial projects, rather than seeking—or building—internal expertise. Contrast this with the high degree of stability in the top ranks of the forty-seven progressive companies, especially as compared with their nonprogressive—and less profitable—counterparts: a much greater likelihood that the officers in 1980 came up through the ranks and had spent most, if not all, of their career with that same company.[10]

Or look at GE Medical Systems, which, like IBM, clearly emphasizes and invests heavily in its own management development. Managers are often sent to company training schools for an extended period of management training, regardless of their level in the organization. As a general rule, except for special projects, Med Systems avoids using outside consultants, whom several managers denigrate with the unfortunate term "prostitutes." Talent, they assume, is home-grown. Med Systems people realize that the company has a history of developing products which have been well received in the market, and although it has not always been the one to originate a design or application, the organization has a tremendous capacity for improving, researching, and expanding upon products that are the creation of others. At virtually all levels and particularly in the middle-management range, there is a notable sense of pride in the way the company is run and the impressive sales of most of the product lines. A high degree of organizational esteem is an important and pervasive element of Med Systems' culture.

Individual esteem plays a role too. There is much evidence that success breeds success. Where there is a "culture of pride," based on high performance in the past, people's feeling of confidence in themselves and others goes up. They are more likely to take risks and also to get positive responses when they request cooperation from others. Mutual respect makes teamwork easier. Thus, high performance may *cause* group cohesion and liking for workmates as well as result from it; pride in the capacity and ability of others makes teamwork possible.[11] In an extension of the "Pygmalion Effect" to the corporation, supervisors who hold high expectations of subordinate's abilities (based on independent evaluations) may enhance that person's productivity.[12] Thus, organizations with "cultures of pride" in the company's achievements and in the achievements and abilities of individuals will find themselves more innovative. This is why formal awards and public recognition make a difference—sometimes less for the person receiving them (who has, after all, *finished* an achievement) than for the observers in the same company, who see that the things they might contribute will be noticed, applauded, and remembered.

It is a self-reinforcing upward cycle—*performance stimulating pride stimulating performance*—and is especially important for innovation. Change requires a leap of faith, and faith is so much more plausible on a foundation of successful prior experiences.[13]

A culture of pride seems to distinguish the innovating companies in a number of fields—success reinforcing an attitude that success is inevitable. Citicorp, for example, the number one U.S. bank holding company in assets in 1983, is known for its innovations in the conservative world of banking: the negotiable certificate of deposit, organization as a one-bank holding company, the floating prime rate, and an automated banking system. While people agree that Citibank is "the most ingenious and imaginative of banks," they also say that "self-doubt has a small place in the typical Citibanker's makeup, and arrogance a large one." Perhaps feelings of superiority on the part of Citibank managers are supported by its typically interesting assignments, rapid promotion, and great deal of internal competition, which makes those who succeed feel that they have been tested under fire—like the rites that weld warrior tribes together with pride in their achievements.[14]

Cultures of pride are undeniably easier to develop and sustain when the company's internal "climate of success" is reinforced by marketplace success, just as in the case of other kinds of organizations whose internal pride is fed by external prestige. As I commented in a *Fortune* article about Tandem Computers, one of the spectacular performers of the last half-dozen years, the company's success can make it feel magical, its leader charismatic, its employees touched by a special "gift."[15] But external success is not the only factor in a culture of pride; there are at least two other important elements reflected in the companies possessing this.

The first is that their progressive practices with respect to people, the people-centered focus of a Honeywell or Hewlett-Packard, make the people inside feel important—not just well treated, but *important*. They are important enough to warrant the time and attention of top management, to be the focus of statements of philosophy, to be the object of interest and concern. This importance that employees feel is reinforced by the second element: the fact that the company's employees are assumed to be capable enough that when there is a problem or a need for change, the company will *look inside first* for new ideas, will provide opportunities for its own membership to become involved. Even an organization in trouble could reinforce the importance of its people in these two ways—and thereby stimulate innovation that might help relieve the trouble. (I keep thinking of Delta Airlines in this regard, one of the forty-seven "progressive" companies and long known for its promote-from-within policies. In appreciation for an 8-percent raise and a no-layoff policy despite operating losses, employees in the fall of 1982 began voluntary contributions to buy Delta a Boeing 767.)

Thus, when Honeywell DMSG managers talk about their pride in the company's, and especially their own groups', people-centered environments, they make it clear that the culture of pride is reflected in people's willingness to cooperate with each other, to help them along. These comments, which arose spontaneously in interviews (we did not ask directly for feelings about the company), are indicative of the views of forty-four managers:

> "This is more humane than other places I've worked, and it's well managed. Other places are more secretive, autocratic, with less room for entrepreneurs."

> "There's a concern with honesty and fairness, from the CEO on down."

> "It runs like a small organization—a little community with its own culture and celebrations."

> "The division has a very open climate. People *want* to cooperate. The strength of an idea will win people over."

> "Managers will let go of employees to let them advance."

And thus, the treatment of people, the attitude toward employees, is an important part of a culture of pride.

Cultures of pride, coupled with cultures of change, provide the context encouraging managers and professional to reach for new, improved ways of operating. But so far I have been emphasizing the structural (job and organization design) and symbolic (culture) rather than the material (rewards). What role, if any, do these more conventional "incentives" play?

The Role of Rewards

The incentives for initiative I have discussed so far are situational rather than material for a good reason. Hope of obtaining conventional rewards seems to play very little role in stimulating innovativeness.

The formal reward system—at least in my research—is a tool used by enterprising managers to encourage others to get on board more often than it is an incentive for them to begin their projects.

Innovators, in their self-reports, were more likely to share rewards of any kind below than to receive them themselves as a direct result of their accomplishment. In the companies with a great deal of entrepreneurial activity, managers frequently could see no relation between eventual rewards like promotions and salary increases and their most significant accomplishments—but, unlike managers in less innovative companies, they also did not seem bothered by this. People tackle innovative projects because they have finally received the go-ahead for a pet idea they have always wanted to try, or they feel honored by the organization's trust in them implied in such a big assignment, or they simply want to solve a problem that will remove a roadblock to something else they want to do, or they take pride in their company and cannot sit still while a problem continues. They do *not* take on this kind of effort because a trinket is dangled in front of them that they can win.

Rewards at Chipco, for example, were theoretically abundant but bore almost no relationship to the size of a manager's accomplishment. As one manager said, "It's possible to get praise and criticism from two different sources at the same time. You have to maintain your objectivity." The managers rewarded most highly (with new job opportunities, good performance reviews, or stock options) generally had minor or marginally successful accomplishments, or good performance on a task clearly called for by the manager's job description. Many accomplishments were not rewarded directly at all. Others were favored with small or informal rewards, or received rewards that may well have been unrelated to the accomplishment. Seven of 34 managers were rewarded with job opportunities, good performance reviews, or stock options; 12 went unrewarded in any formal sense; another 7 received small or informal rewards or received rewards that may well have been unrelated to the accomplishment. Yet the managers, who had usually been promoted several times in their years at Chipco, seemed very well satisfied with the company.

The nature of rewards at Med Systems and the role that they played in facilitating the initiation of accomplishments is also confusing, if not paradoxical. Consistently, the comment was freely made by managers that rewards offered by the organization were typically *not* commensurate with the accomplishment. Frequently, the rewards were delayed or were substantially more informal than expected. Out of 41 managers, 20 felt they were rewarded formally and 19 said they were rewarded informally. Formal rewards included increased salary or a promotion, in combination with a positive evaluation/appraisal in connection with the accomplishment. Informal rewards included a positive evaluation without other action, or an oral comment of praise or "pat on the back." In two or three cases the reward was considered to be so informal by the manager interviewed as to not truly represent a reward at all.

In only a few instances were the rewards provided at a high level or on a national scale (e.g., receiving a national management or technical achievement award along with an increased salary). One very successful manager downplayed the role of these kinds of awards, commenting that Med Systems "was like the Army and would give an award for anything, even if a guy walked around for two years without bumping into anything."

It is clear that any form of reward is not expected to be equal to the time and effort put in for it; hence the motivating factors did not come from a conventional reward structure. Instead of "rewarding" people in the traditional sense—after the fact, after a demonstrated achievement—the innovative companies *"invested" in people before they carried out their projects.* The rewards often consisted largely of getting to carry out the project in the first place, and then, at the end, getting the budget or the assignment to start an even bigger one. The manager responsible for the team successfully designing a new computer at Data General likened this to "playing pinball"; the reward is doing well enough to win a free game—to get the chance to play another one.[16]

Thus, the reward system in innovative companies emphasizes *investment* in people and projects rather than payment for past services—for example, moving people into jobs for which they must stretch or giving them resources to tackle projects they define.

The other important conclusion about rewards is that they occur throughout the accomplishment process rather than at the end. In the initiation stage, the incentive is the very chance to "do something" itself. Later, there are rewards from teamwork. And at completion, as we shall see, many of the most meaningful rewards are those which put people in a position to begin the accomplishment cycle again on a larger scale: enhanced reputation or enhanced position.

The conditions I have described in the highly innovating companies are a far cry from the innovation-smothering arrangements in their segmentalist counterparts. Indeed, the freedom to innovate given to managers, the investment in people supported by a culture of pride, contrasts sharply with the production of powerlessness by any number of American companies. The example of companies like Chipco, GE Medical Systems, and Hewlett-Packard, whatever their other problems may be and regardless of how far they still have to go to realize all their values with exactness, provides hope that American companies *do* know how to create the conditions for more innovation. They *can* be more responsive to their people, who, in turn, can create greater responsiveness to change.

Remember that the distinctions I am making between the more innovating and less innovating companies, or between integrative and segmentalist tendencies, is a matter of degree rather than of kind, a matter of relative superiority—somewhat more of this kind of practice, somewhat less of that. And much of this balance is in the hands of those leaders who set the terms and conditions for the work of the managers just below, who set the cultural tone and design the organizational structure. The culture and structure that results, in turn, encourages or discourages managers to reach for higher, more innovative levels of achievement.

But this is just one part of the innovation equation: the desire. The other part is whether the power to innovate will be available when managers need it.

CHAPTER

6

Empowerment

> I think it's impossible to really innovate un-
> less you can deal with all aspects of a prob-
> lem. If you can only deal with yolks or whites,
> it's pretty hard to make an omelette.
>
> —Gene Amdahl,
> founder of Amdahl Corporation
> and Acsys, Ltd.

I HAVE JUST DESCRIBED situations in which people are not told exactly what to do, and do not have full authority to do it anyway—and I have said that those circumstances, when supported by a culture of pride and change, produce more innovation.

This hardly seems like a prescription for corporate success. Among other things, it contradicts every orthodox principle about job clarity. So clearly we have to add a second set of conditions for successful managerial innovation: the factors which ensure that people can indeed get the power they need to innovate.

I showed in *Men and Women of the Corporation* that people performing tasks involving discretion, visibility, and relevance to "critical contingencies" (pressing organizational problems) find it easier to attract the credibility that brings power.[1] My interest then was in demonstrating how differences in where *individuals* stood in the organization, by virtue of their job design and location, affected their access to power. I was not particularly concerned with how available power was in general. Here the problem is slightly different: *What is it about some organizations that makes power more widely accessible?* What conditions help information, support, and resources to circulate rapidly, so that managers can "grab" and use them?

Even to speak of "grabbing" power conjures up the wrong image, I should hasten to say. In innovative companies, entrepreneurial managers use a process of bargaining and negotiation to accumulate enough information, support, and resources to proceed with an innovation. This is not a matter of domination of others—winning over them and cutting them out—or of monopolization of resources, but rather of coalition building to persuade others to contribute what they can to the innovation's launching.

At "Chipco," managers gave this process a name—"buy-in"—and showed how it served as a control against abuses of power. The very possibility that managers could attract power tools from peers served as a check on their use of the power. Several Chipco managers offered these details about buy-in at a training program where the subject came up:

> A working compromise is better than an optimal solution poorly implemented—this is the spirit of the buy-in.... The process itself, which gives access to ideas and knowledge, is beneficial. ... Do your homework.... You must estimate the proper level of effort needed to get buy-in, and must get key groups and key individuals to buy in.... Use dual-authored memos.... Be clear on goals and objectives.... Buy-in is not a barrier.... It's okay to refuse to buy in, as long as your criticism is constructive. You don't need buy-in on everything.... It's hard to tell when you have buy-in, but it's easy to tell when you don't.... If you can't get buy-in of a missing actor, proceed, and be sure to get his agreement later.... Buy-in reduces some risk, but managers must act at some point; time matters.[2]

Buy-in was an unofficial, informal, but well-known pattern at Chipco, part of the folklore.[3] The need to get buy-in was itself a check that was designed to screen out chancy or unneeded ideas. One manager was sanguine about his inability to get others to buy in to a pet idea, explaining that if he couldn't get their agreement, it showed that he didn't deserve to proceed. Elaborations on the buy-in process involved several steps, which could be independent or part of the same power search: "preselling," "tin-cupping," "sanity checks," and "push-back." Each of these steps, as elaborated by my researcher, Ken Farbstein, reveals the checks and balances involved in the circulation of power.

From talks with others the middle manager has learned who fits into his network, who is competent, and who should be bypassed. "Preselling" the appeal of an idea to one or two key people in one-on-one meetings was done by managers in every department. Once the idea is bought into in this stage, the manager widens the coalition with further one-on-one talks with managers whose support is deemed necessary. Then a meeting is held, ostensibly to gather support, but in fact to demonstrate it and test the competence of new players. One manager explained:

After you've been around here, you know people's track records.
But with new players it's harder to determine their abilities. So
there are a helluva lot of meetings, probing in areas you understand to see if their understanding is the same as yours . . . we
live in meetings.

The step after preselling is a "sanity check" with someone older
and wiser, and possibly outside the department. The sanity check is
a highly legitimate test of the reasonableness and relevance of an
idea whose enactment is in progress. The next step, "tin-cupping,"
is the process of knocking on the doors of managers' offices and
seeking money to fund a pet idea. In its purest form, a manager
solicits a product-group organization for money—e.g., from manufacturing to enable a manager in engineering to build prototypes;
money can also be obtained from clients or customers. (Farbstein
commented in his report to me, "As any beggar can attest, tin-cup-
pers have cash-flow problems.") One manager explained:

> Tin-cupping from users brings in money too late to help. Tin-
> cupping is hard with long-range projects. You have to get the
> right people in, and get enough money up front. . . .It makes you
> susceptible to a fire elsewhere. . . . You have to expect change,
> ups and downs. It's like living with variable interest rates—
> payments may change with the prime rate. . . . You have to be
> optimistic. . . . You can go out of business if no one needs your
> product right away. The more centralized you are, the harder it
> is to appeal through tin-cupping. . . .

The difficulty in tin-cupping is aggravated by the combination
of the oral nature of almost all agreements and the constant change
in the players as they are hired and promoted. New managers must
be persuaded to honor the monetary commitments of their predeces-
sors and must be taught to tin-cup for themselves. We were told it
takes six months to a year to learn tin-cupping. Since folk wisdom at
Chipco says managers should move on to a new job after two years,
this provides a small window of competence. Managers using buy-
in did not necessarily need more money to proceed than did other
managers; buy-in is therefore used more to gather support than to
gather money.

"Push-back," another check on the power-acquisition process,
is a generic term for an expression of disagreement, often by a manager needed on the team. Examples of push-back include telling a
manager he's not the right one to bring about a given task, or giving
someone a small amount of seed money and telling him to try to sell
the idea. Another form of push-back, exercised at meetings, is to say
"The CEO says—" or "A senior vice-president thinks—" which
really means, "This group is about to do something I don't want

them to, and by claiming I know that top management thinks they shouldn't, I hope to stop them."

The buy-in cycle at Chipco elaborates an informal, peer-oriented process of power gathering that would be recognizable by any corporate entrepreneur. The climate that makes this work—reaching outside of and beyond the authority of position to develop an idea for a change—is one in which power tools are locally available and those who control them can be persuaded to invest them in an innovative effort.

POWER TOOLS

Organizational power tools consist of supplies of three "basic commodities" that can be invested in action: *information* (data, technical knowledge, political intelligence, expertise); *resources* (funds, materials, space, time); and *support* (endorsement, backing, approval, legitimacy).

To use an economic analogy, it is as though there were three kinds of "markets" in which the individual initiating innovation must compete: a "knowledge market" or "marketplace of ideas" for information; an "economic market" for resources; and a "political market" for support or legitimacy.[4] Each of the "markets" is shaped in different ways by organizational structure and rules (e.g., how openly information is exchanged, how freely executives render support), and each gives the person a different kind of "capital" to invest in a "new venture."

We can hardly speak of "markets" at all, of course, where the formal hierarchy fully defines the allocation of all three commodities —for example, when money and staff time are available *only* through a predetermined budget and specified assignments, when information flows *only* through identified communication channels, and when legitimacy is available *only* through the formal authority vested in specific areas with no support available for stepping beyond official mandates. In companies where there is really no "market" for exchanging or rearranging resources and data, for acquiring support to do something outside the formal structure—because it is tightly controlled either by the hierarchy or by a few people with "monopoly" power—then little innovative behavior is likely, as we shall see later. On the other hand, the "market" for information, resources, and support is not totally free and open in a corporation either, and even corporate entrepreneurs can find some portion of the power tools already attached to their positions, available for investing in an innovative project.

Indeed, managers who believe in their projects eagerly leverage their own staff and budget or even bootleg resources from their subordinates' budgets. But typically, innovations require a search for additional supplies, for additional "capital," elsewhere in the organization—and innovations thus are ultimately integrative, requiring connections beyond predefined categories.

Three broad aspects of the operation of innovating companies aid power circulation and power access. *Open communication systems* help potential entrepreneurs locate information that can be used to shape and sell a project. *Network-forming arrangements* help them be in a position to build a coalition of supporters. And *decentralization of resources* helps them get the resources to use to mobilize for action. These three clusters of structures and processes together create an empowering, integrative environment.

OPEN COMMUNICATION: AIRWAVES FOR INNOVATION

The innovative managers agreed that the most common roadblock they had to overcome in their accomplishment, if they faced any at all, was poor communication with other departments on whom they depended for information; at the same time, more than a quarter of them were directly aided by cooperation from departments other than their own as a critical part of their innovation. Therefore, a communication system, depending on the kind adopted by a given corporation, can either constrain or empower the effort to innovate.

The most entrepreneurial companies I looked at—the Chipco/ Wang rather than the GE/Honeywell end of the innovation continuum—generally encouraged face-to-face information sharing in "real time"—that is, at the moment the issue comes up. Chipco thought this so important that it developed its own transportation system to move people between facilities, on the theory that in-person meetings work best. In other, similar companies, lack of reliance on support staffs for message taking and typing also facilitated live communication. There were departments with only one secretary for the entire twelve-person staff—cutting down on bureaucratic paper generating, if nothing else.

Examples of "open communication" systems from innovating companies stress access across segments. "Open door" policies mean that all levels can, theoretically, have access to anyone to ask questions, even to criticize. At Wang Labs I was told that there is even a policy that all meetings are open, that anyone may attend any

meeting. Such norms acknowledge the extent of interdependence—
that people in all areas need information from each other. Furthermore, parties and other social events, like Tandem Computer's beer
parties by the company pool or Chipco's baseball games, ensure that
people become known to each other outside of their job roles, facilitating contact back on the job. And sometimes the desire for keeping communication live gets absurd. For a while at a division of one
of these companies, there was a paging system rather than intercoms
on phones, so that people who needed to could reach one another
instantly; but this caused so many distractions for "innocent by-
standers" that it was finally replaced by a telephone system, which
was quieter but also made instant communication less possible.

"Open communication" may mean that problems as well as successes cannot be kept secret and that public punishments occur—
often of people who themselves *failed* to communicate. One marketing manager created brochures for a new division and went off to
produce them without consultation with other department members.
The brochure was beautiful, but "out of sync" with company standards. The division general manager called him to account for this
in front of peers and upper-level managers, and his resources were
frozen. The penalties for secrecy or hoarding of information serve as
a warning to others.[5]

"Openness" at such companies is reflected in physical arrangements as well. There may be few "private" offices, and those that do
exist are not very private. One manager had a "real" office enclosed
by chest-high panels with opaque glass, but people dropped by casually, hung over the walls, talked about anything—and looked over
his desk when he was not there. In general, people walk around
freely and talk to each other; meetings and other work are easily
interrupted, and it is hard to define "private" space. They often go
to the library or conference room to "hide" to get things done, especially on "sensitive" matters like budgets.

There are two problems with "open communication" that middle managers frequently encountered, however: "underload" and
overload. The "underload" problem is when people do not keep
information circulating out of ignorance. The existence of such an
informal system encourages people to think that everyone knows
everything, and so they either fail to pass something on (on the
assumption that it is known already) or take authoritative-sounding
misinformation passed on through informal channels as the "truth,"
and do not bother to check out rumors. At Chipco, people who received unwarranted "bad press" for a supposed mistake sometimes
found that the incorrect information kept moving through the system, with others rarely bothering to seek to add the other side of the

story. At a facility that was visited just before an impending move of part of the staff to a new location, practically all employees seemed to have information about when and how this would happen and claimed that their information was confirmed; the only flaw was that their stories contradicted each other.

But perhaps people did not seek to correct "underload" because of information overload. Managers and professionals felt burdened by inessential communications that were simply cast upon the organization rather than targeted only to the people who should get it. "We hold a meeting where a memo would do," a manager complained, in a lament echoed frequently at the more innovating companies. (But complaints about this should be weighed against the better performance that can result from fewer filters eliminating useful data that might be viewed as "noise" for the filterer.)

Still, for all the drawbacks of open communication, from lack of privacy to slippages in its very informality, it served a very important function for the potential entrepreneur. Information and ideas flowed freely and were accessible; technical data and alternative points of view could be gathered with greater ease than in companies without these norms and systems. And thus both the "creative" and the "political" sides of innovation were facilitated.

NETWORK-FORMING DEVICES: ENSURING A SET OF SUPPORTIVE PEERS

Corporate entrepreneurs often have to pull in what they need for their innovation from other departments or areas, from peers over whom they have no authority and who have the choice about whether or not to ante up their knowledge, support, or resources, to invest in and help the innovator.

The frequency with which peers in other areas cooperate readily, in the highly innovating companies, contrasts sharply with the absence of horizontal cooperation at "Southern Insurance" and other segmentalist organizations. For many innovators, relationships with people in other areas are essential to their success.

This was certainly true for "Bill Golden's" dramatic achievements. Working within severely constrained budgets, Golden, a marketing manager at Honeywell, met the company's sales goal in his first year and tripled it in his third year, achieving the year's goal by March. He did this by restructuring personnel and reorganizing work to separate marketing and sales, while instituting weekly staff meetings to tie the functions together. Not only were the numbers

impressive, so was Golden's endurance. Four managers he characterized as "exceptionally talented" had held this job before him, and
none had lasted more than a year; Golden was now in his third year.

When Golden took over, the sales people were all expected to
do marketing chores, including generating new product lines, taking
care of market planning, and handling constant paperwork. As a
result, he said, "the operation was on its knees. Sales were off by
fifty percent, and the competition was eating us alive." Golden had
to manage the initial turnaround with the same people and fewer
dollars, because of the financial constraints stemming from the
recession; and in midyear, his budget was cut further. But he did it.
And in his second year, because of the credibility of Golden's unit,
he got additional resources and had the best year ever. But he never
could have done it without unusual amounts of support.

Topping Golden's list of supporters was his boss, the officer who
had recommended him for the job. Golden's boss backed him when
he asked for it, gave him latitude, and, perhaps most important,
wanted him to succeed. Golden also needed the support of the product line managers, which he got because of long-standing relationships. "They knew I would never end-run them; they had respect
for me, knew I was a team player." The product line managers were
initially critical of Golden's efforts. They "all wanted to get into the
sales game and have client contact. Two of the three felt they knew
how to sell better than I." But Golden kept stressing the fact that all
of them had common problems and common goals, and he won them
over. He wrote a memo defining responsibilities, establishing sales
people as the first line of client contact and product line managers
as the second, to which they "bought in."

Many innovators besides Golden draw on long-standing relationships with people in other areas to aid their accomplishment.
One long-service GE manager turned around a production area
using a new team approach that reduced both the production time
and costs for his product by 50 percent. But to do this, he needed
both capital funding for the introduction of modernized equipment
and the freedom to move poor performers to areas where their skills
could be better used without jeopardizing the project. Securing the
funds turned out to be easier than getting good people as replacements, and so he used his extensive contacts with peers in other
production facilities and the personnel department to speed up the
process—and speed up his results.

There are four principal kinds of integrative devices that aid
network formation in innovating companies: frequent mobility, including lateral moves; employment security; extensive use of formal
team mechanisms; and complex ties permitting crosscutting access.

Mobility Across Jobs

Mobility—circulation of people across jobs—is a first network-facilitating condition.[6] In the more entrepreneurial companies, managers change jobs frequently, even if the moves are lateral rather than vertical. At Chipco, about two years is an average job tenure, and managers often begin to get ready for another move after eighteen months. At GE Med Systems "unusual" career moves across functions are not uncommon: e.g., from headquarters staff to district sales manager, from finance to manufacturing, from personnel to operations manager for a key product—without a technical background.

Too rapid mobility has some negative consequences for accomplishment completion, as I will demonstrate shortly, but a reasonable pace is one of the keys to the circulation of information and support. The constant moving around of parts of Chipco, for example, meant that people rather than formal mechanisms were the principal carriers of information and integrative links between parts of the system. Communication networks were facilitated, and people came to rely on a strong information flow from peers. As people moved around, they took with them the potential to establish another information node and support base for a particular network in a different corner of the organization. Knowledge about the operations of neighboring functions was often conveyed through the movements of people into and out of the jobs in those functions. As a set of managers or professionals dispersed, these people took with them to different parts of the organization their "intelligence," as well as the potential for the members to draw on each other for support in a variety of new roles.

It does not take very many series of moves for a group that has worked together to be spread around in such a way that each person in it now has a close colleague in any part of the organization to call on for information or backing. The more frequent the moves and the more widely dispersed the original group, the more widely information and support can potentially circulate.

Several things are noteworthy about mobility as a network-forming vehicle and thus an admission ticket to the power centers. Those who move clearly have an advantage in terms of network breadth over those who don't. The moving have both previous and current ties; those who remain in one place while others flow in and out may begin with the same number of direct personal ties, but they soon fall behind unless everyone in their area turns over frequently. Thus, the organization's *opportunity* structure—who moves, out of what jobs, and how often—has a direct bearing on its *power* structure because of the impact on networks.[7] Those who lack the opportunity to move at all or are confined to moving within a narrow space (as are those with nontechnical backgrounds at Chipco) or enter later, when others at their level have already made several moves, pile up handicaps with respect to network access. I am thinking particularly of women in this regard, but the condition could hold true for anyone.

If one thinks of each co-worker or colleague group as a "graduating class" spreading itself over many sectors of the organization and picking up new connections along the way, it is easy to see how the links between people created by moving people through jobs over a variety of areas can be a valuable tool in organizational communication, integration, and empowerment.* Points of view are likely to be more cosmopolitan; segmentalism is less possible. But this kind of change can also create instabilities. It is thus an opportunity, rather than a threat, to people only if it is coupled with basic overall security. And so long-term employment is the second part of the network-building apparatus that keeps power circulating in innovating organizations.

Employment Security

There is both a past and a future dimension to the employment-security issue. Looking back, one can see that over time, relationships form among mobile employees which facilitate the exchange of favors, the willingness to back one another, the agreement to commit to one another's projects. Many entrepreneurial managers are aware of the ways in which their long-standing relationships, and perhaps cashing in on a favor done in the dim past, eased communication, melted opposition, or gave them unusual access for someone in their position.

But this phenomenon is true of less innovating companies as well as the enterprising ones. It is in looking forward that the differences become apparent. The security that comes from an expectation of *continued* "place" in the organization is what creates not only an innovation-embracing outlook—higher flexibility and lower resistance to change [8]—but also a willingness to invest in the future.

Furthermore, it is not only one's own continued employment that seems certain but also that of others. If people fear tackling

---

* This is useful, however, only if the people are moving across parts of the organization that have to work together or that have useful intelligence for each other, as generally occurred at Chipco. At GE, some of the moves took people outside, to other divisions, or brought them in from other sectors, with wide variation in net gain in information for Med Systems itself.

innovative projects that do not guarantee short-term results because they might not last, they also fear investing their support in projects of their colleagues for the same reason.

Thus, Med Systems' and Hewlett-Packard's variants on "permanent employment" helped promote cooperation; providing individual security, we can see, also fosters teamwork. At Polaroid, a company which, until recent setbacks, had made employment security a centerpiece of its people-centered policies and which was small enough for mobility to occur within a narrow geographic area, middle-manager projects generally had the longest time orientation of any of the companies I examined. Those in the middle as well as at the top could afford to take a long view.

Teams

A third network-forming device is more explicit: the frequent use of integrative team mechanisms at middle and upper levels. These both encourage the immediate exchange of support and information and create contacts to be drawn on in the future. The organizational chart with its hierarchy of reporting relationships and accountabilities reflects only one reality; the "other structure," not generally shown on the charts, is an overlay of flexible, ad hoc problem-solving teams, task forces, joint planning groups, and information-spreading councils.

It is common at innovating, entrepreneurial companies to make the assignments with the most critical change implications to teams across areas rather than to individuals or segmented units: e.g., a team of mixed functional managers creating a five-year production and marketing plan for a new product. Such formal teams, not incidentally, served as models of the method that top management endorsed for carrying out major tasks and projects. Indeed, one of the managerial accomplishments at GE Med Systems involved developing a system to measure team performance and to make comparisons across teams. At Chipco, the establishment of formal interdepartmental or cross-functional committees was a common way managers sought to improve the performance of their own unit.

Collaborative and consultative rather than unilateral decisions were the expressed norm. It was expected that all managers—even those not tied into a matrixed situation—would not generally reach decisions alone, without consulting others, and company philosophies explicitly encouraged teamwork. Furthermore, because of an unexpressed but strong "norm of modesty," it seemed easier to get credit and recognition for others than for oneself. Thus, the implication was that it was beneficial to get onto other people's teams as well as getting them onto yours.

Teamwork is not just a high-tech touch. In older industries as well, practice in the use of integrative team mechanisms may account for successful problem solving and innovation. In 1981, for example, a terrible year for retailing, J. C. Penney stood out for its superior financial performance (an earnings rise of 44 percent on a mere 4.5-percent increase in sales). Success was attributed by top executives to a "new management style involving teamwork" and "creating developmental opportunities by helping people understand what happens at different levels of the organization," as the chairman put it. The mechanisms for teamwork involved a set of overlapping permanent and ad hoc committees, the groundwork for which dates at least back to the early 1970s, but which flowered fully a decade later: a management committee of the fourteen top officers; seven permanent subcommittees on key issues such as strategic planning, transitional planning, personnel, and economic affairs; and task forces on operating problems composed so that diverse points of view would be available. The leaders appeared aware that such team vehicles not only paid off in immediate problem solving but laid the foundation for a more informed, versatile, and integrated management.[9]

Note that it is not just *any* team that aids innovation but a tradition of drawing members from a diversity of sources, a variety of areas. Innovating companies seem to deliberately create a "marketplace of ideas," recognizing that a multiplicity of points of view need to be brought to bear on a problem.[10] It is not the "caution of committees" that is sought—reducing risk by spreading responsibility—but the better idea that comes from a clash and an integration of perspectives.

A General Electric Med Systems manager, for example, told me some of the reasons for this diversity on teams in his slightly rueful comment that "It is now impossible to know everything about one of our products, or to have someone work for me who does." So he accepted—he supposed—the need for so many teams and task forces. It was not clear whether specialists or generalists dominated at Med Systems—one self-described generalist said there were "too many Ph.D.s and engineers running around" and another felt like a maverick for being a generalist, but some top executives said that broad-based managers rather than technical experts were the wave of the future. Regardless, both groups were pulled together on teams to solve complex problems or make changes.

Complex Ties and Crosscutting Access

A formal structure acknowledging complex ties also forces a great deal of interunit contact between managers. Although managers nearly everywhere complained about excessive meetings, such occasions provided them with opportunities to develop formal and informal working relationships with persons from many other functions or disciplines. This encouraged coalition formation, helping managers to mobilize support or resources to complete an accomplishment. And it discouraged segmentalist overidentification with one area or divisive, polarizing politics.[11]

"Matrixed" managers at GE Med Systems typically reported to both the function where the manager had traditionally worked (generally the stronger and more direct tie) and a connected department, or the office where tasks would be integrated for a particular product (often a less direct "dotted line" tie). Although the "dotted line" relationship was often ill-defined, it benefited the manager by providing him or her with access to another powerful upper-level manager (generally a department head). Managers were thus able to use their dotted-line reports to secure support, resources, or information, gaining an additional route to vital organizational commodities. There were also important implications for sponsorship; if a manager was unable to obtain sufficient backing from his direct superior, then there was an alternative in the dotted-line boss.

Finally, the legitimacy of crosscutting access promoted the circulation of all three power commodities: resources, information, and support. By this I mean managers could go across formal lines and levels in the organization to find what they needed—vertically, horizontally, or diagonally—without feeling that they were violating protocol. They could skip a level or two without penalty. Indeed, at Chipco managers were frequently counseled that direct access was better than going through channels. At Honeywell's Defense and Marine Group, upper management was viewed by middle managers as "very open, looking for and encouraging innovating approaches," so that "anything that makes sense will be listened to by them," providing the sponsorship to bypass the usual formal channels to make an unusual move. (Six of Honeywell's twenty-seven innovators benefited from a clear, explicit top-management sponsor's making it possible for them to open needed doors quickly.)

Matrix designs, though not essential for crosscutting access, can be helpful in legitimizing it, for the organization chart shows a number of links from each position to others. There is no "one boss" to be angered if a subordinate manager goes over his head or around to another area; it is taken for granted that people move across the organization in many directions; and there are alternative sources of power. Similarly, formal cross-area and cross-hierarchy teams, as in Honeywell's parallel organization, may provide the occasion and the legitimacy for reaching across the organization chart for direct access. (Still, effective managers were careful not to violate protocol too often or to exclude authorities like their own bosses completely, even if they temporarily bypassed them, as I show in Chapter 8.)

DECENTRALIZATION OF RESOURCES

The last broad condition for power circulation is local access to resources.

The existence of multiple sources of loosely committed funds at local levels makes it easier for managers in innovating companies to find the money, the staff, the materials, or the space to proceed with an entrepreneurial idea. Because no one center has a monopoly on resources, there is little incentive to hoard them as a weapon; instead, a resource holder can have more influence by being one of those to *fund* an innovative accomplishment than by being a naysayer. Thus, managers at Chipco could go "tin-cupping" to the heads of the various product lines in their facility who had big budgets, collecting a promise of a little bit of funding from many people. This process reduced the risk on the part of all "donors" at the same time that it helped maintain the "donee's" independence.

Sheer availability of resources helps, of course. Typical organizational characteristics in other research on innovation match those of Med Systems and Chipco: rapid growth, resource abundance, absence of "distress." Richer and more successful organizations innovate more than poorer and less successful ones, especially in technology.[12] At Chipco, for example, money never seemed to be tight; only one manager—in a financial area, to be sure—commented that there was not enough money to fund every good idea.

There are a variety of ways that innovating companies make resources accessible locally or give middle-level people alternatives to tap when seeking money or materials for projects. One is to have formal mechanisms for distributing funds outside the hierarchy. Chipco had a corporate research-and-development committee which heard proposals from any part of the company; at the time I started to meet with Chipco people, the committee had just decided to seek proposals for organizational and work process innovations as well as technical innovations. 3M has put in place "innovation banks" to make "venture capital" available internally for development projects. Honeywell's DMSG divisions have top-management steering committees guiding their organizational-change activities. The original steering committee solicited proposals quarterly from any employee for the formation of a problem-solving task team; the teams may receive a small working budget as needed.

Decentralization itself keeps operating units small and ensures that they have the resources with which to act—and thus makes it more likely that managers can find the extra they need for an innovation locally. Business analysts have commented that Chipco has avoided the problems of large companies despite its large size because it has operated as an aggregate of many small groups. Hewlett-Packard and 3M are among the companies that find a wide variety of virtues in small-scale divisions, creating new ones when existing ones get too large. 3M's gains from this strategy are impressive: in the 1970s sales and earnings rose by almost 44 percent, with a much smaller increase in employment. An emphasis on small size showed up at every level. In 1982, the company's manufacturing plants had a mean size of 270 people and a median of 115; only five had more than 1,000 employees. This enabled, in turn, a variety of other integrative mechanisms to work well: problem-solving quality circles, regular work-crew and management-group meetings, task flexibility, and lots of informal communication. It was also said that work at 3M was thought of in terms of "projects" with a usual size limit of a dozen managers and professionals.[13]

Despite employment of about 60,000, Hewlett-Packard managers often say: "We feel like a small company" and "We are responsible for our own destiny." Corporate philosophy holds that about 2,000 people and/or $100 million reaches the limit of manageability for a division, after that becoming impersonal and procedure-bound, limiting personal growth and innovation. In response, as a division reaches this ceiling, an elaborate "cloning" process takes place. A division will split along some subdivision of product lines (sometimes into two, sometimes into three divisions), moving employees into new plants, each self-contained, each more in line with company guidelines for size. This ensures the existence of small-scale, decentralized, but also highly integrated units as well as the continuity and coherence of the corporate management style and philosophy. And it keeps resources available locally.

Of course, a number of the issues with which managers deal can be handled without money at all. Only about half of the 234 accomplishments in the six-company study required new financial resources. Instead, the most common resource sought was staff time. This was also decentralized in the form of "slack" and local control: people locally available with uncommitted time, or with time that they could decide to withdraw from other endeavors to be attached to an appealing project. Because mid-level personnel, professionals, and staff experts had more control over the use of their time in the more frequently innovating companies, it was easier to find people to assist in a project, or to mobilize subordinates for a particular activity without needing constant clearances from higher-level, non-local bosses.

Most innovating managers at Chipco and Med Systems perceived a great deal of "running room" (freedom) in the course of completing their accomplishments. Of course, it is possible that these were individuals viewed by their superiors as extremely competent managers and subsequently given autonomy by virtue of some outstanding personal characteristics. But the extent of the pattern suggests instead that the structure of these companies diffused the authority of superiors; bosses had to understand that their managerial subordinates must also work actively with other functions, must rely on the participation and contributions of others, and hence were not entirely in control of the outcomes of their own tasks. Knowing this, bosses seemed more likely to grant those under them more autonomy, realizing that their subordinates must have more freedom to take any action that will translate into significant accomplishment because of their connections to other managers and other areas. We can call this a system of "passive sponsorship": superiors may be less likely to take direct action to assist in a manager's accomplishment, but they may offer a general mandate such as "Do what has to be done." Thus, time as a resource is also decentralized and under local control.

In innovating companies, then, a number of aspects of the organization's culture and structure make the potential for power—for the acquisition of information, support, and resources—available to middle-level managers and professionals.

THE CIRCULATION VERSUS THE FOCUSING OF POWER: A QUESTION OF BALANCE

Unlimited circulation of power in an organization without focus would mean that no one would ever get anything done beyond a small range of actions that people can carry out by themselves.[14] Besides, the very idea of infinite power circulation sounds to some of us like a system out of control, unguided, in which anybody can start nearly anything. (And probably finish almost nothing.)

Thus, the last key to successful middle-management innovation is to see how power gets pulled out of circulation and focused long enough to permit project completion. But here we find an organizational dilemma. Some of the focusing conditions are contrary to the circulating conditions, almost by definition. This is what makes organizational innovation so tricky; an organization has to constantly balance the circulation and the concentration of power.

While measures of complexity and diversity in an organization are positively related to initial *development* of innovations, they are often negatively related to eventual *adoption* of the innovation by the organization. Diversity gives the individual more latitude for discovery, but may make it difficult to get agreement on which of many proposals or demonstration projects should be implemented on a wider scale. Similarly, innovation is aided by *low* formalization at the initiation stage, when freedom to pursue untried possibilities is required, and by *high* formalization at the implementation stage, when singleness of organizational purpose is required.[15]

What all this means in practice is something else again. A corporation cannot shift from "decentralization" to "centralization" at a moment's notice. Even if it could, different projects would be at different stages, requiring opposite conditions. The issue is balance: never to disperse power fully but merely to temporarily loosen it, to allow units autonomy and single-minded focus when they need it while preventing segmentalism from setting in.*

General Electric Medical Systems is a good example of a balanced system, in which centralized planning mechanisms ensure sufficient focus for innovations to be developed which can and will be implemented.

First, Med Systems' headquarters retained control of large expenditures and remained central in giving guidance to the company by setting overall strategic direction. "Headquarters" often seemed to managers to speak with one voice, unlike the sense of temporary-but-detachable coalition one gets at the top of a looser system like Chipco. Some areas within the organization attracted more attention than others. A "hot" new product line had received considerable attention for three to four years and was still a focal point. One reason for this, at least initially, was that the division vice-president put most of the organization's resources into this product in order to develop it more quickly and capture a share of the market. This generated a tremendous amount of enthusiasm within this product-line organization but also jealousy in other departments. Diagnostics, for instance, the mainstay and oldest product line in Med Systems, seemed to have suffered the most from the attention given to

---

* This is my answer to a question raised by a colleague about whether there could ever be "creative segmentalism," as when an R&D unit in the throes of an invention has to wall itself off. My answer is no. Temporary protection of activity boundaries or projects—something I show in Chapter 8 to be important in the dynamics of innovation—is not the same as segmentation, which implies the erection of permanent walls.

the new area, and there was vocal resentment from several diagnostics managers.

The issue of product quality was another one on which top management focused attention. For a while there was high turnover of managers of this area until a handpicked "star" took over and remained in the position for about a year to a year and a half. Since then there had been more emphasis upon quality, not simply in the manufacturing stage but in design and engineering as well.

Second, Med Systems emphasized both short- and long-range planning at the middle-manager level as well as higher. Planning was done both for the immediate year ahead and for a five-year period, modified every year. Although several managers complained that Med Systems does not think well for the long run, middle managers' participation in creating a five-year plan enabled them to conceptualize accomplishments and changes that would benefit the organization for many years and might take years to consummate. Vertical communication around these plans served to guide the choice of projects in organizationally useful and connected directions.

Third, a clear financial "results orientation" limited requests for major projects. The ability to obtain essential resources or information was basically contingent upon the manager's promise, generally implicit, to produce something of benefit to the organization or the unit or person for whom one had negotiated. Sometimes this was a formal process—particularly when a manager used a routine approach for requesting large-scale financial backing for a significant task, and in such cases would invariably be asked to guarantee a 20-percent return on the money that was provided to him. There were two different kinds of time frames around which managers organized their tasks and planning. Some projects were linked to the long-term planning process or had a long time orientation. But many managers were also assigned "fast-track projects," with a great deal of attention from top management and easy access to vital resources to complete their tasks. These projects were often described with images conveying intensely high pressure and risk.

Fourth, higher-level managers did not abdicate in the face of all the delegation and teamwork emphasis in the company. They used their power to bend rules, to isolate a project from the matrix to give it sufficient autonomy to pursue an important task to completion. This was most likely to happen in connection with new-product development.

Fifth, there was stability and more formality in lower-level assignments. The company had a set of "permanent" subordinates over whom middle managers had legitimate authority; they were mobilizable for the managers' projects. In short, labor did not float freely; labor was clearly attached, in relatively fixed proportions, to different areas. The matrix stopped just below middle-manager level, and so did other power-circulating conditions such as cross-cutting access. This may seem like an obvious point, but it is worth remembering. While the hierarchy was loosened at middle-manager levels by the matrix, it remained in full force below—perhaps why middle managers mentioned that some of their managerial reports behaved more "dictatorially" then they did. Most managerial projects were accomplished because the manager had a pool of clearly subordinate subordinates, even if he or she behaved toward them in participatory fashion.

Chipco, in contrast, had somewhat less overall concentration and focus, which showed up in a certain air of creative disorder, duplication, and waste accompanying the entrepreneurial spirit there. This relative "chaos" was both cause and effect of Chipco's innovativeness on my measures. As I mentioned in the last chapter, one of the reasons for the large amount of structure/method innovation at Chipco was simply the need to invent mechanisms to cope with the lack of rationalization and routinization in the corporation.

Because so much power, or potential power, circulated so freely at Chipco, it was hard for some people to accumulate enough to sustain a significant project. Since nearly everyone at managerial levels seemed to have the potential to start something, there was often an air of many people running off in all directions. What was the inducement for all of these potential entrepreneurs to join some-one else's team—in the sense of real work, not just verbal support? Or if they did sign on, what would actually get them to meetings if they had other investments in other directions? (Chipco-ers were notorious for arriving at meetings late or missing as many as they attended.) Teamwork, despite its value, was sometimes hard at Chipco because of the "entrepreneurial," not the "bureaucratic" trap; as an observer put it, "The motivation and drive are to seek individual stardom. Besides, people are confident; they feel they *can* do it themselves if let alone." If bureaucrats isolate themselves to protect territory, entrepreneurs may do it to prove their worth.

High-level sponsorship or support for a project counted for *something* at Chipco, in terms of focusing people's attention, but much less than at Med Systems. Similarly, Chipco had a hierarchy, with levels of subordinates fanning out below middle managers, but it did not instill the same automatic respect as at Med Systems, and besides, there were more staff managers without assigned subordinates at Chipco. Furthermore, it was more difficult to isolate projects at Chipco than at Med Systems (perhaps because their issue *was* the organization), harder to buffer them against the effects of external changes.

The need for peer collaboration at Chipco and the infrequency of top-down authorization, compared with Med Systems, made the selling process—"buy-in"—a lengthy, frustrating effort that was hard to sustain over time, as many managers complained. Because agreement in this complex, rapidly changing environment was usually oral, ill-defined, and situation-specific, a "yes—I'll support you" was often not a "yes" a few months later when new players and circumstances entered. Indeed, a safe way to say no was to say "yes" and not mean it. Everything took longer than expected. One manager groused, "Previously, it took me an hour to write a memo. Now it takes a hundred and twenty people-hours—thirty people in a meeting for four hours!"

In such cases, schedules slipped by several months, so that "buy-in" sometimes meant only, "you hang in there so long that there's agreement by default." (Twenty-one months for an average accomplishment involving peers may be too much for a company in the electronics industry; Med Systems, in contrast, often "fast-tracked" important projects.) A variant of the process, which Ken Farbstein called "railroad buy-in," was described by several managers in different departments as a way to speed things up. An idea and a "departure date" (sometimes an obligatory first meeting) would be proposed. Supposedly the train was leaving on that date, with one passenger or a full load. (But the initiating managers deemed it prudent in those cases to wait for the "boarding" of "important passengers.") In each of these cases, well over half the needed supporters got on board early enough, and each of the accomplishments was significant and successful.

Impatience with the process may be the reason one manager enlisted the support of a vice-president who could order people (his subordinates several levels lower) to help the manager search for information; but he still felt the need to define this as "buy-in," calling it "top-down buy-in."[16] Frequent jokes were told about the need to get "buy-in" even to go get coffee, go to the bathroom, or empty the wastebasket, reflecting impatience with the process.

The strong interpersonal communication networks and reliance on peer approval at Chipco sometimes caused competent people to be ignored or humbled. New on the job, one manager asked, "Whom should I talk to about X?" and was told, "Don't talk to Jim: talk to Bob Anderson instead." The manager did so and then explained, "You bypass someone you never met, and assume he's incompetent. People get slaughtered by innuendo. . . . If you're associated with a failing product, you're not to be spoken to." The need to maintain face explains the norm of never surprising, and hence embarrassing, another manager in a meeting. One manager commented, "A person may have some good ideas, but if he's not a good advocate who can present his case persuasively in a meeting with lots of critics, he won't succeed."

Newcomers, especially those from hierarchical firms, typically needed six months to a year to become comfortable with this, and often left in frustration. One manager, who had joined Chipco after years in a more traditional and hierarchical company, was given an award by co-workers for finally learning to master the lateral-communication process, years after arriving at Chipco. Meanwhile, he occasionally longed for the good old days, when life was simpler, even if less exciting and more constrained.

Decentralization at Chipco seemed both cause and effect of what some managers saw as an inability to make long-range plans. Where there is a sense of vast opportunity, there may be little perceived need for planning. The large number of people bargaining for resources and attention in a rapidly growing company made plans little more than goals, and long-range planning science fiction. Planners sometimes said they felt like "voices in the wilderness" with their plans a "necessary evil, subject to change at a moment's notice." This explained the high number of assigned tasks at Chipco which, despite their authorization by a boss, were still not strongly supported by top management. Occasionally, when higher-level management saw a problem but had no time to develop a solution, it would announce a problem and a haphazard solution which would be put into effect unless someone came up with a better one.

Not everything at Chipco worked this loosely, of course. Some areas were more structured than others: finance, manufacturing, and lower-level, nonexempt jobs in general. Still, Chipco had fewer focusing mechanisms than Med Systems. Except for product innovations that were likely to require a top-determined concentration of resources, many Chipco management innovations were vulnerable to replacement later. Despite a large number of innovations, most involved technological or work methods that often did not outlast the innovating manager.

Thus, a company can stimulate a great deal of initiative and enterprise without necessarily getting maximum payoff from managers' projects; power must be *focused* as well as available. In the ideal situation, there is a "marketplace" in which power circulates at middle levels of the organization, guided by a hierarchy above and serviced by a hierarchy below. The innovating organization, in effect, has a kind of quasi-free "market" sandwiched between two "hierarchies."

A Culture for Enterprise and Innovation

Some of the innovating companies, such as Chipco, GE Med Systems, and others in high-tech industries, have an automatic advantage in the innovation game: they are younger, growing faster, in more "modern" industries, in highly competitive markets, and have cultures of change. These are features associated with greater enterprise and innovativeness across a wide variety of organizations and societies. But I can also find examples of the same kinds of innovation-enhancing practices in older firms in more traditional industries characterized by integrative practices and cultures; Procter and Gamble is a frequently cited illustration.

Innovating companies in high-technology fields have also clearly benefited from the impact of location in certain geographic areas where they are surrounded by company-spawning institutions that both keep inventions flowing and also create competitive pressures on companies to offer the kinds of opportunities that attract and hold talent. The connections between companies because of common origins and exchange of personnel also serve to keep their practices current and ensure that they remain innovative.

In Boston, for example, three organization-creating organizations were particularly important to the growth of high-technology companies in the area: the Massachusetts Institute of Technology; Arthur D. Little, the world's largest technical consulting firm; and ARD, a publicly held venture-capital firm established in 1946. Lincoln Laboratories at MIT, established in 1951 to develop an air defense system for the United States, had spawned about fifty companies by 1966, according to Edward Roberts of MIT, who sat on the boards of a number of them; MIT's Research Laboratory for Electronics led to about fourteen companies; and its Instrumentation Lab, about thirty. In 1972, MIT even incorporated a development foundation to launch new companies that would use MIT technologies.[17] Then companies "spun off" from others, as ex-employees formed their own. Sylvania's electronics division in Boston was the "parent" of perhaps thirty-nine companies; Data General was one of a number of companies founded by former Digital Equipment managers; and high-level executives carried expertise from one to another, like the vice-president of manufacturing at Wang Labs, who came from Digital. In Silicon Valley (Santa Clara County, California), Hewlett-Packard set the style, and Fairchild was the company spawner.

In other, smaller high-tech centers, there were similar patterns: Control Data was founded by an engineer from Sperry Univac; Cray

Research by a former Control Data employee. Entrepreneurs often left one firm to start another in a chain still replicated today; for example, James Treybig was a marketing manager at Hewlett-Packard just before he founded Tandem Computers in 1976, and recently a new company, Stratus, was started by former Tandem employees.

It is thus the whole context, the whole system at the more innovating companies—and indeed, norms within their industries—that generates the enabling conditions for managerial enterprise, rather than a set of discretely separable features. Out of the design and structure of the organization arises a set of patterns of behavior and cultural expectations that guide what people in the system consider appropriate modes of operating. At Chipco, managers feel compelled to demonstrate their entrepreneurial spirit and to look for it in others, resisting "stifling" it with too much "bureaucracy." They know they are expected to invent or develop or plan something; this is the cultural image of success. At Med Systems, managers need to haul out their teamwork credentials and show how they use a team to get something done that improves on an already smooth-running operation. Such expectations or cultural "norms" guide behavior in a holistic sense; managers are not responding to a specific set of incentives or a concrete "program to stimulate innovation." This is perhaps why the matrix form of organization has not worked well in companies that do not support it with other culture changes.

The highest proportion of entrepreneurial accomplishments is found in the companies that are least segmented and segmentalist, companies that instead have integrative structures and cultures emphasizing pride, commitment, collaboration, and teamwork. The companies producing more managerial entrepreneurs have more complex structures that link people in multiple ways and encourage them to "do what needs to be done," within strategically guided limits, rather than confining themselves to the letter of their job. They are encouraged to take initiative and to behave cooperatively.

Ironically, though power seeking is a necessity for managers in innovation-producing companies, raw "power politics" seems much more common in the heavily segmented and bureaucratic companies like Southern and Meridian than in more entrepreneurial settings which encourage managerial innovation. The specific, delimited authority characteristic of more bureaucratized organizations not only creates an incentive for territorial protection and fighting across groups, but also creates the illusion that managers can, indeed, act alone, maximizing the value of their own areas without having to take the needs and concerns of others into account. In the more entrepreneurial settings, the very ambiguity surrounding the managers' areas and the absence of clear possession of all the resources, coupled with the nature of the issues that are being tackled, means that managers are impelled to behave more cooperatively in order to survive.

Thus, even though the system in innovating companies is more "politicized" in one sense—with managers having to capture power that they are not directly given in order to get anything done—it is also more "civil," at least on the surface. "Opponents" are won over by persistent, persuasive arguments; open communication is used to resolve debates, not back-stabbing. Perhaps the very publicness and openness of the battlegrounds—if that word even seems appropriate —makes "reason" prevail. It is hard for back-room bargaining or displays of unilateral power to occur when issues are debated in group settings. Public meetings require that concerns be translated into *specific* criticisms, each of which can then be countered with data or well-mounted arguments. And the heavy reliance on informal communication networks as a source of reputation places a check on dirty dealing. "Bad press" would ensure that such a person gets frozen out. An innovating company, then, begins to substitute a control system based on debate among peers for one based on top-down authority.

Life is by no means perfect in the innovating companies. Certainly not all managers have the same access to power; and systems promoting innovativeness also bring with them a new set of problems of managing participation, ambiguity, and complexity. But if life is not perfect, at least the tools exist for individuals to use to make corrective changes.

CHAPTER

7

Energizing the Grass Roots: Employee Involvement in Innovation and Change

> We're a proposal-oriented company. Even the
> janitor can make a proposal and it will be se-
> riously considered.
>
> —"Chipco" manager

THE FREEDOM FOR MANAGERS in innovating companies to experiment and invent does not automatically extend to the people at the bottom. The foot soldiers of the organization are more likely to occupy humdrum turn-it-out-on-schedule jobs that stress reliability, not creativity.

We are so accustomed to the existence of such jobs—on assembly lines, next to bookkeeping ledgers, behind typewriters—that it seems faintly ridiculous to wonder whether innovation is being stifled or the entrepreneurial spirit dampened. After all, even if people had the potential to come up with innovative ideas, when would they develop them, where would they express them, and who would listen, if their jobs are to keep the machinery of the organization running?

It is not hard to see how segmentalism could develop around the production and support ranks of the organization, even if there is a free-flowing integrative atmosphere at the top and middle. By necessity, for efficiency reasons, some jobs have a high component of routine, repetitive, do-it-as-ordered action.[1] The problem for innovation and change is not the *existence* of such tasks but the *confinement* of some people within them. Segmentalism sets in when people are never given the chance to think beyond the limits of their job, to see it in a larger context, to contribute what they know from doing it to the search for even better ways. The hardening of organizational arteries represented by segmentation occurs when job definitions become prison walls and when the people in the more constrained jobs become viewed as a different and lesser breed.

But how, when, and why could a clerk, an assembler, a technician, a bookkeeper appropriately become involved in innovation and change? The usual rationales for increasing employee involvement or participation in problem solving are the high costs of friction between segments (e.g., labor conflict) and the need for higher motivation for production and quality. These conventional rationales, however, are oriented toward doing what the organization is already doing better, faster, and with lower costs, rather than helping it learn to do something new. The mobilization of the grass roots for *change* is a less common phenomenon. Yet an innovative organization cannot afford to ignore any potential source of new ideas, any informant with information about a possible problem or a possible solution— the rationale behind Motorola's "Participative Management Program" or Data General's "pride teams" or Honeywell's "quality circles" and "positive action teams."[2]

An innovating organization needs a work force at all levels that has not become so stuck in the rhythm of routine jobs that it cannot easily adapt to a new drumbeat. For change to be a way of life rather than an occasional traumatizing shock, the "Indians" as well as the "chiefs" have to be engaged in change making and change mastery —while still doing their necessary jobs.

That is the challenge innovative organizations face: to combine the necessity for routine jobs with the possiblity for employee participation beyond those jobs.[3] They need to give individuals the same chance to contribute to innovation at the bottom that they get at the middle and top. They need to encourage an integrative culture that includes all levels, rather than segmenting off the production and support ranks. And they need to make sure that the *idea* of participation does not itself suffer from segmentalist treatment—i.e., as a special "quality-of-work-life program" to "be nice to the workers" but not affecting any other areas of operations, and certainly not part of a broader strategy to produce innovation and change.

"Chipco" decided to take on this challenge, years before the idea of "copying Japanese-style quality circles" became faddish. In a set of innovation-stimulating changes at its "Chestnut Ridge" assembly plant, Chipco developed a model of grass-roots employee involvement for the rest of the company. Because Chipco made it easy for managers to innovate—the resources, information, and support were available—an entrepreneurial woman manager was able to introduce mechanisms in the plant that empowered grass-roots employees to contribute to innovation themselves.

The way it happened at Chipco is similar to the ways other innovating companies with which I have been involved, including two divisions at Honeywell, are creating new vehicles—"parallel organizations"—for energizing the grass roots. And it builds on the best parts of what happened in "Petrocorp's" Marketing Services Department.

EMPOWERING THE PEOPLE AT CHESTNUT RIDGE

The story begins a few years ago, and it has no real ending because events are still unfolding four years later. In fact, like all dramas of change, it is really several stories.

The first one stars "Roberta Briggs," as I'll call her, an entrepreneurial middle manager who was convinced that current and anticipated production problems could be tackled by the production people themselves; she had to sell the idea (overcoming stereotypes about the production staff in the process), get resources, build a coalition, and shepherd the pilot efforts. So Briggs's story has a dash of politics and a fair amount of behind-the-scenes diplomacy in it. And it shows her changing roles: from innovator at the initiation to "prime mover" and pusher as events are set in motion. Every successful change effort has at least one committed pusher.

The second drama includes a much larger cast: all of the shop-floor men and women mobilized by Briggs's project. They spent a year working in teams analyzing the problems of their part of the plant and devising new solutions. As each team finished its work, a new team came forward. The teams themselves are the stars, but a few "character actors" turn in distinguished performances: for example, a technical supervisor—a former Hungarian freedom fighter—whose English was said to be so bad that no one could understand him, who gets so turned on by his team's efforts—and they by his—that the language problems disappear; or the supervisor who turns down a promotion to another plant so he can finish his team's work. This story begins slowly, held back by worker skepticism and passivity, and then builds to a crescendo of enthusiastic effort as the teams begin to see that "management really means it," that they really *are* empowered.

The third saga surrounds the other two; it is a managerial story of how the innovating teams become linked to a permanent set of integrative vehicles at Chestnut Ridge for involving the grass roots in change while still maintaining a smoothly functioning routine production organization. And throughout this drama we see management's frozen countenances—their skepticism matching that of the workers—gradually thawing into high approval.

Here is how it all began.

It was early 1977, and Chipco management was starting to worry. Although Chipco was viewed inside and out as a fast-growth company with almost unlimited potential, market conditions suggested to management in 1977 that some reorientation and evaluation of traditional manufacturing practices might be in order. Chipco's larger size had made it increasingly difficult to manage traditionally—which, in Chipco's case, meant "informally." Increased competition and market pressure for some products necessitated tighter control in the growth and operation of production facilities. And continuing rapid developments in technology suggested a need for more responsiveness on the part of manufacturing personnel, to anticipate otherwise sudden product changes.

On my initial visit to Chestnut Ridge, I was struck first by its woodsy surburban environment, complete with river and wild flowers in the woods, and only second by its size—more than 1,000 people easily disappeared into a squat building sprawling over many acres. Because Chestnut Ridge was one of Chipco's largest plants, it was also among those most affected by the changes facing the company, and so these general issues translated into more immediate concerns. Chestnut Ridge was faced with the need to control growth; moreover, there was strong pressure to reduce costs and increase productivity to maintain profit targets. But plant management regarded its work force, particularly the front-line production supervisors, as insufficiently competent to deal effectively with these needs. Indeed, Chestnut Ridge was widely regarded as having one of the least skilled work forces in Chipco.

Chestnut Ridge was also a "lead plant" for a manufacturing group; this meant that groups of trained managers and technical personnel were drawn off from time to time as the nucleus of a new facility. A temporary hiring freeze made it impossible to bring in new talent. Yet plant management regarded most of those who would normally be promoted as a result of transfers and promote-from-within policies as unpromotable.

New technology was already an issue at Chestnut Ridge: some of the traditional small-batch manufacturing operations were being converted to assembly lines, with considerable uneasiness resulting. There was a rumored move afoot to centralize production under one manager, rather than continuing to have seven separate production groups, one for each set of product lines, each with its own business manager reporting to the plant manager.

The impending reorganization was viewed with concern by some production supervisors. They worried about less group identity for their work units. They worried about the impact of pressures for higher volume, creating less chance for identity with the product. They were concerned that changes at higher levels would mean that they would not be dealing with the same upper managers as before. And they thought that greater specialization of direct production jobs would limit the skill level in each, making it harder to attract or keep good people and downgrading the quality of jobs.

The supervisors worried, in short, that the coming reorganization was one more step in depersonalizing Chipco and removing the familylike atmosphere of earlier years when there was greater contact between levels and an active company social life, complete with almost-legendary baseball teams. The ball teams were one way that people kept in touch across areas and made informal contacts that helped their careers; for a while, people felt it was hard to get promoted at Chipco without company baseball experience, because so many strong bonds were forged on the playing field.

The sports teams also symbolized the youthful exuberance and informality common in the industry—like Tandem Computer's Friday-afternoon beer parties around the company swimming pool. No one would seriously attribute Tandem's phenomenal record—worker productivity estimated to be twice the industry average, a yearly doubling of earnings since its founding in 1976—to the pool parties, but getting too big to continue them might start to dilute integrative network building and camaraderie that certainly contribute to innovation. So Wang Labs, for example, maintains its employee picnics and recreation facilities, to help 10,000 people feel as if they are still part of a "small" company. And when a Honeywell operation in California launched a new marketing campaign by inviting all its employees onto the parking lot to hear executives dressed in rock-band costumes and wigs sing a campaign theme song, it was momentarily reviving that kind of community spirit, leveling status differences, and providing a chance to have fun together. But at Chestnut Ridge, the plant's size and high proportion of new employees were making it hard to maintain plantwide team spirit, even with the help of its sports leagues. Anyway, after the coming reorganization, the task distance between levels might be too great even if the social distance could be bridged. The production supervisors were right to fear the loss of community at Chestnut Ridge.

Furthermore, to pile problem upon problem, the production organization was still adjusting to a recent tightening of distinctions between supervisors and nonexempt work-group leaders, eventually called "work coordinators," who performed some supervisorlike tasks but were entitled to earn overtime. Chestnut Ridge had more of these quasi-supervisors with ambiguous status than any other Chipco manufacturing facility; 90 of 126 group leaders would need to be upgraded to supervisors according to Wage and Hour Law distinctions. But in this case an "upgrade" was not always well received, since the overtime earned by work coordinators often gave them higher total compensation than was currently offered to supervisors. There were rumblings of discontent from the coordinators. Thus, Chestnut Ridge staff were wondering if the old spirit of open communication, teamwork, and loose boundaries was in jeopardy. Could segmentation be creeping in?

As a consequence of these pressures, Roberta Briggs, then manager of personnel staff functions at Chestnut Ridge, suggested to plant management the possibility of developing a project that would address these issues, simply titled the "Production Project." After months of discussion, she received funding through one of those innovation-stimulating vehicles at Chipco: a central corporate R&D committee newly interested in people matters. Money was provided on a step-by-step basis, with funding for each additional step contingent on satisfactory results from those preceding. The process of defining the project and gathering support began in February 1977 and culminated with a proposed project plan, approved by the R&D committee in July.

The Production Project to be launched at Chestnut Ridge had two very different overall objectives. One was developmental—to increase the effectiveness and promotability of lower-level production supervisors and those nonexempt workers who might replace them. The second was an organizational objective—to increase the production organization's capacity to innovate and to manage changes with minimum disruption, such as the move to more efficient assembly lines.

Four more immediate goals were also identified: to improve management and effectiveness of production organizations; to give supervisors and managers tools—information and understanding—to manage their groups better; to produce concrete changes in the structure and function of the production groups so that opportunity and influence were improved; and to develop a model that could be used in other plants and organizations for improvement projects.

The vehicles were to be twofold: a plantwide *steering committee* composed of key managers, which would monitor the gathering of information and receive proposals from supervisory groups for action projects; and a set of *task forces, or action groups,* reporting to the steering committee, composed of supervisors and those below, that would tackle concrete problems, identified via data gathering, to increase opportunity and power. No one at Chestnut Ridge used the term "quality circle" in 1977, but that was one function the action groups served. But unlike the way quality circles are used in most companies, the Chestnut Ridge action groups could come together from across the plant to tackle plantwide problems, and they could have direct access to management and to management resources through the steering committee.

The activities at the Chestnut Ridge plant proceeded through five overlapping stages: initial education and support building; information gathering and diagnosis of needs; team formation and action planning; implementation; and integration and diffusion of results within the system.

Building a Broad Base of Support: Education

Money and a tentative okay from her boss in hand, Briggs could start moving the Production Project from "hers" to "ours." The goal of the first stage was to educate and persuade: to get support for the project and its theoretical framework on the part of corporate and plant management and staff. Briggs, her staff, and her consultants conducted a set of individual interviews, followed by several group discussions and seminars presenting the project's theoretical framework and seeking refinement of the project plan.

The theory behind the project was derived from *Men and Women of the Corporation*, which Briggs had recently read: that the relatively ineffective behavior and lack of promotability of many of the first-line supervisors were the result of insufficient opportunity and power available to them in the existing structure.[4]

The same increased growth that had led to Chipco's success had also rendered less effective the informal processes through which both opportunity and power had been accessible to the first few Chipco generations—a problem at all growing companies, and why Hewlett-Packard, 3M, and others seek to maintain small-size units. This was further exacerbated at Chipco by the increased pressure toward control and centralization represented by the assembly-line changeover. Thus, development and installation of a "parallel participative organization"[5] of cross-hierarchical task forces reporting to a steering committee would increase participants' effectiveness and recognition by providing new sources of opportunity and power and would shift management's perception of their competence and promotability, while serving as the necessary flexible integrative structure to respond appropriately to the emerging changes.

The proposed action seems straightforward enough, especially in retrospect, but for Chipco in 1977 it took a great deal of selling. Recall some facts about a corporation like Chipco that would make

Briggs's reasoning atypical. First, the emphasis was on individual competence and the entrepreneurial spirit; the idea that "anyone who is good can rise around here," coupled with almost-legendary stories about people who had shot up from the shop floor to the officer ranks in the early days, made it hard to do anything but assume that individuals were responsible for their own success or failure.

Thus, many Chipco people saw the problems of production inefficiencies and poor supervision at Chestnut Ridge as reflecting incompetent individuals who should be either replaced or given enough training that they would shape up. Sometimes the term "animals" was used to refer to the workers at Chestnut Ridge, a term given currency by such acts as the time some exuberant workers threw a computer into the river running by the plant. Of course, all this is relative; compared with a typical old-fashioned auto factory or steel mill, Chestnut Ridge was a model of cooperation, decorum, and cleanliness, and its work force was generally well educated.

This first assumption is often made in companies other than rapidly growing entrepreneurial ones, of course; it is typical American corporate thinking to blame poor performances on individual failings. But a related aspect of the Chipco culture was less typical, especially in contrast to its innovative counterparts like GE Med Systems or IBM: an antibureaucracy bias that made it difficult to look for solutions in setting up an organizational mechanism, as opposed to just training the individuals and expecting them to change as a result.

The other issue at Chipco was the discrepancy between the enormous opportunity and power open to people in the middle and upper levels and the much smaller amounts for those at the bottom, from supervisors through the direct-labor ranks.

While Chipco's labor policies and benefits were highly progressive and overall turnover rates or other signs of dissatisfaction low even for its industry, it shared this gap with other progressive fast-growth, high-technology firms; even manufacturing managers at Hewlett-Packard, often touted as the ideal people-conscious company, recognize problems with opportunity and power on the shop floor. Certain jobs, such as those in direct production, were more structured and routinized than the relatively freewheeling positions in the middle, and the expectations for performance were more concrete and quantitative.

The Chipco atmosphere also had differential impact on lower levels. Some of the same constant change, informality, and decision-by-networks-and-coalitions that provided middle managers with freedom and challenge could appear to the rank-and-file as confusion, chaos, and favoritism. While middle managers could easily trade higher salaries elsewhere for Chipco's wide career-growth opportunitites, this trade-off was not so meaningful to the lowest-echelon employees, whose compensation was lower to begin with and who had fewer chances to move up—especially if Chestnut Ridge management did not see them as having potential.

The discrepancy between the situations of middle and bottom made it hard for some managers, looking at their own experience, to see *why* the bottom could possibly be unhappy or need a change in structure. Some staff wrote off the dissatisfied rumblings of workers in the plant as part of a general, to-be-expected "noise level." One relatively new manager, fresh from business school, taking a visitor around the plant, mentioned his "brainstorm" about how to improve attitudes, stimulated by the lines painted on the floor that helped color-code and separate departments: "Why not make a T-shirt with the colored lines and patterns on them and the motto, 'Trust the system'"? Farther along, he showed the visitor the new assembly operation, where the workers were boxed in by conveyor belts, commenting that they looked like "caged animals." His feelings that this is "better than any place like it, the best place in the world to be in terms of opportunity" seemed to the visitor to reflect a "dream world of management" far separated from the reality of life on the shop floor.

When Briggs introduced her Production Project, the plant manager was lukewarm and kept his distance. But because Roberta Briggs could get independent funding and had the backing of key corporate and local managers, she had her own power base; so the plant manager, rather than overtly oppose her, simply stayed away.

One of her primary in-plant supporters was "Bob Lizzeri," a business manager for the largest product line, known as a rising star and slated to take over the new centralized production-manager position; it is often those with opportunity and power themselves who are readier to support change. Roberta's coalition of supporters beyond Chestnut Ridge included the general manager for the manufacturing group to which Chestnut Ridge belonged, a manufacturing vice-president, a corporate personnel director on the central R&D committee, and her counterparts in two other plants who were very much interested in launching a similar project in their facilities. She invited them all to seminars. She also drew on the services of an external consulting group, with Barry Stein and, later, Daniel Isenberg playing leading roles.

Some of the senior people invited to in-plant and offsite seminars expressed doubts about the feasibility of a project that seemed to rely excessively upon the as-yet-untested skills of those at the lower levels of the organization. Other concerns were appropriate managerial ones similar to those I hear repeatedly at other companies: Would there be an impact on compensation? Weren't the proposed activities things that the production manager should be doing? Wasn't the project challenging a company "god"—its new mechanized assembly line?

Roberta Briggs described Bob Lizzeri's approval of the project this way: "He turned green, swallowed, clutched his heart, slid under the table—but said, Go ahead."

Thus, doubts were dealt with openly, the rationale and hypotheses were discussed, and measurable objectives set. The meetings and seminars were polished and effective. Furthermore, Roberta Briggs was smart, and she was persuasive. Her own credibility with manufacturing managers was high, because she had helped them on a number of personnel and planning problems. She was a shrewd businesswoman with an M.B.A who could discuss facts and figures, but her low-key humor-filled personal style fitted the informality and youthful culture of Chipco. The R&D label on the Production Project made it easier for some people to swallow it as a "test" rather than a change of assumptions. She had already raised the money to do it. And there were enough managers close to the shop-floor experience, like Bob Lizzeri, who knew *exactly* what Briggs was talking about. They understood, as Chipco people like to put it, the "pain level" in the production area, and they were willing to do something to change the organization as a route to improving the performance of their people. They were aware of the risks, but liked the underlying theory. It made sense to them. And so they "bought in."

By the fall, then, despite discomfort over the possibility of failure, the exploratory sessions and formal seminars led to formation of a steering committee for the project, consisting of Briggs, Lizzeri, business managers of two production groups, the plant personnel manager, and a member of the employee-relations staff; all were volunteers. One of the consultants was an *ex officio* member.

A project advisory group was also created, composed of senior manufacturing and personnel managers from within *and* beyond Chestnut Ridge in order to provide formal support and power for the activities to be undertaken—another integrative mechanism. This supplied knowledgeable counsel for decisions needed for implementation; legitimacy for plant people to participate; sources of visibility, recognition, and reward; and high-level linkages to prevent the project from either floating unconnected to the rest of Chestnut Ridge and Chipco or appearing mysterious to nonparticipants. Chestnut Ridge was not to be an organizational deviant or a cultural island the way the Petrocorp Marketing Services Department was; Chipco cared about reaping results and using them elsewhere.

In October 1977 the steering committee held its first meeting and approved the official "launching" of the project itself, with an in-plant survey.

Moving the Project Downward: Data Gathering and Communication

The second stage began to involve the production employees themselves; data were gathered from them and fed back between November 1977 and February 1978. A formal questionnaire was designed, with advisory-committee and steering-committee help, to provide measures of the information, flexibility, opportunities, and problems that people had in their jobs. These were filled out anonymously by 25 percent of the workers and 66 percent of the first- and second-level supervisors, meeting in business-group units in a conference room in the plant.

A memo was sent to everyone in production announcing the survey sessions:

> You may be asked to participate in a meeting to gather information about the production organization. We're writing to tell you what it's about.
>
> Chipco has always taken pride in offering opportunities for people to grow. But frankly, we've gotten pretty big, and it's harder to find out what the opportunities are and where they are. We have to figure out how to flag the ones that exist and how to make more of them. We need everybody's help to make sure we do the right thing.
>
> We're starting just by trying to find out what's really going on. We need to stop guessing and start checking. That's where you come in.
>
> With a little help from our friends, we've put together a questionnaire to get the kinds of answers we need. It will take less than an hour to fill out, and after that, we'd like to get your comments and reactions. It's easier to do this in small groups— that keeps the hassles down. You'll be getting the word soon about a time and a place. If you're willing to help, and if you want to get your ideas into the pot, this is a chance. But it's up to you—*this is strictly voluntary*.
>
> We'll get back to you with results and next steps.

The problems surrounding production workers in the plant became clear even at the first survey session, punctuated by anxiety, mistrust, and distancing. One of the questionnaire administrators noted:

The first group was supposed to begin at 8:30. At 9:00 Tom Smith decided "I better go find out what's happening." He managed to round up some people, who came in at about 9:20 to 9:30. Not one of these people knew why they were being brought together. They had not read the memo about the questionnaire that was supposed to have been circulated. Everyone was totally in the dark. I think this created some amount of apprehension in some of the groups (especially the direct labor, who felt out of place, and were somewhat intimidated by us).

Then they entered the room tentatively, seeking direction. "I don't belong here." Off their home territory. Nervously expectant. They don't know us, but they know authority when they see it. Nervousness shows . . . speaking in whispers, afraid to get coffee which has been offered. In this group the older people seem more relaxed, verbal. (Less at stake?) Most approach the questionnaire seriously. (Want to do a "good" job?)

Question asked . . . which supervisor? immediate or the highest one? Seems to be some question as to who has real authority here.

Point brought up about question 16—overtime. One possible answer says "This is out of my hands. I'm never asked." Overtime is out of their hands, but they are *always* asked. Takes much time and effort for us to explain what should be done.

They help each other on the questionnaire. Jokes about some points concerning factors in promotion. "Good press" gets a good round of laughter. "Hard work" gets even more. An older woman says, "I don't think they even notice." All the jokes and laughs at this point surround the issue of powerlessness at getting raises and promotions. "Promotions are based on who you know or who likes who."

More shoptalk; mostly about anger over increases in vending machine prices (which in truth are very low). Real strong complaints about this. Everything is out of their control.

Discussion after questionnaire:

Feel treated like they are in grade school; demeaning; changes made, no one knows who or why. Not informed of a possible change until after it has taken place.

Specifics: Work coordinators were eliminated in this group. All seem unhappy with this. The WCs worked alongside the people, understood their functions and jobs. The supervisors are out of touch with the workers. Supervisors are smug, disliked, don't know floor jobs. . . . Workers are not consulted about production changes.

Changes in general: considered superficial. Day-to-day there are no changes: only hypothetical. Immediate production is all that is important. Goals of change are not met because management wants immediate results in production.

Problems mentioned: Hiring practices—workers have little

control over who gets hired. Jobs get posted, management has already filled the positions. Management refuses to follow current workers' recommendations about who to fill what job with. Another major issue is parts; hire more people in parts, just creates more paperwork. More people working on parts, the less they are available. . . . Can't get parts from another department unless you know someone. Only other way is to follow the red-tape chain; break one link and you never get there. System is getting more complicated, less productive.

The initial memo promised that workers and supervisors would see the results of the survey, and this commitment was repeated in the survey session. But nearly everyone was quite skeptical about ever seeing anything. One worker became quite interested in the project and approached the questionnaire administrators as they were having lunch. He wanted a copy so he would be able to remember what had been asked. They told him he would get feedback the next month; he was sure he never would. They told him they had a commitment from plant management; he said that didn't matter. The administrator commented:

If things go through Chipco channels, people at lower levels never expect to hear what is going on. No matter how much reassurance we gave him, he simply said, "I doubt it."

But the prompt feedback of results just a few weeks later to everyone in the production organization at Chestnut Ridge was one of the first tangible signs that this project was serious, and that serious change was intended. This started to convince people that the project represented something potentially quite useful for them; management *was* providing an opportunity.

Findings were presented through several short memos distributed to employees, voluntary discussion sessions, and access for everyone to copies of the whole questionnaire, with overall response patterns indicated. Then in January 1978, meetings were held with each production group, on a voluntary basis, to discuss the specific results for that group. A memo was circulated to members beforehand, along with a copy of the questionnaire with group results indicated. Data were presented not as fixed phenomena but as suggesting questions for discussion. For example, the memo to one business group looked like this:

One way we've been using the information that you provided us through the questionnaires is to take a closer look at each of the business groups and the differences among them. Following are some highlights about your group and some of the questions they suggest. . . .

Supervisors

The more flexibility you have in your job, the fewer the difficult headaches to take care of. Flexibility seems helpful. What kinds of flexibility particularly help in this connection? In what ways? —If this is true, how can jobs be made more flexible?

You generally feel that there are many different jobs available for you to advance to, but you don't seem completely sure what they are. What might be done to help you to get more of this information?

Direct Labor

The more time spent at Chipco, the greater the number of things that might hurt your chances for promotion, raises, advancement, etc. Why is this? Does this mean that the more in the organization, the more these things become known, or is there a real change? . . .

Training and learning around your jobs are very important —some of you said you don't have skills, training, and knowledge that you need or at least as much of it as you would like. Some of you said that you have skills, training, and knowledge that don't get used in your jobs. How can this be changed?

Many of you are somewhat dissatisfied about your opportunities for advancement. They seem more limited, less clear than you would like. Is there something that might be done about this?

One business manager sent his own memo to his boss and to the product managers for his product line:

SUBJECT: PRODUCTION SUPERVISORY PROJECT

My people and I have been participating in a project with Roberta geared to developing our production supervisors here in CR on a plantwide basis. As a first, we have obtained definitive feedback from our troops concerning the ins and outs of a floor person's job. Attached for your perusal are the results of a questionnaire completed by 13 of our direct labor people and 4 of our work coordinators. I think you'll find the results interesting.

In summary: The majority of our people learned their job by talking to and observing others. Personal friendships and the environment are the high points of the job; they'd like to see lots more training. The biggest headaches are overtime and getting parts/supplies to do their jobs. Work coordinators have an average amount of influence and rarely discuss performance. Poor quality and not delivering are most likely to cause problems on the job. Rumor control is the prime source of info re what's going on outside the group; they feel least informed re what's happening outside the plant.

The feedback meetings discussing the data provided confirmation of the initial opportunity/power framework and reinforced the commitment of project staff. They also served to bring many of the work groups together for the first time, to open dialogue about solving their work problems together. Dan Isenberg wrote, after meeting with three of the business groups:

> The idea of opportunity as a variable affecting motivation is very clear to the supervisors. They feel frustrated and helpless in the face of fewer job openings and entry level jobs from which people increasingly cannot move. This is true of the "touch-up" area—the supervisor says there is little that he can do to make the job less boring and less monotonous. Other people in his group are more fortunate, including the senior supervisor, because he supervises two groups and thus can move people from the entry level touch-up area into a more challenging and rewarded task. The other problem with moving people out of touch-up is that it is getting to be impossible to replace them given the hiring freeze.
>
> People understood the concepts of power as capacity and could identify with the frustration that comes from not having the capacity to get the job done, especially because of a lack of physical resources such as supplies. I think that managers at Chipco *do* have a fair amount of flexibility, and the rules and regulations—the red tape quotient—are fairly new. One said: "The inventory sits there. It affects the power in your job. You feel pretty powerless if you have a million dollars' worth of inventory sitting behind you on the line and can't get a particular part. That results in frustration."
>
> I had a conversation with a supervisor from the other group when we were looking over the questions about the data that we had brought to the meeting. John is very interested in getting his own people to answer the questions, perhaps by filling out the questionnaire. He said that since volunteers came to the data collection sessions, there was trouble getting enough volunteers toward the end so that some whole groups were "volunteered" by their managers. He does not feel that the data are therefore representative of his group. But what was interesting was he said he had been thinking for a while that he really ought to get his people together to discuss some of these issues (he called it their myths and beliefs) but just hadn't gotten around to doing it. So this project gave him the impetus.

Communication was being opened up in the production organization. This was the "breadth" phase of the Production Project: all employees were involved in a discussion about the patterns and the problems in their particular work unit. Next, the project moved to its "depth" phase: identifying areas for concentrated, in-depth problem solving. And now Briggs and the consultants became supporting actors as the production people themselves moved to center stage.

Structure Building and Action Planning

The next step was to create a structure for problem solving and action outside of, but complementary to, the existing hierarchy and assignment of jobs. The steering committee would manage a series of time-limited action groups that would carry out specific projects. This task-force/steering-committee structure, "parallel" to the hierarchy, represented the expansion of opportunity and power through a chance to get training that was not strictly technical; a chance to have impact on the company in ways other than through the immediate job; a way to detour around bureaucratic structures that might not be working, to see and solve their problems; and a mechanism for managing new activities that exist outside of people's jobs.

The strategy, in short, was to create a setting for experimental action and career development that was loose enough to allow for flexibility and some trial-and-error, yet connected enough to the existing organization that the lessons learned could easily be seen as relevant to the larger setting and ultimately incorporated into ongoing operations.

At a large meeting following the feedback session, this participative structure was described to supervisors, and ultimately three groups volunteered to be the first action groups—to tackle problems they cared about and to demonstrate that the parallel organization could work. One of the pilot action groups was an existing group, one was a mixture of many levels within a business area, and one was composed of representatives of several areas. The steering committee made it clear to supervisors that the time required for the action groups was legitimate and appropriate, subject to the continuing requirement to meet agreed-on business targets. Furthermore, a representative from each pilot group—chosen by the group in any way it wished—now became part of the steering committee.

Each of the three pilot action groups met four or five times over two months to devise plans. They were given a boost by a series of brief educational events: workshops in meeting skills and in planning. (But the training was not nearly as extensive as the twenty or more hours involved at Honeywell's printed-circuit-board plant or at General Motors' QWL plants; it is not clear that so much elaborate training is necessary in more progressive companies that already have good relations with the workers and a congenial atmosphere for participation.) These groups got short seminars as they needed them and moved into action fast.

The teams quickly presented options to the steering committee, and the committee helped refine the plans to make them appropriate and workable. Then each group drew up a proposal and plan for an action project relevant to its particular business unit which its people would present to the corporate R&D committee for its own funding.

The first action group, the one most involved in the shift to an assembly line, elected to tackle the big worry: to redesign the organization of the line itself, to reduce the disempowerment and isolation that the data indicated, and to increase productivity. The second group proposed to develop a mechanism for bringing new people on board at Chestnut Ridge quickly and effectively: they were concerned about the finding that people had little help in learning their jobs or career opportunities. The third group prepared two projects. The first was to systematically attack operational inconsistencies among the seven production groups, inconsistencies that were responsible for many problems in production; the survey data had shown that intergroup transactions and flow of materials were big "headaches." Their second plan was to develop, with the plant training staff, more sharply focused and timely training programs for supervisors, because survey responses made the existing ones look inadequate.

The steering committee was delighted. It felt that all these projects had significant business payoffs in addition to their quality-of-work-life implications and was happy to endorse them. The groups could then present their proposed projects directly to the R&D committee for approval.

The presentations to the corporate committee were made by teams of direct-labor people and first-line supervisors—men and women seven or eight levels lower in the hierarchy than those to whom they were presenting. All proposals were accepted—enthusiastically. Members of the corporate group were "astonished," they said, at what they had seen; the reappraisal of Chestnut Ridge people had begun.

The action groups left that meeting with their commitment to their projects even firmer—ironically, not because of the praise and funding, it seemed, but because of the implied insult in the committee's surprise at their competence. "We'll show them we're not a bunch of turkeys" was the attitude.

Action

Now the pilot action groups began carrying out their tasks. The groups' own initiative and commitment were critical; they were no longer being pushed solely by Briggs and the consultants.

The groups were free to make the key decisions about how to organize themselves, how to involve and inform other direct-labor people, and how to maintain current production levels. They managed small budgets authorized by the steering committee. They developed specific plans and objectives, educated themselves about the topic of their particular projects (often rapidly outstripping their "expert" resources), visited other companies, gathered more information about Chestnut Ridge relevant to their projects, explored other resources available at Chipco, and coordinated with each other and the steering committee as they sought to follow through on their proposals. The steering committee met regularly to hear reports and offer guidance.

This act of the drama comes to a swift curtain because once the vehicles are in place, action is often surprisingly fast. In this case, all four action-group activities were completed within six months, by September 1978, a year after the Production Project began.

The assembly-line-redesign group developed a strategy to increase job effectiveness on the line through use of flexible, horizontally integrated teams. This team-assembly concept was especially well received and quickly implemented. It provided for choice in whether direct labor would work in teams or in the conventional, individual assembly mode, since the action group had found that not all workers *want* to do their jobs in teams (about a third did not). Shortly thereafter, another assembly-line group at Chestnut Ridge approached the pilot action group about setting up a similar activity, and pilot-group members, including direct labor, acted as their consultants. A few years later, a leading business publication featured the results of the team-assembly option in a major news story:

The Chestnut Ridge plant started a four-member team in 1977 to assemble standup equipment. The team is made up of volunteers from four different skill areas. Each taught the others his job, and now they can switch jobs at random.

A production manager says the team uses four times less space to assemble a machine than the assembly line process. Cycle time improved by 60 percent, and output per employee increased "slightly," he says. The team manages its own workplace without close supervision, and its members seem to like the responsibility that comes with directing their own work. "It's better to work under these circumstances because you have a feeling of companionship," says one. "Team assembly helps workers get to know each other and help each other."

The other groups also produced innovative new methods for improving Chestnut Ridge's effectiveness. For example, the plant could put to immediate use a report on procedural inconsistencies across the production groups that impeded work coordination, along with suggestions and options for action, and a recommendation for establishment of a supervisors' forum as a continuing mechanism.[6]

Institutionalization of Participation

When the first set of action groups were finished with their work, they disbanded as planned. Management was extremely impressed with the quality of the products, and the recommendations began to be implemented by plant staff, with the occasional involvement of action-group members.

The steering committee liked the impact of the project on the plant, even beyond the action-group projects. Members especially noted the markedly improved "managerial" capacity (of both supervisors *and* direct labor), including an increase in planning skills, less narrow concern for one's own job and more for the whole organization, increased motivation on the job, decrease in the strength of stereotypic views by both labor and management personnel, better communication, ability to analyze data, and better use of plant resources. Bob Lizzeri was particularly struck by the way the project aided innovation and change: the supervisory-driven job redesign, the bottom-up participation, the willingness of the rank-and-file to change and to take charge of their own future.

In short, people in the factory who participated in the project were said to have gained important new skills, become more productive, and in the process, grown more satisfied with their jobs. They were not an unusual group, either; statistical tests on background characteristics of project participants and nonparticipants showed no significant differences.

The new "energy" was palpable when I visited the plant. No one stood still or worked in isolation for long anymore. There were frequent gatherings in bunches, animated discussions, and huddles during the lunch period. At one work station, the "red line," the mountains of parts I had seen before, had dwindled to a molehill because the group had reduced in-process inventory by their team methods. But there were new piles around here and there—of the books the supervisor had purchased for his people; the work area was getting to be known as the "Chestnut Ridge management library." At a break, I heard employees discussing their plans to tour other Chipco and outside plants. People at the next station, the "blue line," couldn't help observing that something was different next door, and they started using "red line" workers as consultants so that they too could make changes.

But changes in some of the people impressed me the most. The level of communication was so good everywhere that the Hungarian freedom fighter with a heavy accent was not only understood, he was a fountain of ideas in action-group meetings. And a rather mousy young woman supervisor blossomed because of her learning of the project, becoming more poised, forceful, and effective with her people, because she now knew how to involve them and get things done for them.

The increase in skills did not go unnoticed by more senior managers, and suddenly the plant that had "lacked promotable people" found workers from its Production Project in demand. Several pilot-group members got raises of from 5 to 15 percent, based on their project performance; they would otherwise have had no raise. A few workers and supervisors were reappraised as more promotable. One supervisor, who had been ranked in the middle of all the production supervisors, was moved to the top of the list. Several people who had been thought flatly unpromotable were in fact promoted, in at least one case into an entirely different function than would normally have been considered. There was a new, positive cycle: opportunity improving performance enough to bring more opportunity.[7] But one supervisor also turned down a promotion to stay with his project through completion.

The steering committee took these results and made presentations to others at Chestnut Ridge, including the plant manager and his staff, to inform them of the Production Project achievements. Action-group participation was formally incorporated into all new job descriptions and made an integral part of performance evaluations (linking the "parallel" and regular organizations). Thus, people would get *job credit* for work in the parallel organization—a big step for Chipco at that time. There is nothing like the "bottom line" to show that management intends this to continue, of course—the reason the Nashua Corporation, one of the early users of quality circles, makes up to 50 percent of a manager's bonus contingent on participation in the program.

The parallel organization embodied in the steering committee and the action groups became a permanent part of Chestnut Ridge. A plant charter was written by the steering committee late in 1978 and then adopted by plant management in 1979. It states that the steering committee represents "an opportunity for people of all levels within Chestnut Ridge to participate in the experimentation of managing change. It is also a forum in which ideas toward improving processes are developed." Thus, the parallel organization at Chestnut Ridge remains an ongoing mechanism to solve problems. Employees can submit proposals to the steering committee to start projects using their ideas. Indeed, Chestnut Ridge was featured in the national business press in 1981 as a successful example of the new forms of worker involvement, and visitors often come to the plant to see how the parallel organization works. This external visibility is not incidental to Chestnut Ridge management's continuing commitment.[8]

THE PARALLEL PARTICIPATIVE ORGANIZATION AS AN INNOVATION AND CHANGE TOOL

The Chestnut Ridge experience, and others like it, shows that worker involvement *works*.

What it works for is not merely gains in immediate productivity and quality, or reductions in friction. These short-term benefits may occur, as they did in small measure at Chestnut Ridge, but they are usually difficult to attribute directly to any one event; too much goes on at any one time in a dynamic organization. The more important benefits are longer-term, those associated with innovation, with increased organizational capacity: new problems tackled, new designs for more effective work systems and personnel programs, and a larger repertoire of skills on the part of workers and managers, who also know each other better and communicate better. Even if seen purely as a *training* device, employee involvement at the grass roots has extra payoffs because problems are solved while people are educated.

To some corporate managers at companies more traditional than Chipco, "quality of work life" is synonymous with "play"—turning the company into a country club and losing work time when people return late from their lunch-hour jog. But what I have seen at Chestnut Ridge and elsewhere was just the opposite: action groups using their lunch hours to hold a team meeting, or getting together after hours on their own to study a new book on their project areas. It was not "play" but "work" that drew the action groups to the project; the personal gains to them included increased knowledge and skill as well as organizational contributions that might result in greater recognition. It was not "free time" from their jobs they were seeking so much as a chance to do even more—if the tasks were significant and likely to be rewarded.

At the outset of the Production Project, supervisors at Chestnut

Ridge were worried about losing the baseball-team spirit that came
from a smaller plant in which people played together. But the new
spirit that came from *working* together seemed to some of them even
greater, because the accomplishments mattered to the company and
made a difference to their work lives. It is precisely the fact that
strong bonds can be forged on a joint task—maybe stronger than in
social situations—that leads to the fear of some spouses about the
appeal of their partners' opposite-sex workmates or to concern of
proponents of workplace participation that absorbing work on team
projects may pull people away from their families. Work teams that
feel they are pioneering and innovating can develop such special
feelings that they may choose to symbolize it in badges of identity:
signs, symbols, and names. The beaming faces of Honeywell quality-circle members pictured in a main hallway display stand just
above their group names: "99.9 percent," "star tech-ers," and other
playful, self-chosen ways of expressing team-ness. Thus, from the
shared work itself may come the greatest flowering of communal
spirit for project teams.

In a surprising number of instances, the human energy is there,
waiting to be plugged in. Managers in a first-time participation effort
often worry about whether there will be enough volunteers, but
nearly everywhere I have been, there are always substantially more
people than management expected ready to invest time and effort in
a project—for Northwestern Bell's "coming issues" task forces, for
the General Dynamics Convair plant's human-resource council, for
a Honeywell division finance department's problem-solving task
teams, just to name a few. Indeed, the Honeywell finance-depart-
ment head had wondered if there would be enough people to staff
four teams—and then half the department volunteered, requiring
twice the expected number to be squeezed in.

Certainly the motives for volunteering may be mixed: looking
good to the managers pushing this, for example. But some of the
results of people's efforts to free time for innovative projects are
impressive. Repeatedly people get their own work done more effi-
ciently when they have a place to put the freed-up time. At the
Lordstown Vega plant after the famous strikes in the early 1970s,
workers doubled up, performing the work of two for fifteen minutes
in order to earn a fifteen-minute break. A Cummins Engine plant
gave everyone the equivalent of two hours' training a day (bunched
at intervals) on the theory that there is enough slack because "people work only six hours a day anyway." Productivity more than met
expectations, a former plant manager reported. At a Honeywell facility in Arizona manufacturing printed circuit boards, output per
week rose from 1,800 pieces in December 1980 to 4,500 in November 1981, with scrap almost disappearing, while employment

dropped from 120 to 85 people as a result in part of the efforts of problem-solving teams on the shop floor. One week the employees were told they could go home when they reached the week's quota, in the expectation that this would happen about 2 P.M. Friday; instead, it happened by midday *Thursday*.

So the time can be found; the question is whether there is anything worthwhile to do with it. At the Honeywell circuit-board plant, did workers go home because they had nothing else to do? There was speculation in May 1982 at the Honeywell plant that boredom was setting in because employees had nothing like the Chestnut Ridge Production Project into which they could pour their energy; once shop-floor problems were solved, there was no vehicle yet for them to tackle larger, plantwide issues.

I know that this may sound utopian to conservatives and exploitative to radicals. But my response is the same to critics on both ends of the political spectrum. It appears that when it is in the interests of the people involved, and they are given genuine opportunity and power, they can be committed to finding the time to contribute to solving organizational problems. It is, I recognize, easier to make the negative case: that *lack* of opportunity and *lack* of power lower performance by dampening motivation and preventing access to tools for accomplishment. Suffice it to say here that in the companies I have observed, projects involving grass-roots participation have not suffered from a lack of willing volunteers who find or make slack time for their involvement.

Chestnut Ridge's success drew on the kinds of more generalizable conditions that support innovation via grass-roots participation. First, it was part of an organization providing resources, support, and assignments leading potential "entrepreneurs" to seek innovations; specifically, the availability of a corporate R&D budget helped set the stage. But all this would have been latent without a "prime mover," an entrepreneurial manager willing to push for change— and able to find a coalition of others willing to support it. The system was carefully prepared, especially through education of top managers and key stakeholders in the plant. Coalitions were built— integrative teams of backers as well as workers. Critical supporters were brought together formally as a steering committee, including top managers with line authority who could act on worker recommendations and link them to the rest of the organization. Then formal data collection with feedback provided a shared information pool available to all employees, but it raised questions rather than drawing conclusions, so as to permit ample employee input. A "breadth" strategy of giving all supervisors information and encouragement to use it with their people was combined with a "depth" strategy of picking volunteer groups to develop high-payoff projects.

There was more power available to the actors, too: educational support and team building for the action groups; company data and company funds; access to high-level decision makers and company experts—in short, information, resources, and support. Continual communication between the steering committee and the action groups (eventually represented on the committee) ensured that the action groups were on track and relevant to the business, making it possible for the first wave of projects to end with a commitment to implement action-group recommendations and reward the teams. As results accumulated, the parallel organization was institutionalized in a plant charter, job descriptions, and performance appraisals.

The immediate gains of the Production Project were applauded throughout Chipco. But perhaps the most important thing that Chestnut Ridge gained from the project was a way to stay ahead of change—to prepare for and anticipate and create change—by being able to flexibly engage the talents of its workers in the parallel participative organization, and then apply what was learned in the parallel organization (individual skills or organizational systems) to the workings of the hierarchy. The parallel organization was an innovation-producing mechanism at the same time that it was a tool for energizing the grass roots, for improving their competence and productivity.

Employees can be energized—engaged in problem solving and mobilized for change—by their involvement in a participative structure that permits them to venture beyond their normal work roles to tackle meaningful issues. They gain an experience of the *communitas* of teamwork on a special project—to use Victor Turner's term for dramas of high involvement—which lifts them out of the humdrum, repetitive routines of their place in the ongoing structure.[9]

In ongoing operations, with some form of orderly structure dominating, specialization, differentiation, and status differences prevail. But in the parallel organization, workers and managers are involved in more egalitarian teams, where status distinctions are leveled and all struggle together for a joint solution. Indeed, involvement in the parallel organization—membership in a task force or action team—may be the closest to an experience of "community" or total commitment for many workers, a dramatic, exciting, and almost communal process brought to the corporation.

What imbues this with meaning for employees is not just the sense of being part of a group, but the *significance* of the tasks taken on: the feeling of pride and accomplishment at building—and building something relevant to the larger organization. There is a "high" in parallel-organization tasks unlike what most people feel in their jobs in the hierarchy. As one Honeywell manager said to me, "I've worked here for thirty years, but the only thing I'll really remember and that I'll be remembered for is the new system my task force designed."

There has typically been some confusion for management analysts between the creation of a parallel organization and its results in short-term gains. Specific changes—e.g., team assembly methods, new communication vehicles, or more open career access—are often equated with "work reform" itself. But the more critical point is the addition of a problem-solving structure that cuts across the hierarchy. For example, the Volvo auto factories' experience with quality of work life has often been described in terms of a new technology and team assembly, whereas those are just two specific manifestations following an extensive set of organizational structural changes that involved decentralization and creation of parallel problem-solving and decision-making vehicles.[10]

The parallel-organization concept also suggests some revisions in organization theory. The Chestnut Ridge experience and that of similar projects shows that it is possible for a "mechanistic" production hierarchy and an "organic" participative organization to exist side by side, carrying out different but complementary kinds of tasks.[11] These two organization types are not necessarily opposites, but different mechanisms for involving people in organizational tasks. Of course, to be effective the two organizations must also be linked, through devices like a management steering committee.

The hierarchy is the maintenance-oriented structure for routine operations: it defines job titles, pay grades, a set of relatively fixed reporting relationships, and related formal tasks. In the hierarchy, opportunity tends to be limited to formal promotion paths, and power flows from the contacts and resources inherent in a defined position. Its main function is the maintenance of production and the system that supports it; that is, the continuing routinization of useful procedures.

The problem-solving participative organization, on the other hand, is change-oriented and embodied in the parallel structure; it is ultimately integrative, making segmentalism impossible. People can be grouped temporarily in a number of different ways as appropriate to the problem-solving tasks at hand, not limited by position in the hierarchy. A different set of decision-making channels and "reporting relationships" is in operation, and the organization as a whole is flexible and flat. In this parallel, more fluid structure, opportunity and power can be expanded far beyond what is available in the regular hierarchical organization. The main task function of the parallel organization is the continued reexamination of routines;

exploration of new options; and development of new tools, procedures, and approaches—that is, the institutionalization of change.
Those *new* routines, as their utility is demonstrated, can be transferred into the line organization for maintenance and integration.

The simultaneous availability and operation of these two structures and their interconnections provide a basis for the efficient operation of each, because both are *equally formal* structures, able to
carry out their specialized functions directly. An integrative culture
connects both, and thus, even differentiated activities cannot turn
into segmentation. It is potentially possible for people to be simultaneously involved in a work organization in two different ways and
through two different mechanisms.[12]

An innovating organization needs at least two organizations, two
ways of arraying and using its people. It needs a hierarchy with
specified tasks and functional groupings for carrying out what it
already knows how to do, that it can anticipate will be the same in
the future. But it also needs a set of flexible vehicles for figuring out
how to do what it does not yet know—for encouraging entrepreneurs
and engaging the grass roots as well as the elite in the mastery of
innovation and change.

PART IV

MANAGING IN THE INNOVATING ORGANIZATION: SKILLS FOR CHANGE MASTERS

Power Skills in Use: Corporate Entrepreneurs in Action

> The twentieth-century scene contains huge and powerful units which compete not so much with one another but as a totality with the consuming public and sometimes with certain segments of government. The new entrepreneur represents the old go-getting competition in the new setting. The general milieu of this new species of entrepreneur is those areas that are still uncertain and unroutinized. . . . The areas open to the new entrepreneur . . . are those of great uncertainties and new beginnings. . . . The new entrepreneur makes a zig-zag pattern upward within and between established bureaucracies. . . . He serves by "fixing things."
>
> —C. Wright Mills,
> *White Collar*

WHAT IT TAKES to get the innovating organization up and running is essentially the same two things all vehicles need: a person in the driver's seat and a source of power.

The acts of myriad individuals drive the innovating organization. There would be no innovation without someone, somewhere deciding to shape and push an idea until it takes usable form as a new product or management system or work method. And that process of pushing and shaping requires power sources and power skills.

Thus, the first step in change mastery is understanding how individuals can exert leverage in an organization—the skills, strate-

gies, power tools, and power tactics successful corporate entrepreneurs use to turn ideas into innovations. Getting a promising new
idea through the system—or pushing others to do it—is the way in
which corporate citizens with an entrepreneurial spirit make a difference for their organizations.

THE QUIET ENTREPRENEURS

"Corporate entrepreneurs" are the people who test limits and
create new possibilities for organizational action by pushing and
directing the innovation process. They may exercise their power
skills in a number of realms—not only those which are defined as
"responsible for innovation," like product development or design
engineering.

I have found corporate entrepreneurs in every function, bringing about a variety of changes appropriate for their own territories.
Some were *system builders* (e.g., designers of new market-research
departments in insurance companies, of long-range financial-plan-
ning and budgeting systems in rapidly growing computer firms).
Others were *loss cutters* (e.g., prime movers behind getting founder-
ing products into production faster, behind replacing obsolete quality-control systems in record time). Still others were *socially
conscious pioneers* (e.g., developers of task forces to reduce the turnover of women in sales, of new structures to engage employees in
solving productivity problems). And there were *sensitive readers of
cues about the need for strategy shifts* (e.g., fighters for reduced
manufacturing of favored products because of anticipated market
decline, for culling out the losers among 1,200 product options offered to customers).

These "new entrepreneurs" do not start businesses; they improve them. They push the creation of new products, lead the development of new production technology, or experiment with new,
more humanly responsive work practices.

Occasionally, they become "heroes" publicly identified with a
successful innovation—in a small way like the accounting manager
at a Honeywell facility pictured prominently in the employee newspaper because he led a group that discovered ways the company
could save a few million under the new tax laws; or in a large way
like Thomas West, the engineering manager at Data General who
led a product-development effort that designed an important new
computer in record time and became the "star" of a best-selling
book.[1] But more often, few of their projects make headlines. And if
they do, it is unlikely that many of the middle-level corporate entre-

preneurs who push the innovations would be mentioned, for by the time that most of these achievements find their way into company product announcements or balance sheets, it is hard to attribute individual responsibility or to single out anyone who deserves the most credit for what turned into a joint effort. Even Tom West would not have been publicly recognized if he had not been immortalized in a book; by the time the computer was announced to the public, he was invisible, and even the team of design engineers he marshaled was not sure exactly what West had done.

Thus, the quiet, sometimes local innovations that corporate entrepreneurs recounted to me are not always by themselves "big changes" or "big events." Only in retrospect can a company point to a particular decision or specific event as one that brought it into a new state, and by the time this happens, large numbers of other people will have to have been involved. But if these innovators are not alone or individually the architects of sweeping reconstructions of a company, they are cumulatively a major force for change. The hundreds of small improvements, subtle readjustments, and visibly successful new techniques they initiate can, in the aggregate, gradually move or improve a company. The first step in "macrochange," or major reorientations, as I will argue later, is often the "microchanges" created by numerous corporate innovators.

As we have seen, there are marked differences in how much the entrepreneurial spirit flourishes in different environments, especially when I compare the 115 innovations in my six-company study with the rest of the 234 managerial accomplishments examined in detail, and add to these the other accounts of productive change from Wang, Hewlett-Packard, and elsewhere. There are almost twice as many well-regarded managers and professionals in any functional area carrying out innovative projects in firms characterized by integrative practices as in those exhibiting too much segmentalism, as I reported earlier. The environment, more than the person, makes the biggest difference in the level of innovative managerial activity. Individual characteristics play a role only when the company's environment discourages initiative and innovation; then younger managers, newcomers, and—interestingly—women are more likely to risk testing limits.[2] (I found many fewer women than men in positions high enough to define projects resulting in innovations, but many more innovators among the fewer women up there.)

It would be appealing to say that corporate entrepreneurs are idealists captivated by the idea itself and eager to show its value; but they are human like the rest of us and driven by the same mixes of "pure" and "impure" human emotions and needs. I saw people whose jobs were too large and who *had* to define an innovation in order to have a concrete goal to attain, and some whose jobs were too small and who were straining at limits. I found people who were looking for their chance in order to advance their careers, and others who were reluctant entrepreneurs, rising to the occasion because of a fervently held value. Some were angry at the system and wanted to prove themselves as a form of "revenge"; others were simply having fun, because creative change can be enormously captivating and energizing.

But once they begin on the route toward innovation, the logic of the process takes over, and individual differences fade into the use of predictable skills to travel over familiar pathways.

Though innovators are diverse people in diverse circumstance, they share an integrative mode of operating which produces innovation: seeing problems not within limited categories but in terms larger than received wisdom; they make new connections, both intellectual and organizational; and they work across boundaries, reaching beyond the limits of their own jobs-as-given. They are not rugged individualists—as in the classic stereotype of an entrepreneur—but good builders and users of teams, as even classic business creators have to be. And so they are aided in their quest for innovation by an integrative environment, in which ideas flow freely, resources are attainable rather than locked in budgetary boxes, and support and teamwork across areas are the norm.

However differently they start, corporate entrepreneurs soon find that they have something in common: the need to exercise skills in obtaining and using power in order to accomplish innovation.

POWER AND INNOVATION

Innovative accomplishments stretch beyond the established definition of a "job" to bring new learning or capacity to the organization. They involve *change*, a disruption of existing activities, a redirection of organizational energies that may result in new strategies, products, market opportunities, work methods, technical processes, or structures. And change, no matter how desired or desirable, requires that new agreements be negotiated and tools for action be found beyond what it takes to do the routine job, to maintain already established strategies and processes.

To initiate and implement an innovation, people need that extra bit of power to move the system off the course in which it was heading automatically, like the extra muscle thrust to turn a boat or the extra engine power to turn an airplane. As long as people are merely the custodians of already determined routines and directions, they too can operate automatically, staying within their segment, working with the resources or information handed down to them. But innovation, in contrast, requires that the innovators get enough power to mobilize people and resources to get something *nonroutine* done.

Innovative accomplishments differ from merely doing one's basic job—even if the person does it well—not only in scope and long-run impact but also in what it takes to carry them out. And this is why power is so important. Power, as I define it, is intimately connected with the ability to produce; it is the capacity to mobilize people and resources to get things done. But people who are "just doing their jobs" do not need to "mobilize" anyone. Little is problematic. They have a job to do; they are told in detail how to go about it or they already know, from past experience with identical assignments; they use existing budget or staff; they do not need to gather or share much information outside their unit; and they encounter little or no opposition. So they can act on their own authority and do not need to seek or use additional power. At the same time, they do not produce innovation for their companies; they repeat what is already known, though sometimes a little faster or better. And often they get so good at repeating it that they find it hard to stop even when conditions change.

The accomplishments nonentrepreneurial managers recount, for example, generally involve beating their own record, advancing their own careers, smoothing friction with subordinates, or adding incrementally to an ongoing process, rather than introducing anything, changing anything, or redirecting or reorienting their area. They rarely describe orientation to a specific goal, a drive toward a concrete achievement or a tangible project. There is only an ongoing stream of organizational events in which these managers have a clearly delineated territory and rather unchangeable activities; sometimes they cannot even single out achievements from this ceaseless flow.

It is also for this reason that noninnovators lack the excitement about their work characteristic of the more entrepreneurial managers, and probably why I found their interviews so dull and lifeless. Routine jobs, after all, lack the adventure associated with carrying out specific projects and watching results pile up—a clear "something" where there was "nothing," order out of chaos. There is none of the drama and even romance inherent in overcoming obstacles, in proving something, in jumping hurdles. So managers whose accomplishments were basic ones coming directly out of routine job charters had little to say about what they had done; their achievements were modest, and there was no real story. My interviews with them were not only dull, they were short:

QUESTION: (to a division sales manager): What resources did you need to proceed with your idea for a sales training program?

ANSWER: None. I was in command in my division; I just wrote out the ideas in a training process.

QUESTION: What about information?

ANSWER: I don't know what you mean. I had an idea, and we developed the training.

QUESTION: Was anyone opposed or critical?

ANSWER: The manager of a neighboring department.

QUESTION: How did you handle this opposition?

ANSWER: I told that manager that this was not his business.

QUESTION: Did you win over the "critics"?

ANSWER: No.

QUESTION: Did you hit any roadblocks or low points when it looked as if the whole thing would flop?

ANSWER: I never thought about a flop.

Thus, nonentrepreneurs, when they do achieve, tend to produce only a narrow range of accomplishments: those clearly identified with the specific mandate in their job, such as incremental improvements in individual or unit performance, or importing and implementing a well-known practice. They stay within their identified segment and define problems segmentally—as small, isolated, bounded pieces. They are likely already to possess nearly everything they need to carry out the related activities. And so they tend to act alone. Their ability to act unilaterally is, of course, a function of these two features, which make the accomplishments low-risk as well as, generally, low-gain.[3] Being low-risk, they are also not particularly threatening to others, arousing little opposition. And what opposition occurs can often be handled merely by avoidance.

But innovations are not safe, bounded, or easy; nor can innovators generally just "command" their subordinates or ignore their critics. Innovative accomplishments, when I compare them directly with basic nonentrepreneurial ones using statistical tests, are perceived by those involved as riskier, and they are more controversial —they generate stronger feelings around the organization both pro and con. (Thirty-five percent of the innovations, compared with only 15.9 percent of basic managerial accomplishments, were rated as involving some risk.) They require greater dollar investment (25 percent of innovations versus 12.6 percent of basics requiring large amounts of resources); they involve working across boundaries of work units and getting peer support (35.8 percent of innovations versus 21.6 percent of basics involved work outside the immediate work area); and they benefit from higher management sponsorship (61.7 percent of the innovations versus 37.5 percent basics involved this). They also are more complex, take more time, and have larger dollar payoffs, where this can be measured.[4]

Even in the (apparently) most straightforward technical projects, like new-product designs, managers had to convince others that *this* was the move that made sense and then gather the action tools to try something new and different. I had somehow romanticized research-and-development departments, for example, as noncontroversial havens for innovation, places where talented people were given the job of creating innovation, and they could simply run with their ideas. But even in research labs, success in getting the instruments of power are as important to innovating as getting the technical instruments; indeed, the two were connected. "Programmable innovations"—which I should really call "improvements" because they were just incremental advances in already known processes—are much like basic, nonentrepreneurial accomplishments in being easy to carry out within given arrangements. But "breakthrough" developments—true innovations—require seeking more information than the manager routinely has, perhaps getting budget increases or redirecting resources away from where they had been allocated, and forging alliances with officials outside of the R&D department who have to be convinced that the particular project the manager is pushing is what the organization needs most at that moment.

Sometimes political fences have to be mended along the way, too. An engineering manager at Polaroid had the task of directing the effort to solve an operating problem in a complex camera; his team's analysis indicated that a basic redesign was called for. This delighted the manufacturing people, who had been blamed for the defect, but it ran up against the designers, who insisted he go back to look at the manufacturing process. No action was possible until this conflict between production and design staffs was resolved and they could agree on the definition of the task.

If such issues were involved even in technical areas, then it is not surprising that they are even more common around innovations in areas with fewer "objective measures" or "right answers," such as changes affecting people, policies, and structures.

Only in retrospect, I saw, does a successful innovation look "inevitable," like the right thing to do all along. Up to that point, just about all innovating has a "political" dimension, even though, in some of the companies I studied, the use of the term "political" was unpopular, and managers liked to act as if there were not a "politi-

cal" side to innovation. But I am using "political" not in the negative
sense of backroom deal making but in the positive sense that it
requires campaigning, lobbying, bargaining, negotiating, caucusing,
collaborating, and winning votes. That is, an idea must be sold,
resources must be acquired or rearranged, and some variable numbers of other people must agree to changes in their own areas—for
innovations generally cut across existing areas and have wider organizational ripples, like dropping pebbles into a pond.

The enterprise required of innovating managers and professionals, then, is not the creative spark of genius that invents a new idea,
but rather the skill with which they move outside the formal bonds
of their job, maneuvering through and around the organization in
sometimes risky, unique, and novel ways. This is what the corporate
entrepreneur has in common with the classic definition of an entrepreneur. Organizational genius is 10 percent inspiration and 90 percent acquisition—acquisition of power to move beyond a formal job
charter and to influence others.

Here is where the environment—the organization's structure
and culture—enters the picture. All the enterprise, initiative, and
bright ideas of a creative potential innovator may go nowhere if he
or she cannot get the power to turn ideas into action.

Recall that organizational power derives from supplies of three
"basic commodities" that can be invested in action: *information*
(data, technical knowledge, political intelligence, expertise); *re-*
sources (funds, materials, space, staff, time); and *support* (endorsement, backing, approval, legitimacy).* Even though corporate
entrepreneurs can find some portion of these power tools already
attached to their positions, especially in the more empowering companies, it is more typical that innovations require a search for additional supplies, for additional "capital," elsewhere in the
organization. Thus, a great deal of the innovation process consists of
a search for power.

In listening to corporate entrepreneurs recount their innovative
accomplishments, I heard a surprisingly universal rhythm to their
activities, a symphonylike structure encompassing many variations,
playing back and forth between main themes and subthemes, but
within a familiar pattern. Despite the differences in the innovations
themselves, the composition of an innovation seemed to follow a
common logic, one in which the finding and investing of power tools
figured as a dominant motif. The acquisition of power tools can be
the longest and most difficult theme in an innovative accomplishment, depending on how easy the organization makes it to tap

* See Chapter 6.

sources of power and depending on how many technical bugs the project has.

A prototypical innovation, led by a corporate entrepreneur, has three identifiable waves of activity, occuring in sequence or as successive iterations and reiterations: first, *problem definition*—the acquisition and application of information to shape a feasible, focused project; second, *coalition building*—the development of a network of backers who agree to provide resources and/or support;[5] and third, *mobilization*—the investment of the acquired resources, information, and support in the project itself, including activation of the project's working team to bring the innovation from idea to use.

In more complex innovations, multiple entrepreneurs and multiple teams move through these phases in parallel, with their individual notes sometimes joined in chords.[6]

PROBLEM DEFINITION: GATHERING INFORMATION FOR SALABLE INNOVATIONS

Innovations inside large organizations are not generally the result of a lone operator's deciding to take initiative on his or her own, in the absence of any cues from the environment.

Innovators may be visionary and self-directed, but they are only occasionally fully self-starting, in my experience across companies. Even entrepreneurial managers are frequently stimulated by the assignments they are handed, though these assignments, if they result in innovation, generally leave room for the individual to guide the creation of something new in the course of fulfilling the assignment.

The slightly smaller number of self-initiated projects I saw (45 percent of the total innovations) also did not occur in a vacuum; the entrepreneur listened to a stream of communication from superiors, peers, subordinates, users, or customers and identified a perceived need—a "sore spot"—a missing piece for a high-priority area, a hole in the system. (Indeed, it is well known that a large number of important industrial innovations were first suggested by users of the products.[7]) One engineering manager who led innovations in production methods on his own initiative got the idea to work on this from a casual conversation about current problems over lunch in the cafeteria with the head of manufacturing; the manager mulled it over, then persuaded his boss to let him take a next step of seeing whether a concrete project should and could be developed in this area.

Thus, active listening to the information circulating in the neighborhood is really the first step in the generation of an innovative accomplishment, and information is the first power tool. In the beginning, it seems, the more available sources, the better. In order to translate a set of vaguely expressed needs or a broadly worded assignment into an opportunity for concrete action that produces innovation, the entrepreneur benefits from moving outside received wisdom and established categories. This may be done by examining the issue from a variety of different viewpoints, perspectives, or frameworks to get a new angle on it, particularly through informal discussions with those in other functions with their own perspectives on the matter. More formal means, like task forces, also help when more than the usual information is required, and many areas have to contribute. Multiple perspectives are key.

In the early stages of project definition, the prime mover of an innovation may spend more time talking with people outside his or her own function than with subordinates or bosses inside; an R&D manager said he "hung out" with product designers while trying to get a handle on the best way to formulate a new process development project. "Roberta Briggs" of "Chipco's" "Chestnut Ridge" plant was known for walking through the plant almost every day, spending her time actively listening to the cues about coming problems, in Chipco's equivalent of what Hewlett-Packard calls "management by walking around." What she heard led to the successful Production Project in Chapter 7.

While gathering information, entrepreneurs can also be "planting seeds"—leaving the kernel of an idea behind and letting it germinate and blossom so that it begins to float around the system from many sources other than the innovator.

Problem identification often precedes project definition, for there may be many conflicting views in the organization about the best method of reaching the goals. Discovering the basis for these conflicting perspectives while gathering hard technical data is critical at this stage. In one case, the information popularly circulated in the company about the original design of a part was inaccurate. The innovator needed his own information in order to show that the product problem he was about to tackle was not a manufacturing problem but a basic design flaw. So the early information an entrepreneur acquires establishes the basis for the entire project, not only technically but politically; and this information is generally located in large part outside the manager's own work unit.

A great deal of technical data bearing on the manager's area is in the hands of suppliers, users, or other points in the production-to-sales chain. For new kinds of initiatives, this information has often not been collected previously and must be pulled together from many fragmentary sources. Such technical data are necessary not only for project activities themselves but also to establish professionalism and expertise (knowing more about the area than anyone else) as an aid in selling the project every step of the way.

"Doris Randall," for example, was the new head of a backwater purchasing department that she feared would join personnel and public relations as the "three Ps" of women's ghettoized job assignments in the electronics industry. But she eventually parlayed technical information from users of the department's services into an agreement from her boss to allow her to make the first wave of changes. No one in her position had ever had such close contacts with users before, and Randall found this a potent basis for reorganizing her unit into a set of user-oriented specialties, with each staff member concentrating on a particular user need. Once the system was in place, and hers acknowledged to be functioning as the best purchasing department in the region, she went on to expand this kind of reorganization into the other two purchasing departments in the division.

A second kind of early information need is "political": information about the existing stakes in the issue and needs of other areas that could be tied to the project to help sell it and support it. Even the most hard-nosed, technically focused engineers I saw acknowledged the need for as much political as technical information in early stages of an entrepreneurial project; without it, they felt, a project would never get beyond the proposal stage. "Taking time to understand how the system works" was the way one young entrepreneur defined his strategy.

A third kind of information involves data to demonstrate need and to make a convincing case that the chosen method can address the need. Some enterprising managers conduct formal surveys of users of their area's output or seek models of similar approaches used successfully elsewhere. Thus, the Production Project initiated by Briggs at Chipco took real shape only after two-thirds of the managers and a quarter of the workers in the plant responded to a questionnaire—itself based on the needs apparent from Briggs's earlier information-gathering efforts. In the course of formal surveys or informal conversations, the entrepreneurial manager is also, of course, beginning to lobby for and line up supporters for his or her own view of how best to get the desired results. These kinds of information, then, serve dual goals of providing data that can shape project activities for maximum utility and of creating a tool that can be used to persuade resource holders to back the project because of the numbers of potential supporters already identified.

At this point, all the fragments of information, from initial vague assignment through technical and political data, begin to be focused and directed on a particular target. Entrepreneurial managers now use their intuition to take imaginative leaps into unknown territory in which they have to assume that what is still uncertain in the minds of others is possible. They develop personal immunity to the comments of others that the kind of thing they have in mind "can't be done, has never been done, was tried and didn't work" or other expressions of reluctance to make commitments. The manager now has in mind a specific project that embodies a vision of possible results, in bottom-line terms where possible, but also encompassing the interests of the other stakeholders.

"Heidi Wilson's" experience draws these threads together. In her newly created job as a head of a Chipco manufacturing planning unit, Wilson's assignment was to improve the cost efficiency of operations and thereby boost the company's price competitiveness on its computers. Her boss told her she could spend six months "saying nothing and just observing, getting to know what's really going on." One of the first things she noticed was that the flow of goods through the company was organized in an overly complicated, time-consuming, and expensive fashion.

The assignment gave Wilson the mandate to seek information but not to carry out any particular activities. So Wilson set out to gather organizational, technical, and political information in order to translate her ambiguous task into a concrete project. She followed goods through the company to determine what the process was and how it could be changed. She sought ideas and impressions from manufacturing line managers, at the same time learning the location of vested interests and where other patches of organizational quicksand lurked. She compiled data, refined her approach, and packaged and repackaged her ideas until she believed she could "prove to people that I knew more about the company than they did."

Wilson's next step was "to do a number of punchy presentations with pictures and graphs and charts." At the presentations, she got two kinds of response: "Gee, we thought there was a problem but we never saw it outlined like this before" and "Aren't there better things to worry about?" To handle the critics, she "simply came back over and over again with information, more information than anyone else had." When she had gathered the data and received the feedback, Wilson was ready to formulate a project and sell it to her boss. Ultimately, her project was approved, and it netted big cost savings.

A specific project has now taken shape and must be sold—a necessity despite the fact that the manager may initially have been handed this area as an assignment. It must be sold because the initial

assignment, though bearing some legitimacy, may contain no promises about the availability of resources or support required to do
something of greater magnitude than routine activities.

The most salable projects are likely to be *trial-able* (can be
demonstrated on a pilot basis); *reversible* (allowing the organization
to go back to preproject status if it doesn't work); *divisible* (can be
done in steps or phases); consistent with sunk costs (builds on prior
resource commitments); *concrete* (tangible, discrete); *familiar* (consistent with a successful past experience); *congruent* (fits the organization's direction); and *with publicity value* (visibility potential if
it works). But when these features are not present—as they are unlikely to be in more "radical" innovations—then projects are likely
to move ahead if they are either *marginal* (appear off-to-the-side-
lines so they can slip in unnoticed) or *idiosyncratic* (can be accepted
by a few people with power without requiring much additional support).[8]

It is not hard to see why some organizations reap the rewards of
many more innovations than others, because even getting to the
point of defining a salable project can be stimulated or discouraged
by the environment. Where integrative connections help information flow freely and people to talk across organizational boundaries,
then more potential entrepreneurs are likely to see opportunities or
to get the information to move beyond an assignment to propose a
new option for the organization. Ideas for innovations begin to take
shape in companies in which the first essential power tool—information—is available, and exchange of ideas is encouraged.

COALITION BUILDING:
FROM CHEERLEADING PEERS
TO BLESSINGS FROM THE TOP

Having defined the project, entrepreneurial managers next need
to pull in the support and resources to make it work. The broader
the ramifications of the issues involved in the proposed innovation,
and the greater the attendant uncertainties, the larger the coalition
of supporters needs to be if the idea for innovation is to result in
productive action.[9]

So innovators have to be team creators as well as team users,
skilled at getting others to "buy in" or "sign on." They may build a
loose network of allies, they may "play the pivot" between groups,
or they may gather or work through a formal task force or team. What
is critical at this point is not only the quantity but the source of the
power tools that are gathered, and this is why broad coalitions are

important: whereas resources like money may be universal and usable regardless of their origins, support or legitimacy for a project involving change has "currency" only when it comes from the right sources, the people who currently control pieces of the territory the innovation crosses.

Successful innovations are carried out and produce results in situations where a number of people from a number of different areas have a chance to make contributions, as a kind of "check and balance" system on activity that is otherwise nonroutine and therefore not subject to the usual controls. The need for coalition building before much project activity can be under way also ensures that appropriate support will be available to keep up momentum and guarantee implementation.

The order of coalition building generally resembles a zigzag more than an orderly progression up the chain of command.

"Clearing the Investment"

The first step typically involves "clearing the investment" with the immediate boss or bosses, first explaining the project idea and later keeping the boss informed about its status. The boss's overt support may not be sought at this point—unless he or she was the source of an assignment stimulating the innovation—but merely passive acceptance of the innovator's attempt to seek wider support. In a few cases, bosses indicated disapproval at this point, but the entrepreneurs went ahead anyway, in one instance because he had another reporting relationship where he got a better hearing, and in others because the managers were willing to take the risk once their investments were "cleared" by at least putting the boss on notice; later, they reasoned, they would come back with evidence to convince the boss.

"Preselling" and "Making Cheerleaders"

The most common pattern of support seeking, after approaching the boss, was to go lower, then higher in the organization. Early supporters of an innovative project, I found, were generally at much lower organizational ranks than last supporters, because higher executives often wanted evidence that a project was backed by peers before committing themselves to it. (Last supporters were 2.2 times as likely as first supporters to be above the immediate boss.)

This step was "making cheerleaders"—a process of "preselling" that lined up supporters in advance of formal approval from higher levels. Innovators used a variety of metaphors for this, revealing subtleties in their companies' cultures: at Chipco, the computer firm, support seeking was sometimes called "tin-cupping"—begging for involvement; at GE, a heavy-equipment manufacturer, it was referred to as "loading the gun"—lining up the ammunition.

Thus, peers, managers of related functions, stakeholders in the issue, potential collaborators, and sometimes even customers would be approached individually, in one-on-one meetings that gave people a chance to influence the project and the innovator the maximum opportunity to sell it. Seeing them alone and on their territory was important; the rule was to act as if each person were *the* most important one for the project's success.

It was also important to avoid putting uncommitted people into a room together where they might discover their common opposition or worries about the project. This was related to the *rule of "no surprises"*—not to hold any meetings where people would hear something they had not been prepared for, or eased into accepting, in advance, especially because of the potentially threatening nature of innovations. A financial manager proposing to develop a new budgeting process told me such individual premeetings were critical in launching the innovation; because of technical concerns about the mathematics involved, one-on-one conversations gave potential supporters confidence in him when critical questions were raised at formal meetings, and they did not have to mask their ignorance of the statistics with carping criticisms of him or his idea.

"Horse Trading"

Now "horse trading" can begin: offering promises of payoffs from the project in exchange for the support or even time and money that peers and others might contribute.

One product manager in a computer company was particularly good at "trading up" every time he met with a potential coalition member; he leveraged what he learned in each conversation and managed to exchange the information he was gathering for support. For example, he learned the logistics group's side of the story about a product problem, which he could then convey to the engineering manager, eventually taking back a new point of view to logistics; "Getting them all to sign on for my project was easy after that," he said.

"Steve Talbot's" trades, used eventually to raise market share dramatically and create a new product line, were perhaps the most creative. As an "operations manager" at Chipco with no staff or budget to call his own, he could do nothing without working through the functional managers. So he worked out a series of "deals": if the product managers, who were generally rich, would provide the funds, he would persuade the sales department to reorganize to put a single sales person in charge of each product line, thereby giving the products the personalized touch the product managers felt they deserved. With this agreement in hand, he approached sales; "I have the money to get you new sales people," he told them. It was not hard to get their support for a reorganization plan which was part of Talbot's master scheme for improving the region's market share —and he did the hiring himself.

Sometimes the entrepreneur is in a position to make direct horse trades, or promises of something tangible in exchange for chipping in resources or lending support, like usable information or more funding at the end or an operational problem solved. But more often the exchange agreement involves something less concrete: your support now in exchange for my support later, or support now in exchange for a piece of the credit, recognition, or visibility to higher management. When this is the case, and the innovator cannot pull the dollars or the horse out of his or her pocket, supporters need reassurance that the horse they are backing is not in danger of being a dead one; they need evidence that the project will work, that their investment will offer a return.

The innovator has two forms of reassurance to offer. One is his or her own skill in persuasion by having done enough homework in the project-definition stage to have the right kinds of answers, to look like the expert he or she has become. The second kind of reassurance is *spreading the risk:* making sure that there are enough supporters lined up that their very presence both indicates the likelihood of success and distributes the responsibility among a larger group. So each party approached may look for evidence that others are involved and withhold his or her final commitment until all the pieces are in place. Sometimes peers would be asked only for "pledges" of money, staff, or backing, to be cashed in after higher management approved and authorized the investments.

"George Putnam," for example, a product-testing manager at Polaroid wanted to prove that a better model should be displayed at a major tourist site than the one then favored by the marketing department. He had to be a good trader and involve a large group, because he needed to get support for doing something not in his charter for which he needed a wide variety of tools. Building on a long-term relationship with the people in corporate quality control and a good alliance with his boss, Putnam sought the tools he needed: the blessing of the vice-president of engineering (his boss's boss), special materials for testing from the materials division, a budget from corporate quality control, and staff from his own units to carry out the tests.

Putnam did it, he said, by showing each manager how much the others were chipping in. Then Putnam met informally with the key marketing staffer to learn what it would take to convince him. As the test results emerged, Putnam took them to his peers in marketing, engineering, and quality control so they could feed them to their superiors. (The accumulating support persuaded the decision makers to let Putnam proceed; they eventually adopted Putnam's choice of a model, which did indeed become a stronger moneymaker.)

Securing "Blessings"

The coalition is still an unformed and tentative one until the final step in the process takes place: getting the "blessing" of relevant higher-level officials. The mixing of metaphors seems right; at the level of peers, the "cheerleaders" and "horse trades" are part of a game, but the matter becomes very serious when it is time for the more "religious" matter of official sanction from higher levels.

At this point, the innovators are likely to skip levels and go up as high as they have to to gather the ultimate support (and information or resources) that will make the project ready to go. Top management sometimes directly sponsors or champions a project, especially where large sums need to be expended. But more often, the role played by the top is to offer those signs of generalized endorsement which help convert potential supporters into a solid team. The issue may be simply, as it was in one case, persuading top managers to show up at a meeting where the proposal is being discussed; without this, their "pocket veto" (unofficial tabling of the issue by nonattendance) can give other people excuses for holding off on a decision.

The enterprising manager has to know who at the top executive level has power on this issue (including material resources or key initial approval power) and then negotiate for their support via polished formal presentations. One engineering manager said that he came armed with slides and lists of potential project contributors even though he was meeting alone with an executive with whom he regularly played squash. The *rule of specificity* holds in convincing the top: the more specific the request, the easier to get support. "Blank checks" are rare, except in cases where the innovator is expected to be the "miracle worker" rescuing the company from a sure disaster.

Whereas peers and stakeholders are often sold on the basis of their own self-interest and the promise that the manager knows enough about the issue that he will not get them into trouble, top executives may need more guarantees about both the technical and the "political" adequacy of the project. In the case of an innovative new budgeting process, a key vice-president approved enthusiastically, but other vice-presidents had their staffs take the proposal apart to make sure the mathematics worked; it was a good thing both for the enterprising manager and for his high-level supporter that the numbers were right. Key executives are often evaluating a proposal in terms of its salability to *their* constituencies; sometimes the manager arms them with materials or rehearses them for their own presentations to other top approval points such as an executive committee or the board, if the proposed innovation has big stakes.

The right approach to top management and skill at coalition building were critical in one important innovation at Polaroid. A technical-department head whom I'll call "Arthur Drumm" led the team that developed a new measuring instrument which dramatically improved the company's product quality. In retrospect, he did just what someone in his position should be doing. But at the time, he was the only one arguing for the utility of this approach; those around him were not sure it was needed or would pay off. So Drumm spent months developing the data to show the need, eventually persuading several of his bosses two levels up to kick in a few hundred thousand dollars for this development. He pulled together a task force with representatives from all manufacturing sites to advise on the process and ensure that the instrument would fit in with existing operations. Then there were arguments about the budget line from which the money would come: R&D? Engineering? Early opposition from one high-level manager was balanced by backing from two others who were coached by Drumm for an officer-level meeting at which they presented his proposal.

Drumm felt that it was his skill at managing organizational relationships more than technical know-how which eventually netted the company this extremely valuable new technique.

Presentations of proposals to high-level managers, then, can result in securing promises of supplies or help in obtaining them. Sometimes enterprising managers walk out with the promise of a large capital expenditure. They may get help in obtaining staff or space. But most often—since many of the projects that occur in the middle are not high-budget operations, or the budget can be found lower down in a decentralized organization—high-level managers provide a general expression of support that makes it possible for the enterprising middle manager to sell his or her own staff on the project. The fact that top management has conferred attention on this activity makes it possible to sell others on investing *their* time.

The very fact of availability of a key executive for a meeting on this issue is a signal to the rest of the organization. At Honeywell's largest division, for example, the day the general manager agreed to personally chair the steering committee for employee problem-solving activities was the turning point, making it easy to get everyone else's previously only tentative commitment.

And sometimes the results are material, not merely symbolic. The ultimate go-ahead for Chipco's Chestnut Ridge Production Project was given by plant management after Briggs brought in funding from a corporate R&D committee; these resources made higher-level support tangible.

In a number of cases, top management's promise of resources may be contingent on getting others on board, and so the early coalition becomes even more critical. "If you can raise the money, go ahead with this" is a frequent directive to an enterprising manager. At General Electric, a service manager approached his boss and his boss's boss for a budget for a college recruitment and training program that he had been supporting on his own with funds bootlegged from his staff. The executives told him that they would provide a large budget if he could get his four peers to agree to support the project. Somewhat to their surprise, he came back with this support, having taken his peers offsite for three days for a round of negotiation and planning. In these cases, top management is both hedging its bets and using the ability to secure peer support as a control mechanism on otherwise uncertain or risky ideas.

Formalizing the Coalition

With promises of resources and support in hand, the enterprising manager can now go back to his or her boss or bosses to make specific plans for going ahead with the project. In most cases, the bosses were simply waiting for this tangible sign of power to continue the authorization for project activities, but in other cases, the bosses may not have been fully committed and may be sold only because of the higher-level support acquired.

At the end of the coalition-building phase, some managers may wind up with a formal task force, team, or defined mechanisms such as advisory committees for continuing to involve and inform their key supporters.

Sometimes the new alliances forged are useful in and of themselves. "Frank Jones," for example, a chemical-process manager at Polaroid who was waiting for his chance to upgrade the department and develop new processes, gathered data about production which persuaded the plant manager as well as Jones's second boss, a chemical-operations manager, to help him sell top management on the large capital outlay required. He developed staff support from a number of the functions in the plant, especially the technical specialists and production people. In the course of making it possible to speed up the development of emulsions and help in the marketplace, Jones also was central in creating a new set of working relationships at the plant level which had other payoffs in productivity.

But regardless of the formality of the mechanisms, continuing involvement of the coalition is important, because once action begins, the promised support or the promised resources will be drawn on to sustain the project.

The network of supporters, of course, has not merely been passively involved during the coalition-building phase; their comments, criticisms, and expressed objectives have helped redefine and shape the project into one that is more likely to succeed. So the third result of the coalition-building phase is a better information base—a set of reality checks which ensure that projects unlikely to succeed will go no further. In this sense, the corporate entrepreneur is in part the vehicle through which the organization and its members realize their own ends, in part the agent of others, guided by their interests: in effect, the chicken serving as the egg's way of making another egg.

If corporate entrepreneurs have to be skilled at building coalitions, it is equally true that the company environment in which they operate has to help clear the way. Where segmentalism drives wedges between departments, between levels, and simply between people, then coalition building is discouraged, and innovators are cut off from some sources of power tools. But where there is a history of teamwork and cooperation, where multiple centers of resources exist and are eager to invest them, and where integrative sentiments prevail over segmentalism, then the efforts of innovators, in the stage where they seek the support to move from idea to action, are more likely to succeed.

The structure and culture of a company helps determine how many good ideas ever get far enough along to begin to take shape in an innovation-producing project. The few innovators at companies like "Southern" and "Meridian" went through exactly the same definition and coalition-building phases as those at the high-innovation companies, and they needed the same power tools of information, resources, and support, but they found these much scarcer and difficult to acquire.

MOBILIZATION AND COMPLETION:
KEEPING THE ACTION PHASE ACTIVE

As the action phase begins, the entrepreneur now switches roles, from composer, solo artist, or leading actor to "prime mover" or conductor.[10] A large number of additional players may become involved at this point; when the action of implementing the innovative idea begins, the stage may be quite crowded, as the project team workers are collected and forged into an operating entity.

Most often, the action players are subordinates. The manager or team leader brings them together, gives them briefings and assignments, pumps them up for the extra effort that is going to be required, seeks their ideas and suggestions as a way of both involving them and further refining the project, and promises them a share in rewards. When the action team consists of peers rather than subordinates, the details change, but the operating principle is the same. "I made it work by more selling than telling," one manager put it.

With the team onstage ready to carry out the specific steps needed to make the innovation a reality, the entrepreneur's job is primarily to manage the rest of the world, to handle external relations so that nothing gets in the way of action and completion. The project leader's part involves maintaining the boundaries and integrity of the project by handling interference and opposition, maintaining momentum and continuity of team effort, making any other organizational changes needed to ensure the project's success, and seeing that external communication paves the way for a successful reception of the results. The entrepreneur continues to be more politician and PR agent than technical expert, although he or she may contribute a crucial technical solution at just those moments when the team is losing faith.

These external, boundary-oriented tasks are sometimes just invisible enough to the team that they may think that they are doing all the work that needs to be done, that the technical designs or prototype building or guideline writing *are* the project. At Data General the engineers for a new computer were deliberately kept uninformed about how much work Tom West, the initiator and manager, was doing in keeping the rest of the company from interfering with the project:

> From time to time, Rasala and Alsing would tell some of their troops that West was acting as a buffer between them and the company bureaucrats, but the two managers didn't go into details. To do so would have violated West's unspoken order—"an

unspoken agreement," said Alsing, "that we won't pass on the garbage and the politics." They wished sometimes that the rest of the team could get a glimpse of West when some other manager dared to criticize the Eclipse Group or one of its members. West, notoriously, kept a double standard in such matters. He criticized other groups but would brook no criticism directed toward his own. He could carry this policy to absurd lengths, Rasala thought. Sometimes West would simply ignore criticism directed toward his team. Sometimes he would answer it with questions in this vein: "Are your people working sixty hour weeks?"[11]

Handling Opposition and Blocking Interference

The first action-phase task is to handle criticism or opposition that may jeopardize the project.

It is striking how little overt opposition is encountered by entrepreneurial managers—perhaps because their success at coalition building determines whether a project starts at all. (Only a quarter of the 234 accomplishments systematically examined encountered direct opposition from peers inside or outside the work unit, but practically all of what opposition there was occurred for innovative rather than basic accomplishments.) Opposition or resistance seems to take a more passive form: criticism of specific details of the plan, foot-dragging, low response to requests, unavailability, or arguments for preferential allocation of scarce time and resources to other pet projects. In the last-cited case, it is not so much that the enterprising manager's project is argued *against* as that arguments are mounted *for* other activities.

Perhaps the lack of overt opposition comes from the strength of the initial assignment and signs of backing by key authority figures, as well as the political skill of the enterprising managers. But another explanation might lie in the extent to which overt conflict is dangerous to those engaging in it. For just as the entrepreneurial manager needs their support in this instance, so might they need his/her or his/her advocate's support in the future. Furthermore, it is also hard to oppose those who have been winners before; a good reputation is a potent shield.

Early opposition is likely to take the form of skepticism and therefore reluctance to commit time or resources. Later opposition is likely to take the form of direct challenge to specific details of the plan that is unfolding. But in general, opposition is more likely to arise at the action phase of the project than at any other. A few key supporters may be all it takes to get a project launched, but the voices of the critics and skeptics get louder once it appears that action is actually occurring. It is sometimes surprising to managers that the critics had kept so quiet before. One manufacturing manager involved in gearing up for a new-product production had approached a large number of executives in other areas while doing the cost estimates; they appeared positive about his efforts. But later, when he began organizing the manufacturing process itself, objections began to surface from those same people.

The nature of the opposition becomes clearer at later points in an accomplishment's history for two reasons: because the impending development or change has become more concrete; and because the very act of contacting others in the course of bringing an accomplishment into being may mobilize what would otherwise have been latent or unorganized opposition. Each action designed to develop a broad base of support or solicit suggestions also bears the risk of arousing opponents or eliciting criticisms. Ironically, the very way in which the effective enterprising managers handled later-stage criticism—by open discussion in which they provided a carefully thought-through answer to each criticism—can also potentially run the risk of publicizing the arguments against their plan. The managers who deal with this most easily, of course, are those who have authorization to *implement* as well as develop a plan, because as long as they can keep their opposition at arm's length, not interfering with or disrupting their activities, they can eventually counter all arguments with proof of results.

I could identify a number of tactics that innovators used to disarm opponents: *waiting it out* (when the entrepreneur had no tools with which to directly counter the opposition); *wearing them down* (continuing to repeat the same arguments and not giving ground); *appealing to larger principles* (tying the innovation to an unassailable value or person); *inviting them in* (finding a way that opponents could share the "spoils" of the innovation); *sending emissaries to smooth the way and plead the case* (picking diplomats on the project team to periodically visit critics and present them with information); *displaying support* (asking sponsors for a visible demonstration of backing); *reducing the stakes* (deescalating the number of losses or changes implied by the innovation); and *warning the critics* (letting them know they would be challenged at an important meeting—with top management, for example).

Thus, the enterprising manager may now be spending as much time in meetings, both formal and individual, as he or she did to get the project launched. Equally thorough preparation for these meetings is necessary, because skepticism and objections can be countered with clear facts, persuasion, and reminders of the benefits that can accrue to meeting the objectives. In most cases, this is enough; at Honeywell there was a prevalent view that the "strength of ideas" would win people over. But this did not always work, and sometimes the power base reflected in the coalition of supporters had to be activated. One high-level champion of an innovator's project had to tell an opposed manager to back down—that the project was going ahead anyway, and his carping was getting annoying.

In total, I saw remarkably little backbiting and undermining surrounding the innovative accomplishments, and by and large rationality seems to prevail—once the initial political homework has been done. After all, one of the circumstances in companies with a great deal of innovation is the realization on the part of managers that they must continue to work together over time, and that cooperation is an essential ingredient in accomplishment. This reduces acrimony and overt rancorous conflict.

Depending on the kind of project, subordinates are sometimes the source of direct opposition. This is more likely in reorganizations, or the attempt to develop and introduce a new method, than it is in new-product or market initiatives, because natural resistance to change in their work sets in. When this occurs, innovators sometimes rely on team members to bring others into line. One manager, attempting to reorganize technical and maintenance functions to separate the two specialties, was unable to get subordinates moving. Then one person who was leaving the function—and could then admit that the reorganization was a sensible idea—agreed to persuade his peers.

For practically all the innovations I saw, a manager could not simply *order* subordinates to get involved; doing something above and beyond that requires creativity and cooperation means that subordinates have to be fully committed or it will not work.

Maintaining Momentum: Team Building in Action

The second action-phase task is maintaining momentum and continuity of effort. Here "opposition" or "interference" is internal rather than external.

Foot-dragging or inactivity is in many ways more potent than overt opposition, for several reasons. First, it is the one "weapon" even the most powerless people possess: simply withholding effort. Secondly, a "pocket veto" avoids the discomfort of the confrontation of saying no directly, and it buys time. It may be a while before the resistance is discovered, or there may be an acceptable waiting period during which the manager cannot legitimately repeat his or her request in the absence of a reply or a sign of action. This is why the manager's team-building skills are so important: to ensure enough feelings of ownership and involvement on the part of all project supporters, actors, and suppliers that this form of hidden opposition does not occur.

But even where passive resistance is not a problem, foot-dragging may still occur in the most widely supported projects. Both enterprising managers and their team members complained constantly about limited time. The biggest problem for many of them was the tendency for day-to-day routine activities to take precedence over special projects. In addition, it is easier to whip up excitement over a vision at start-up than to keep it in mind when facing the practical tedium of work. In these cases, sustaining enthusiasm over the long haul is a matter of both the personal persistence of the managers and the strength of their support: when key authorities are clearly waiting for the results, there is impetus for persistence.

Particularly where subordinates or team members have a number of other responsibilities they are carrying out, it is critical for the prime mover to be the force that keeps the project going. A manager might pull in some of his or her bosses or supporters at this point—without compromising the autonomy of the project—to get people back on board when foot-dragging occurs. For example, the manager might arrange a well-timed visit from the big boss, or a letter reminding everyone how important the project is.

Keeping up momentum can be important even where subordinates officially devote all their time to the project. One manager at Chipco, whose assignment was a full-time project to develop new, more efficient methods of producing a product ingredient, held daily meetings with the core team, met frequently with operations managers and members of a task force he had formed, put out weekly status reports, and made frequent presentations to top management. Or leaders might nurture crises to the point where the team can get "charged up" about putting out "brush fires" and find their spirits revived.[12]

The comments of one Honeywell innovator involved in new-product development reflect the mix of pushing individual accountability and constant team building that helped him maintain effort for a major technological breakthrough:

Most of my time during this project was spent making sure that goals were clear: identifying problems and resources or providing a focus on getting the problem fixed. . . . I called in everyone on my team to my office when we got the designs and got them to agree on ownership. I said that each person's ass was tied to it, and we had two days to sign the production plans indicating that they were okay, or to make quick changes if they weren't. . . . The most productive use of my time was communication, giving rewards. . . . It is very important to give information to members of the team—the troops. Too often reports are written just for the boss, which results in a misguided emphasis. . . . I also met on a regular basis with small groups. It was not typical to do this, and it was painful and time-consuming. However, it did have real benefits. It kept ownership alive. It kept everybody focused on the importance of our efforts.

"Secondary Redesign": Rule Changing, Bending, and Breaking

A third task of corporate entrepreneurs in the action phase is to engage in whatever "secondary redesign" of systems, structures, or methods is necessary to keep the project going.

Entrepreneurial accomplishments often end up with such secondary results—a number of other changes made in order to support the central change. For example, a General Electric manager whose team was setting up a novel computerized information bank held weekly team meetings to define tactics; one of the fallouts of these meetings was a set of new awards and a new performance-appraisal system for team members and their subordinates. As necessary, new arrangements might be introduced in conjunction with the core tasks. Methods and structure may be reviewed, and when it seems that a project is bogging down because everything possible has been done and no more results are on the horizon, then a change of structure or approach can result in a redoubling of efforts and a renewed attack on the problem.

Attempts to limit interference and operate freely to "do what needs to be done," as some managers put it, may also stimulate attempts to change organizational rules and arrangements. Whereas some kinds of opposition—direct challenges and criticism that might result in cutting off the flow of power supplies—need to be countered directly, other kinds of potential interference simply need to be kept outside the boundaries of the project. In effect, the manager and his or her group define a protected area for their work—erect a fortress. The manager goes outside this area to head off critics and to keep people or rules now defined as external from disrupting project tasks. Then the manager patrols the boundaries. Or the manager may act as an "interference filter" to block interference from the top or from any formal rules that have to be bypassed to make the project work, so as to permit the action team to do whatever is required to complete the project successfully.

Furthermore, managers may bend rules, transfer funds illicitly from one budget line to another, put designs into production before they receive official approval, bootleg resources from subordinates' budgets, develop their own special rewards or incentive systems which offer bonuses above company pay rates, hold offsite meetings at resorts against company policy, or make sure that superiors stay away unless needed. Important features of the IBM 360 were developed by scientists who ignored the company's command to stop a particular line of research.

One extremely successful marketing manager, for example, took a variety of risks with his budget, using it in unusual ways, but with high payoff. He bought demos for customers from the production budget. A few times he used money to have product features developed. He spent advertising money on product development, getting some new products out of it; for trade shows, he sometimes traded off or shared budgets with product-line management. He also required his people to take at least one course a year, skimming the dollars for this from other budgets. Because of limited R&D funds, he had his people seek and obtain customer contracts which supported in-house product development without loss of proprietary rights to development.

As the leader of an important product-development team commented, "A lot of things we did were unique to that environment. It's clear they weren't always the way things should be done. We all enjoyed it. Anytime you do anything on the sly, it's always more interesting than if you do it up front."[13] Of course, as I showed in Chapter 3, breaking rules was almost a necessity for the few entrepreneurs who emerged in the uncongenial environments of Southern Insurance and the like.

External Communication: "Managing the Press" and Delivering on Promises

The last set of tasks of the action phase brings the accomplishment full circle. The project began with the gathering of information, and now it is important to continue to send information outward.

The innovator and his or her team members both move out and begin to allow others in. They may attend meetings, make presentations, and put together shows that emissaries can take on the road. At the same time, when the project nears completion and there are things to see, they may begin to bring important people in to view the activities—starting to "make cheerleaders" for the presentation of results, to "sign up the rest of the corporation." Even more now than at other times, successful innovators "manage the press," working to create favorable and up-to-date impressions in the minds of peers and key supporters. They do not exactly hide bad news, though they may gloss over it, but they try to send out their own information broadcasts in advance of the grapevine so that outsiders hear the team's version first.

Delivering on promises is also important now. The team needs to meet deadlines as much as humanly possible, to deliver early benefits to investors, and to keep supporters supplied with information—all of which maintains the credibility of the project and its innovator even before concrete results can be shown. And information needs to be widely shared with the team and the coalition as well. The manager may remind people periodically of what they stand to gain from the accomplishment, may hold meetings to give concrete feedback to project actors and to stimulate pride in the project, may make a point of congratulating individual staff on their efforts, and may invent awards and ceremonies and celebrations. I found that a high proportion of all innovators do this, even when they do not get direct rewards themselves (40 percent did not).

One issue is whether company norms make this standard practice. At Southern Insurance, when "Bob Smith," a district manager, took up a collection to give the clerical workers on his successful team a $2,000 bonus—while receiving a slap on the wrist from his management for being a norm breaker—this kind of generosity stood out as highly unusual. But at Honeywell, by contrast, trophies and dinners and plaques and breakfasts and ceremonies in the parking lot—for everyone from technical whizzes to secretaries—are common ways managers recognize their teams.

As completion draws near, then, the sharing of information about the contributions of all contributors is increasingly important; after all, many people have invested power in innovators or joined their efforts because of the promise that, as a Chipco manager put it, "everyone will get to be a hero."

The extent to which "heroes" emerge, of course, is a function not only of the skills of the manager and his or her team but also of the quality of the corporate environment: whether the company makes power tools easily obtainable or less obtainable, smooths the way for innovative projects or puts roadblocks and obstacles in their paths. Thus, it is clear that some companies regularly produce both more innovations and more heroes than others.

Innovation and the Participative/Collaborative Style

Corporate entrepreneurs acquire their power through mobilizing others (and being mobilized by them) as collaborators; they are not "solo artists." They secure information, support, and resources by building an actual or implicit "team" of people who will maximize goals of their own through their involvement. The "power skills" that produce innovation are unlike the image many of us hold of organizational politics: staking out a proprietary claim to an area that can be manipulated at whim and without interference.

Instead, corporate entrepreneurs produce innovative achievements by working in collaborative/participative fashion: *persuading much more than ordering* (although threats, direct orders, or pressure from a bigger boss might be used as a last resort); *team building*, including creation of formal task forces or committees, frequent staff meetings, frequent sharing of information, use of regular brainstorming sessions; *seeking input from others,* including needs of users, suggestions from subordinates, review by peers; *showing "political" sensitivity* to the interests of others, their stake or potential stake in the project; and, last but not least, *willingness to share rewards and recognition.*

A participative/collaborative style, in short, means that the leader interacts and listens; it does not imply that the person is "Mr. or Ms. Nice Guy." One highly successful innovator at GE, who excelled at building teams for dramatic turnarounds, was described by peers in a training program as "unemotional, a hard driver, action-oriented." "Steve Talbot," the Chipco operations manager who built his own base to open new market opportunities, did not hesitate to use threats or have his own boss pull rank to get other functions on board for his efforts. A Honeywell entrepreneur who worked in a highly collaborative fashion, sharing information freely and soliciting staff opinions in regular team meetings, also didn't hesitate to "tie each person's ass" to the plans he or she agreed to. Tom West, the prime mover behind the new Data General computer, was aloof and distant from the design engineers two levels below and sometimes short-tempered with the managers directly under him—in reviewing his achievement, he thought that maybe next time around he'd learn to say "please"—but the collaborative team spirit he engendered was remarkable: about thirty people routinely working eighty-hour weeks for no extra pay.

Thus, corporate innovators may *empower* subordinates, involve them and give them latitude, but they still set tough standards. "Participative management is not permissive management," one Honeywell manager said.

Whatever benefits participative styles have for the carrying out of routine work, they are essential for producing innovations. After all, routine accomplishments can also work with a more traditional autocratic style: doing what is assigned, so not needing external support; having all the tools to do it, so not needing to get anyone else involved; and simply ordering subordinates to do what is required. For example, my statistical comparisons show that basic managerial performance within the boundaries of the job-as-given, and with no new learning emerging (what I called "basic accomplishments"), was much more likely than innovations to involve: *exclusive use of formal channels*—e.g., preexisting budgets (25 percent of the basic accomplishments versus only 12 percent of the innovative ones); *little need for extensive resources and support* (29.6 percent of the basics had low or nonexistent resource needs versus only 9.2 percent of the innovations, and 29.6 percent of the basics had low need for peer and top-management support, versus only 13.3 percent of the innovations); *less movement or activity outside the immediate work unit* (63.6 percent of the basics had little extra-unit activity, versus 55 percent of the innovations); and *little peer collaboration or exchange* (50 percent of the basics involved little peer support, versus only 31.7 percent of the innovations).[14] And in general, they were more likely to involve an authoritarian style.

But for innovative accomplishments, participation, collaboration, and persuasion are necessary, because *others* control the information, resources, and support—the latent power tools—the entrepreneur needs to acquire and invest in his or her efforts. The innovations produced by corporate entrepreneurs often involve the need to seek funds or staff from outside the work unit; set up or work through team mechanisms; attend a large number of meetings of those with a stake in the project area, and make a large number of convincing and reassuring presentations; seek information from outside the department—"political" as well as technical; require "above and beyond" effort from staff. Thus, one of my most robust statistical findings was the association of corporate innovation with a participative/collaborative style.

Seen in this light, the corporate entrepreneur has to understand the uses of *participation* in order to realize *individual* visions.

At the same time, the participative/collaborative processes required to get the power for the project serve as risk reducers, guarding against failure and encouraging completion of the accomplishment. The involvement of others serves as a check and balance, reshaping the project to make it come ever closer to a "sure thing." The involvement of others also puts pressure on the entrepreneur to follow through; once others have invested their information, resources, or support in a project, they have a more active stake in whether results in fact emerge. Thus, what begins as a "risk" because an individual puts an untried idea forward and stakes his or her credibility on its ultimate success is gradually converted into a joint effort in which risk recedes as participation grows. In fact, the few ultimately unsuccessful attempts at innovation in my research can be attributed to a failure of participation—a failure to build a coalition of supporters and collaborators.

The organizational context and political processes surrounding the activities of entrepreneurial middle managers act to eliminate failures early. The very looseness of authority that gives corporate entrepreneurs both the incentive and the freedom to act also serves as a control on their actions. The necessity of seeking legitimation, forming coalitions, generating persuasive information, counting on other people for resources, and so forth—all this means that few projects that could fail will go very far, and it also means that many others along the way develop a stake in successful completion. Of course, there *are* bad initial ideas and dead-end tracks, especially in more technical areas. But these will, via persistence and politics, be converted into a more success-prone direction.

Tolerance for uncertainty is more important than tolerance for risk. It is far riskier to attempt to innovate in an uncongenial environment than in one supporting experimentation; companies that seek more innovation would be better off changing the corporate culture and structure than trying to select or train "risk takers." But what corporate entrepreneurs do show, compared with their counterparts, is a longer time horizon, conviction in an idea, no need for immediate results or measure, and a willingness to convey a vision of something that might come out a little different when finished. And moreover, they have *fun*. Tracy Kidder commented about Tom West's new computer project:

> Adopting a remote, managerial point of view, you could say that the Eagle project was a case where a local system of management worked as it should: competition for resources creating within a team inside a company an entrepreneurial spirit, which was channeled in the right direction by constraints sent down from the top. But it seems more accurate to say that a group of engineers got excited about building a computer. Whether it arose by corporate bungling or by design, the opportunity had to be grasped. In this sense the initiative belonged entirely to West and the members of his team. What's more, they did the work, and with uncommon spirit.[15]

Entrepreneurs are, above all, visionaries. They are willing to continue single-minded pursuit of a clearly articulated vision, even when the line of least effort or resistance would make it easy to give up. But they also need other people to contribute to and participate in their efforts. And so, as circumstances create opportunities for corporate entrepreneurs to move beyond their jobs to act boldly and creatively—to the great benefit of their organizations—the collabo-

240

rative process through which they acquire the power to act also serves as a check on their actions. In an ironic sense, freedom and control, individual action and teamwork are roughly equilibrated in innovating organizations. As individuals with an entrepreneurial bent find the power tools to initiate innovation, they create and work through participative teams. And those teams make it possible for other potential entrepreneurs to step forward with useful new ideas.

CHAPTER 9

Dilemmas of Participation

> "It's a quagmire."
>
> "That's not a very attractive image."
>
> "Well . . . it's a lobster trap. Once you're in it, it's hard to get out. I'm saying, it's worth it, but look at the risks realistically."
>
> —Interview
> with a manufacturing manager
> at Hewlett-Packard

ONE WAY OR ANOTHER, the innovating organization accomplishes a high proportion of its productive changes through participation.

Corporate entrepreneurs—single-minded individuals that they are—still get their projects done by crafting coalitions and building teams of devoted employees who feel a heightened sense of joint involvement and contribution to decisions. The integrative, participative vehicles surrounding innovators—open communication, interdependent responsibilities, frequent team efforts—keep them close to the power sources they need to operate, ensuring access to information, resources, and the support needed for implementation. Involving grass-roots employees on participative teams with control over their own outcomes helps the organization to get and use more ideas to improve performance and increase future skills. Whether called "task forces," "quality circles," "problem-solving groups," or "shared-responsibility teams," such vehicles for greater participation at all levels are an important part of an innovating company.

Masters of change are also masters of the use of participation.

But some of the claims for the effects of greater team participation make it sound like Ms. Lydia Pinkham's Organizational Elixir: millions of dollars in cost savings at a bank! production time cut in half at an instrumentation factory! labor grievances reduced by a factor of nine! "deadwood" transformed into live wires! work-force contentment! energy! productivity!

Many of these claims have solid evidence behind them. Energizing the rank-and-file through participation in team problem solving has indeed produced significant results for many companies. And the effect on the employees themselves does resemble that of a "tonic": the excitement of getting involved and making an impact,[1] especially for those who have never experienced it, for those for whom it is a departure from organizational routine.

But mention "participation" to others, and it may seem like just one more staff or committee meeting. One innovating company's initial definition of participation was "task forces"; getting something done meant working through a task force. So at the drop of an issue, a task force would be formed. There were task forces for major events, and task forces for trivia. There were meetings upon meetings upon meetings, until conference rooms overflowed, and offices seemed empty. And there were also loud groans, after a year or two of this, if anyone dared to suggest that a task force should be convened. Not only were people using task forces where it made more sense to do something else, but they had also neglected to devise any ways to get rid of many of them when they finished their initial work. Some groups were good at finding more things to do, and they were rapidly turning into standing committees, with more and more task forces piling up on top of them. Zealous managers, eager to show that they were in tune with the CEO's push for participative management, started counting their task forces rather than thinking about the substance of what was, or was not, being accomplished. No wonder people at that company were beginning to question the value of participation.[2]

Participation, it is clear, needs to be managed just as carefully as any other organization system, and it creates new problems demanding attention in the course of solving others.

Participation in a team with responsibility for a joint output is not always a preferable process for carrying out tasks; there are circumstances under which authoritative, unilateral decisions or delegation to a single individual makes more sense. Several decades of social-psychological research[3] and the accumulated wisdom of the companies struggling with participation make clear that the use of teams is most appropriate for purposes closely related to staying ahead of change: to gain new sources of expertise and experience; to get collaboration that multiplies a person's effort by providing assistance, backup, or stimulation of better performance; to allow all of those who feel they know something about the subject to get involved; to build consensus on a controversial issue; to allow representatives of those affected by an issue to influence decisions and build commitment to them; to tackle a problem which no one "owns" by virtue of organizational assignment; to allow more wide-ranging or creative discussions/solutions than available by normal means (e.g., to get an unusual group together); to balance or confront vested interests in the face of the need to change; to address conflicting approaches or views; to avoid precipitate action and explore a variety of effects; to create an opportunity and enough time to study a problem in depth; to develop and educate people through their participation: new skills, new information, new contacts.

In short, a great deal of innovation seems to demand participation, especially at the action or implementation stage.

Simply reverse these conditions, and it is clear that there are also times when participation or employee involvement is *not* appropriate: when one person clearly has greater expertise on the subject than all others; when those affected by the decision acknowledge and accept that expertise; when there is a "hip-pocket solution"—the manager or company already knows the "right answer"; when someone has the subject as part of his/her regular job assignment, and it was not *his/her* idea to form the team; when no one really cares all that much about the issue; when no development or learning important to others would be served by their involvement; when there is no time for discussion; when people work more happily and productively alone.[4]

Participation is a way to involve and energize the rank-and-file; it is not a single mechanism or a particular program. And it is certainly not the latest new appliance that can be purchased from a consultant or in a do-it-yourself kit, assembled, plugged in, and expected to run by itself. There are a large number of perils and problems, dilemmas, and decisions, that have to be addressed in managing participation so that it produces the best results for everyone. These revolve around how teamwork is initiated; how it is structured and issues chosen—both the inner workings of teams and their relationship to the rest of the organization; and how both members and onlookers evaluate the whole process.

It is striking that many of the problems of participation arise less sharply and are easier to resolve in integrative environments than in segmentalist ones. A foundation of mutual respect, cooperation, open communication, and crosscutting ties—the empowering conditions described in Chapter 6—can make it easier to add still more teamwork, extending it to lower levels of the organization, providing an important multiplier.

DILEMMAS OF BEGINNING

Participation-by-command: the paradox of initiation. Several years ago I coined an informal definition of participation, to poke fun at the contradictions involved in the launching of most corporate participation programs: *Participation is something the top orders the middle to do for the bottom.*

I found that this definition was true enough to get applause regularly from managers listening to me talk about participation: they were commanded to stop commanding. More often than not, participative activities are initiated because someone at a high level directs others to get involved in task forces, to set up teams, or to treat their subordinates differently—and sometimes tells them, in addition, that they will be measured on how well they do this. In short, lower-level employees are to be given a greater voice, but their managers—who will be rewarded or punished for doing it— have little say. It certainly appears that leaders are not modeling the behavior they want others to adopt.

But how else can participation get launched? Even in situations in which leaders carefully work out all details with the managers below them first, or give employees a say in when and how participation will begin, someone somewhere still has to be the initiator and pusher. In business organizations, grass-roots activism or revolution is neither likely nor desirable as an impetus for participation. *Someone* has to start the ball rolling, and it is hard to prevent at least some others from feeling that participation has been imposed.

This paradox can block action. The "democrats" among us may avoid beginning participative activities until they have complete consensus and voluntary involvement—which at best slows everything down and at worst forestalls any action. After all, people find it difficult to vote in favor of a system they have never experienced, so it may be necessary to push the first experiment in order to get enthusiastic consensus about the second.

Sometimes there are concerns about the contradiction between the command style of top management and the participative activities it is pushing, a concern that arises particularly in segmented companies emphasizing vertical command. This was the case in one major industrial corporation, in which the chief executive foisted the idea of participation on his staff in what some considered an authoritarian manner, and then directed the rest of the organization to set up a variety of participatory structures. The corporate organization-development consultant was bothered by the contradiction and spent most of his time trying to confront the chief executive about his personal style in the hope of changing him. Meanwhile, the organization was indeed gearing up for a number of participative activities that could have used the consultant's help. In a sense, the chief executive's personal style did not matter; the important thing was that he was enabling—or forcing—a number of vehicles for greater employee involvement throughout the rest of the organization. Making those work would eventually have greater impact on the organization than teaching its leader to say "please."

There is an ideal middle ground, of course, that involves key managers and employees in making the decision to go forward by exposing them to the same information the initiator has, by making the initiator's thought processes transparent, and by engaging them in a pilot project for which they help set the ground rules and standards. The first step in many successful employee-involvement efforts, like the one at "Chestnut Ridge," is extensive education and formation of a steering committee. But still, someone pushes, and some others may at first feel forced to go along.

The irony of participation-by-command will eventually fade into historical memory as participation becomes seized and owned by those engaged in it—as long as a second dilemma of initiation is handled well.

"Why aren't they grateful?": the paternalism trap. Participation sometimes appears to be imposed on employees by an organization's leadership in "the employees' own interest," in a liberal "Father knows best" style of organizational change, or what has been called "liberating the masses by beneficent dictate": "Like it or not, you're free."[5] The management expects gratitude for the gift it has given, even though workers may not have sought it and have other concerns. So the thank-you note to management contains resentment and a pouring out of other accumulated complaints, rather than praise for management's enlightenment. Management, in turn, concludes that its employees are ungrateful, and tries to take back its gift.

Even when participation is undertaken in response to perceived interest by employees, leaders can easily fall into the paternalism trap by treating participation as a gift rather than a *right* and as a luxury rather than a results-oriented, task-related organizational tool. Imagine employees' reactions to these two different ways of announcing the start of a participation program:

Announcement A: "Because we at Widgets International care about our people and want all our people to feel included, we are spending some of our 1980 profit to develop a series of em-

ployee task forces and work teams that will give employees a
chance to express their views to management."

Announcement B: "Our business at Widgets International has
been changing over the last few years, and we are increasingly
concerned about improving productivity and quality and developing new production methods in order to stay competitive. We
need to use the talents of all our employees more fully in order
to do this. We are creating a series of employee problem-solving
teams to help us work on these issues."

Announcement A is likely to be greeted with amusement, skepticism, and passive indifference, although a few employees might
indeed feel grateful for their company's paternalism. Announcement
B, on the other hand, is a more straightforward, businesslike statement of what management hopes to gain by tapping its people's
talent.

Employees resent being treated like children for whom managers decide what they *really* need, and this can cast a negative
shadow over early participative activities—even though, ironically,
participation may give workers their first chance to *tell* management
what they really need. Furthermore, treating participation as a management gift makes it vulnerable to removal: "Management giveth,
management taketh away." Rights are protected, and people do not
have to grovel for them. But gifts keep the giver in control and carry
obligations of reciprocity—leaders expecting something in return. A
"Petrocorp" Marketing Services Department director warned his
managers, "You don't give someone opportunity as though it were a
gift. You present them with a *chance* and hope like hell they grab
it."

Furthermore, treating participation as a luxury is insulting to
employees; they do not want to waste their time at something that
management hardly takes seriously and does not consider relevant
to business objectives. Rather than making employees feel that
"we're doing this for them," management should be clear about, and
acknowledge, its own gains from participation, and make sure it is
choosing and designing participative activities to get them: e.g.,
more creative solutions to problems, more worker responsibility for
quality and production, programs better tailored to employee concerns, better "early-warning systems" for communication about
problems, and more ownership of new systems so that they get implemented faster. Instead of hoping for gratitude for how enlightened and giving they are, leaders would instead be engaging
employees' talents in getting something beneficial to the organization.

"Participate or else": the question of voluntariness. Another issue in launching participation is which people get involved, and how. Must participation be obligatory, if not mandatory, in order to deserve the label? That is, must everyone be included? Or does participation by definition require volunteers? Can people choose not to participate or to abdicate their decision-making rights to others—representatives, leaders, or managers? Or should participation be a matter of interest in the issue and willingness to volunteer? There is a clear dilemma here: if participation relies on volunteers, it may not be representative; if it does not, it may be coercive.

A variety of evidence suggests that people differ in their interest in participation,[6] although, as we see later, this often has much to do with the nature of the job. But certainly people care more about some issues than about others, and it is unreasonable to assume that all employees will contribute or involve themselves in the same way. It is also unreasonable to hold off making decisions or taking action unless everyone is involved, just as the search for consensus can often hang up a group. And some kinds of participation, such as problem-solving teams, require the right skill mix. Thus, it is important to establish criteria for action, representation, and membership on teams.

The question of voluntariness of participation is always tricky. On one hand, it is important to handpick people for teams or task forces who have the skills and enthusiasm for carrying out the activities. On the other hand, it is equally important to avoid making participation simply another "job" that people are assigned to without any say in the decision, in true segmentalist, authoritarian fashion. In the Petrocorp Marketing Services Department project, the first members of the Communications Council were appointed by the directors. One appointee, questioning his involvement, said, "*I* don't see any problems with communication in this department," and the appointed chair missed the council's second meeting.

Even if the official "rule" for participative activities is volunteers-only, informal pressures can arise that make participation mandatory *de facto* if not *de jure*. For example, quality circles in U.S. companies generally consist of volunteers from each work unit who want to put in the time and effort. But what happens if those involved get differentially rewarded because of greater visibility or access to information? What if nonparticipants in task forces are labeled "non-team players" because they turn down the opportunity? Thus, it is also important to avoid the kind of peer and management pressure that makes it difficult for workers to say no even if they are asked formally whether they want to get involved. What is important here is balance.

DILEMMAS OF STRUCTURE AND MANAGEMENT

"Escape from freedom": the need for structure. This reflects the importance of clear limits in making an empowering, freedom-generating process like participation work. Erich Fromm's classic book title was intended to convey the ways that people slide into neurotic behavior when given freedoms they cannot handle. True "freedom" is not the absence of structure—letting the employees go off and do whatever they want—but rather a clear structure which enables people to work within established boundaries in an autonomous and creative way. It is important to establish for people, from the beginning, the ground rules and boundary conditions under which they are working: what can they decide, what can't they decide? Without structure, groups often flounder unproductively, and the members then conclude they are merely wasting their time. The fewer the constraints given a team, the more time will be spent defining its structure rather than carrying out its task.

Total freedom, with no limits set, will not occur in a business organization anyway. But the limits can be vague, unclear, contradictory, hidden and subject to guesswork. So the group might make a large number of false starts before it finally learns what is permissible and what is not. It might spend most of its time discussing *how* to decide rather than deciding. Too many choices, too much up for grabs can be frustrating. Anchors are necessary, something to bounce off of, some constraints or criteria or goals.

Thus, turning over a task or an issue to a group of organization members with no guidelines, objectives, constraints, or limits can be extremely ineffective; yet some people (both advocates and detractors) think that this is what participation must mean. Responsible parties (managers or leaders) do not give up all their control or responsibility for results just because they are involving a wider circle of people; nor ought they to leave the participating members to flounder without help. Honeywell's DMSG developed its steering committee for innovative problem-solving "task teams" to deal with just that need. A large number of task forces had been established in the group to create new policies and programs, but until the top-management steering committee was formed, they reported nowhere, had no standards to meet, and received no guidance. The steering committee, which included the vice-president/general manager, quickly brought order and direction, and the task teams went on to produce impressive products.

It is significant in this respect that participation works better where the parties involved in it are strong, and there is clear leadership in the organization. It is a common statement by those experienced in union-management committees that these do not work where either the union or management is weak. The benefits of participation do not seem to occur in stalemate situations where no one has enough power to generate action, or in situations where power is equalized but low. In a comparison of twelve plants, more participation was associated with better communication, better performance, interunit cooperation, top-management support, lower costs, and less work pressure—a better-organized system with better management; but flatter and thus theoretically more equal power distributions alone were not associated with such positive results. However, when participation was coupled with everyone having *more* power (more total influence in the system to get things done), the benefits were even greater. In short, *leadership*—the existence of people with power to mobilize others and to set constraints—is an important ingredient in making participation work.

But structure means giving people full information about the ground rules; it does not imply the imposition of mindless formulas for action—giving people a set of rote motions to go through that have worked somewhere else or have been specified in minute detail. This is the problem with many packaged programs that have been sold to U.S. companies under the "participation-and-" label. (The "and-" generally refers to productivity improvement or quality improvement or morale improvement.) Many American companies are fond of using what Barry Stein has called the "appliance model of organizational change": buy a complete program, like a "quality-circle package," from a dealer, plug it in, and hope that it runs by itself. The opposite extreme from no-structure is overstructuring participative activities with no thought to the appropriateness of the structure for the place where it is being used, and with the elimination of one of the values of participation to employees: the chance to exert more control over work situations.

Quality circles are especially prone to being introduced by overly specific structuring of actions rather than by education in the principles and skills to make this kind of involvement work. The most effective set of work teams handling quality issues at a Hewlett-Packard facility were a natural outgrowth of existing practices, and they were called "quality teams" rather than quality circles. The initial structure consisted of forty hours of training for supervisors, who then devised their own methods. In six months, efficiency was improved by 50 percent by the leading teams, and space was cut by 25 percent. A manufacturing head concluded, "It was not a rote process." In contrast, a neighboring plant that adopted the formula method, complete with one-two-three do-this/do-that structuring, ran into problems with its quality circles. Treated by employees as "a typical fad, a typical campaign," this program was "running out of gas."

Delegation ≠ abdication. Thus, related to the structure issue is a key lesson for managers: delegating responsibility to other people does not mean abdicating managerial responsibilities for monitoring and supporting the process. Some managers assume an either/or world where either they are in complete control or they have given up all control. But delegation—whether by a management team to a set of employee teams or by a single manager to his or her subordinates—means that the manager not only sets the basic conditions but also stays involved, available, to support employees, reviewing results, redirecting or reorienting the team as necessary. Leaders can also help to coordinate activities, centralize record keeping, and serve as points of contact with other departments. Of course, sometimes a manager who simply wants to prove that participation does not work will throw a task at an unprepared team and abdicate all responsibility—thereby setting up the whole thing for failure.

People frequently want leadership and guidance. The problem of authority in organizations in which employees are given a greater chance to contribute and participate is not resolved by its elimination. In one company's experience with autonomous (no-supervisor) work teams which all "evaporated" within two years, a skilled worker involved in one observed:

> Looking at the . . . work project, basically it's a [no-supervisor] structure in order to give the individual more incentive. What I want to say is that it is a waste of time logically and theoretically. . . . I was wondering if QWL couldn't restructure the project; define an ideal lead, and train a person to be that ideal lead. The more I think about it, the structure of manufacturing is authority and supervision; and without it, it would be chaos.[8]

A senior manager agreed that there were problems, but associated them with the failure of higher-level management to provide the right amount of structure and not with the problems of team responsibility:

> The girls out there . . . are really down. . . . They're dissatisfied, because they don't feel they've been getting the things to do the job properly. That's partly our fault. I think we've gone through a bit of a learning process in that we've kind of turned them out on their own. A brand-new product, a different concept of working, and it's floundering. I think, given enough time, we could make it successful, even without authority. There's some natural leadership coming out of this.[9]

"Who cares?": reporting and accountability. One of the reasons for managers to stay involved, even when delegating responsibility

in a participative fashion, is that—ironic though it may seem—the personal concern of the manager for results is the sign to employees of caring. If the manager or the initiator of a team simply walks away from the process once having launched it and never asks for reports, or monitors and measures output, then employees begin to wonder whether this is indeed a high-priority use of their time. They wonder whether anybody really cares about this. Was it simply an empty activity to give them the illusion of participation? But clear accountabilities and reporting relationships are a way of indicating to employees exactly who *does* care and exactly what the value of their activities is.

The twenty-five-hour-day problem. The last issue in launching participation is to find and manage time. Participation in teams and involvement in decisions are time-consuming, or they take time in addition to core jobs, and time is a finite resource. So where will the time come from? Will participation be on company time, or employee time? If it is on company time, is it off the budget of the particular manager, or is it compensated in another way? Members or workers may not always feel that the extra time they must invest in meetings and in informing themselves is justified, particularly where they feel inadequately paid for it. Worker apathy can be a problem; teams can shirk responsibility because they do not feel like putting in the time. So corporations must legitimate participative activities and entitle participants to take time for them as long as they can still do their core jobs effectively.

When time use is not legitimized, inequities can result. People have different amounts of control over their time, and thus greater or lesser opportunities to participate. In Norway, for example, women are less able than men to take advantage of widespread industrial and civic democracy because they must rush home to shop or take care of the children, while the men in this traditional family system are free to put in time after work hours in employee and community councils. In American companies, professionals are likely to control their own time or get easy management approval for task-force involvement, while clerical workers are not, making it harder for them to put in the effort or play as important a role in participative problem solving.

Time negotiation involves more people than simply those directly authorizing or involved in participative activities that extend beyond the immediate job: managers who need to "release" their subordinates' time, peers who wonder if they will have to pick up the slack for a colleague with other commitments. A segmentalist tradition is likely to produce resistance to releasing employees, who are hoarded, as we saw in Chapter 3, just like any other commodity not easily "lent" to another area. For the first "volunteer"-task-force

head in the Petrocorp Marketing Services Department, this created
a number of uneasy situations. There was no clear negotiation with
her managers, and she had to promise the people *on* the task force
that they would not have to spend much time. She described the
rewards and punishments she found in trying to get the task force
moving to department managers who were trying to learn how to
manage participation more effectively:

> This project grew on its own. . . . There was commitment to the
> idea, but the reality of spending time and the money has not
> been borne out. I was not really given the blocked-out time. . . .
> I got people to join the Central Resources Task Force by telling
> them they wouldn't have to do any work—advise only. I've
> been stuck with all the work. I didn't realize I'd be spending
> this much time on this work.
>
> The rewards are sort of nebulous. There's an ego thing in-
> volved. The punishments are quite specific: my boss implying
> that I wasn't doing my regular work. This caused a lot of per-
> sonal pain; what am I supposed to be doing? I took my task-
> force work home to do. My secretary paid the price for helping
> me. She always had to go back to a loaded desk.

Her boss replied:

> Before, this project of Jennifer's came out of the blue. I didn't
> think too much about it at first. After a while I began to see more
> and more of it. Maybe too much?

Everyone listening agreed that top management would need to le-
gitimize task-force and project involvement, letting managers know
that employees were entitled to take the time if they could show that
their core job would be done just as effectively.

DILEMMAS OF ISSUE CHOICE

The "big decision" trap. People are generally paid in corporations to do their jobs. The more time and energy participation appears to take away from this, with no compensating rewards, the
more it will be resisted. But the closer the arena for participation is
to the job territory, the more it will be embraced. Local, task-related
issues seem to work best. Concrete agendas and clear technologies
also make it easier for groups to find the stability to proceed. John
Witte's quantitative pre- and postparticipation survey measures
at "Sound, Incorporated" showed that participation on companywide joint planning committees may have *increased* alienation, but

participation on work teams with control over the immediate job situation *reduced* it.[10] Other observations confirm this. The more distant, broad, and open-ended the territory over which participation extends, the less likely is it to have the desired results without considerable difficulty. This is what I call the "big decision" trap.

A common assumption by managers in debates about participation is that people want to be involved in the "big decisions" about the overall management of their organization or other sweeping concerns. But my observations and a wide variety of evidence indicate that at least initially, employees would rather be involved in local issues reflecting daily annoyances, and that they see some issues as quite appropriately belonging to management. In surveys at two companies about to launch participation programs, workers expressed clear preferences for more involvement in issues such as work procedures and work rates (78/83 percent wanting "some say" or "a lot of say"); quality standards and wages were next in line. But questions of management salaries, hiring and firing, and job assignments held low interest, and the setting of production levels was of interest to only slightly more than half the workers.[11]

It is important not to make assumptions about issues of greatest concern to employees; they should be asked. But secondly, it is striking how often people are most concerned about the "little things" (at least in the eyes of managers) that make a great deal of difference to them in their daily working environment. Involvement in these immediate and local issues is often what they care about most, and they may be content to leave the "bigger" issues to experts and managers.

There is an instructive joke about how a husband and wife divide up their decision territories. He takes all the "big" ones: U.S. foreign policy, the energy crisis, the stock market. . . . And she takes all the "small" ones: which house to buy, school for the children, purchasing decisions . . . Unfortunately, some organizations approach participation in just this way, surveying employees about their reactions to distant aspects of corporate policy but giving them no say in the "small," local matters that may concern them most. It can be frustrating to be just one of many people voicing an opinion about large decisions (the anonymity of the vote) while having no control over local ones (such as the arrangement of equipment or the timing of coffee breaks) that could make a significant difference in work-life quality.

Furthermore, as a place to begin moving toward more participatory systems, local issues for specific, smaller groups make the most sense. People might feel uninformed about macro-issues and thus threatened by having their opinions solicited. Or they might feel that one more voice does not matter anyway. But when it comes to local issues, people can talk from their own experience, respond from their own needs. People are always experts on themselves and on what touches them personally. There is no knowledge gap between workers and managers here.

The "agenda" trap: the need for visible results. These are more likely to be produced by local work teams than by broad, open-ended policy groups anyway. People need to feel that their scarce time is well spent on things that produce something tangible and visible. Therefore it is important, in assigning the tasks for participation, that they be clear, concrete, and likely to produce a solid result.

In the beginning, of course, there may be a certain number of teams or activities that I call "agenda-setting": simply generating issues that will be discussed later. At one leading high-participation company, efforts to increase involvement seemed to be faltering because of disaffection on the part of members of several high-visibility task forces chartered by top management, and the same kind of disaffection was occurring in work-team meetings. It turned out that the method of participation the company had hit upon was to pull together teams to make recommendations about general approaches to a subject. Then each of the recommendations of the first wave of teams would be handed over to a second round for refinement and further specification. Then the second-wave recommendations might go to still a third team or individual to consider implementation issues. In short, each team was setting an agenda for another team rather than engaging in action that would have organizational impact.

At a certain point, then, people become impatient with simply setting agendas and want to be in a position to take action and bring in results.[12] It is better to involve people in teams at the work site or task forces with a specific action mandate than simply to have a number of committees engaging in loose discussion that seems to go nowhere. "Too much talk, too little action" is a common complaint about participative vehicles that do not have concrete tasks to carry out. For this reason, a Hewlett-Packard facility uses its MBO process to prioritize a team's activities; they are encouraged to work on a succession of easy problems before tackling tough ones.

People are skeptical about participation just for show, without any impact on substance: for example, endless surveys of employee opinion without change in the underlying conditions giving rise to that opinion. One company with repeated survey results showing negative attitudes decided that this proved it needed more surveys; several cynical managers said they would "lie" on the next round to give the company the findings it wanted, in the hopes that the time-consuming surveys would end.

Participatory mechanisms that simply set agendas rather than produce action, such as representative councils or advisory bodies, can come to be regarded as part of the control system of the organization, rather than an empowering device, if those mechanisms have no real impact on at least some activities—an attempt to manipulate people by giving them the illusion of a voice.[13] In short, if people are given a chance to talk, other people have to listen—and perhaps modify something as a result. Tangible signs of both the listening and the new outcome are important.

"Is participation its own reward?" There is also another kind of visible result beyond impact on the organization: rewards to participants, compensation, and recognition.

For a while, participation is indeed rewarding in and of itself, especially if it is novel and exciting and provides access to higher-level management as well as the fun of working in a spirited team. But once participation leaves the experimental stage, compensation and recognition have to be more formal. Without this, people can quickly lose enthusiasm, or their efforts will go where the money and the recognition are. There is research evidence that extrinsic rewards for participation are important,[14] that people need to feel that they benefit from contributing more to an organization's effective functioning. In a trucking company, employee stock ownership coupled with participation was associated with more commitment, involvement, and satisfaction than participation alone,[15] indicating the payoff for people who see that they will be personally and individually rewarded for the effort involved in participation. Similarly, workers who participate in productivity-improvement projects need to feel that eventually, they as well as the organization will benefit. At "Chipco's" Chestnut Ridge plant, the solution was to add parallel-organization participation to job descriptions and performance-appraisal forms.

The issue of credit and recognition is important to resolve at the beginning, or there can be problems later. The experience of a major manufacturer with task forces on a variety of employee concerns is instructive. One task force worked particularly well, turning in early its highly professional recommendations for a new communication program, based on extensive hearings with rank-and-file employees. All the recommendations were accepted by the steering committee, and task-force members felt proud of their achievement. So proud, in fact, that the task-force chair carried around with him a diagram of the program, which he showed to anyone he encountered, clearly presenting it as "his" program. He even told a colleague that out of his thirty years of company service, *this* was what he would be remembered for. He was leaving his mark on the company via the task-force project.

All well and good so far. The company was hoping that participation would stimulate just such responses, the moments of energy and excitement punctuating what some employees felt were otherwise grim or routine existences. But the chair was not alone in feeling that he had produced something of value—so did the other team members, especially a professional who was an expert on the program area and felt that he had been particularly influential in the results. For him, it was twenty years of service, but the feeling was the same: *This* is my first above-and-beyond creative contribution. He wanted recognition and credit too. But since the company had not thought about how to recognize task-force members, and since the chair seemed to be monopolizing attention out of his own previous deprivation, it was unclear whether the rest of the task force would ever have any reward beyond their own private knowledge that they contributed.

DILEMMAS OF TEAMWORK

One of the things that help a team work well is the feeling on the part of members that they are an integral, connected part of the group. In a task-oriented team, this means that their contributions are welcome and valued, that they are as important as any other member of the team to the final product. After all, why waste scarce organizational time to feel irrelevant? "Participation" must mean much more than observation and tacit approval of others as *they* do all the important work.

But four kinds of "inequalities" can drive a wedge between individuals and the team.

The seductiveness of the hierarchy. Segmentation cannot be overcome just by sudden definition of "teams." Teams that are pulled together from different external statuses, with the awareness that they will be returning to them, may slip into deference patterns which give those with higher status more air time, give their opinions more weight, and generally provide them with a privileged position in the group. There is a great deal of experimental social-psychological evidence that external status strongly influences the reactions of others in a group discussion. Similarly, stories from marooned military units indicate that people fall into positions determined by their place in the military hierarchy even though external constraints are removed, and the survivors constitute a "community" more than a hierarchy.

So teams, especially in segmentalist environments unaccustomed to the mixing and matching of an integrative culture, may end up duplicating the organizational hierarchy in miniature inside themselves: higher-status people dominating, lower-status people dropping out. Members of one task force in the Petrocorp Marketing Services Department felt that the most important factor inhibiting their participation was the presence of the boss. At two different high-technology companies where I observed cross-hierarchical task forces in operation, the same pattern was repeated: I could guess the relative levels of participants in meetings just by noting who "took over" and whose comments were treated most attentively— regardless of who was officially the chair. In another sense, the chair *was* the highest-status member of the group, and meetings were run as though the rest of the participants were staff that had been assigned to him, instead of partners in a joint task.

The seductiveness of the hierarchy has emotional roots. The emotions that make it easier to reproduce the hierarchy than to operate as partners are principally fear and comfort. The basis for fear is obvious: "crossing" a powerful figure in a group, even if the purpose of the group is to get diversity of opinions, can make people afraid of external retribution later. So the lower-status people hold back, or feel very daring if they contribute. But there is also a comfort factor: it is easier to maintain familiar patterns of relationships and interaction than to experiment with the unfamiliar. Over time, with appropriate support from the higher-status figures, people are more likely to try to act in ways that place them outside the hierarchy. This is what we mean when we say that a group "loosens up."

"Participators are made, not born": the knowledge gap. There are also task-related reasons that hierarchies magically reproduce themselves inside teams designed to "level" status and improve cross-hierarchical communication and cooperation to solve a problem, or that other forms of "inequality" develop. Effective participators are, to an extent, "made, not born." It takes knowledge and information to contribute effectively to task teams, and this has to come from somewhere. Those people with more information about the matters at hand have an advantage over the others, and those in communication-rich integrative companies are all more likely to have the tools for effective participation than those in companies when segmentation significantly reduces information flow.

Often organizational position, with resulting differentials in information access, can create this difference in the team. Wherever there is a knowledge gap that is not closed with information before the team meets, inequalities develop that are often frustrating to the

258 SKILLS FOR CHANGE MASTERS

less well-informed group members, who respond by dropping out or failing to appear at meetings.

Participation *per se* does not always equalize power and may even increase discrepancies. If more poorly informed members sit with the more knowledgeable and skilled in meetings where all of them are theoretically making joint decisions, the less knowledgeable not only may be "shown up" for their lack of knowledge, thus losing power, but may also be forced to endorse, *de facto*, the decisions they supposedly helped make. Their right to complain later is lost.[16] This is one reason that worker participation on boards of directors was found to have mixed, rather than positive, benefits in Europe. People of managerial or equivalent status had an automatic information advantage. (Of course, the same thing is true of inside versus outside directors, leading to the common observation that management's viewpoint dominates board discussions.)

The secretaries in the commercial section of the Petrocorp Marketing Services Department eventually decided *not* to continue their rotating participation in section meetings because of the knowledge gap, which in turn stemmed from the segmentalist culture. Far from making them feel included and giving them a voice, sitting in on section meetings made one of them feel "dumb," since she could not follow the content of all the discussions. A few others felt they had to keep quiet, since they clearly knew less than anyone else. Some were bored and felt it was a waste of time. Special briefings by one of the managers before the meetings were a help, but not enough to compensate for the inequality caused by differential organizational status. This is not a "woman's" problem, either; I have observed the same hesitation and frustration on the part of male production workers in noninnovating companies suddenly thrust into management meetings unprepared.

Differential personal resources. There are also related problems from unequal distribution of skill and personal resources. People bring to groups different levels of personal attractiveness, verbal skill, access to information-bearing networks, and interest in the task.[17] Victor Vroom and other psychologists have argued that there are personality differences among people that make some "fit" better than others in participatory groups requiring responsibility and active involvement, although the evidence for these differences is mixed.[18] It is easy to see, however, that personal characteristics and mutual attraction can play a role in helping people become connected to social networks outside the team that give them an advantage inside it: more informal status, better reputation, earlier gossip. Furthermore, interpersonal attractions among team members can lead to subgroups engaged in "natural" patterns of friendly communication which give people an advantage in team discussions. It is not simply that people may support one another on the basis of friendship, but also that the opportunity for people to discuss issues in smaller units outside the group may mean that they come to the meetings with their thoughts better formulated and their arguments rehearsed.

There are also specific skills involved in articulating opinions, developing arguments, and reaching decisions that are differentially distributed across organizational populations—and not just because individuals are intrinsically different. Hierarchy, or at least the structure of the line organization, intrudes on the participative team again. Development of the kind of skills necessary for effective participation—ability to push a point of view, ability to see issues in context, and so on—is closely associated with job characteristics. The effectiveness of participation in local department committees at British Rail, for example, was a function of the job; jobs with high autonomy (which required working alone and unchecked, as well as initiating action) predisposed people to learn the requirements of decision making. (Thus, the conditions described in Chapter 5 that stimulate innovation, which include broad assignments and the ability to initiate, also help make people more effective team participants.) Then, on top of that, actual involvement in a decision-making process on the job (obtaining and processing information, evaluating outcomes, defining action strategies) tends to teach people to articulate corporate goals. Finally, jobs with more information passing through about local and corporate issues also give their occupants an advantage in effective participation.[19]

The seniority/activity gap. A final source of inequality comes from relative seniority or activity in the team itself. Outsiders or newcomers or those not attending meetings regularly often feel uncomfortable about speaking up, especially if the group has developed its own language, abbreviations, or understandings. Sometimes the group will deliberately close ranks in the presence of a newcomer as an occasion to reinforce its own solidarity. And in what has been termed a "competence multiplier," the most active members may gain a monopoly on the skills required for effective decision making and therefore become even *more* active, beginning an exclusionary cycle.[20]

Thus, in general, the "best" participators who come to dominate team discussions may again turn out to be those already best placed in the hierarchy and in the networks spawned by it. "What kinds of things inhibit your participation?" members of a functional area were asked in an anonymous survey, in reference to their quality-circle–like problem-solving meetings. Among the replies: "Know-it-all centers"; "Owl and Pussy-Cat group glances"—meaning side-long looks exchanged by just a few members; "lack of information on subject under discussion"; "things outside my responsibility and interest"; "fear of being attacked by group"; "older members who make newer members feel insecure."

Overwhelmingly, the element of the meetings that everyone liked best was the presentations by the group's boss—relying on the familiar hierarchy and reducing any feelings of peer competition, because during *his* presentation no one had to feel unequal to a peer. There were also feelings that individual participation would be improved if people did not feel they had to perform and if they all got the same information in advance.

The internal politics of teams. Of course, declaring people a "team" does not automatically make them one, nor does seeking decisions in which many people have a voice ensure that democratic procedures will prevail. A philosophy of participation in no way eliminates jockeying for status or internal competition if people bring self-serving interests into a group, or if they have differential stakes in the outcome, or if they come from segmented organizations whose structure and culture encourages divisiveness and noncooperation across areas. There may be differential advantages to individual members to be gained by pushing particular decisions over others; there may be differential benefits to be reaped outside the group by appearing to be a dominant force in it—like the ambitious young manager who wants to impress his boss with his "leadership" skills. People bring different needs and interests into any kind of group from their location outside it, and these can serve as the origins of team politics.

How much differential needs and interests politicize a team is in large measure a function of how the team is set up in the first place. Group dynamics becomes more competition-centered when rewards or recognition outside the team are scarce, and members are direct competitors for them. There is also more internal politicking when some functions, represented by team members, think they stand to lose by certain decisions of the team, and the representing member is under pressure from colleagues as well as personally concerned. It is a simple psychic-economic calculation: do the gains from dropping certain interests/goals in the name of cooperation outweigh the losses? Cooperation and reduced politicking are more likely to occur when team members are participating in the group as individuals rather than as representatives, because they can make individual deals free of the pressure of a "shadow group" symbolically looking over their shoulders. (Indeed, when teams begin to jell as cooperative entities, even representatives sometimes forget their external group affiliation in favor of team identification—sometimes to the detriment of the constituency supposedly being served by the participation of its representative.)

Beyond the politics of interest maximization, teams are also arenas for the flexing of power muscles in and of themselves. There is often nothing inherently more "democratic" about certain decisions because they were made by teams rather than individual managers. Teams can turn into oligarchies, with a few dominant people taking over and forcing the others to fall into line. There are many examples in history of supposedly representative mechanisms' sliding into oligarchies—e.g., the reputed takeover of some unions by small groups with shady ties. The benign "tyranny" of peers can substitute for the benign "tyranny" of managers, with conformity pressures as strong and sanctions for deviance as impelling. In one highly participative factory, workers complained that they felt too dependent on their teams for evaluation and job security and feared being ostracized by a clique.[21] Members of autonomous work teams in a Cummins Engine plant were likely to be harder on absent members, according to a former plant manager, than management would have dared to be; they would often appear at the doorstep to drag a person in to work if the claimed illness did not satisfy team members. (Of course, they relied on each other's contributions more than in a conventional work situation.) Indeed, management often *counts* on this peer pressure to stay in line as a side benefit of participation.

Finally, teams become politicized when there are historic tensions between members that have not been resolved before the "team" is formed, tensions that are more likely in category-conscious, segmentalist cultures than in integrative ones, where ties cut across levels, functions, and social categories.

These tensions can rise to greater importance when hostile parties are thrown together and forced to interact, especially if they have to rely on each other for reasonable outcomes. This statement challenges a classic social-psychological cliché, based on a famous experiment by Muzafer and Carolyn Sherif, that groups in conflict who suddenly find themselves dependent on each other for survival develop "superordinate goals" which relieve the tensions; they discovered this by fostering group rivalry and then imposing a crisis at a summer camp.[22] But that was summer camp, not a corporation. Experience from joint labor-management participation in problem solving suggests that there are circumstances in which hostility may increase, rather than decrease. If no attempt has been made to create a more integrative system, to resolve tensions and improve communication before the meeting, and if the situation is frustrating—as meetings can easily be—the emotions may rise to the surface, and members of the opposing camps may start blaming each other for team problems. At British Rail, participation by worker representatives in management meetings resulted in increased tension between managers and workers, especially because worker representatives tended to include those more critical of manage-ment.[23]

Successful labor-management committees that seem to belie this do so not because participation automatically created a "team" out of adversaries but because careful groundwork was laid before the parties ever came together to begin joint problem solving. The Sherifs' hypothesis seems to be borne out by the cooperative rela-tionship of the automobile manufacturers and the United Auto Workers, making mutual concessions in a time of crisis, where the "superordinate goal" is survival of the U.S. auto industry. But the two groups got to the point at which joint participation was possible only because of preceding efforts to improve communication, dem-onstrate good faith, and remove irritants.

Thus, "power" and "community" can run at cross-purposes. The more forces there are that fan the political flames within a par-ticipative "team," the more likely it is that members will feel uncom-mitted to the team and unwilling to invest scarce organizational time to make it work. "Political" conflicts and tensions need not be con-scious or overt to be disruptive; there may simply be subtle discom-forts that members can barely articulate which tell them that this is a place to withhold commitment.

The myth of "team." "Inequality" and "politics" in team dis-cussions are not generically so bad. After all, the people we are talking about have learned to live with both in the rest of their service in the corporate hierarchy. Dominance of the "best"—most skilled, most informed—participators seems likely to produce better decisions. "Political" discussions may mean that a variety of in-terests are more accurately reflected in ultimate decisions. So the solution to the problems of lowered commitment that these phenom-ena create should not lie in expecting the skilled and informed to stay out of discussions or those with special needs or interests to forget them. But that "solution"—holding back—is in fact what is fostered by the next dilemma of participation: "team" mythology.

The mythology that surrounds the idea of "team" in many orga-nizations holds that differences among members do not exist—be-cause they are now a "team"—and therefore it is not legitimate to acknowledge them or talk about them. Everyone has to act as if they were all sharing equally in the operations of the group. While inside the team, they have to pretend that they do not see that some are more able than others, or that the highest-level people are dominat-ing, or that the chair is railroading another decision through. Where "team" mythology is strong, only an outsider—a consultant or facil-itator or naive visitor—can open it for examination.

In some organizations (or perhaps it is an American phenome-non), the idea of participation is imbued with a mystique that makes legitimate differentiation among participants difficult. Falling back on the external hierarchy is easier for a group than developing internal rankings, because the hierarchy was created by someone else and does not force the group members to confront their own differences or inadequacies. Even though implicit "rankings" are manifest in practice, as the group carries out its deliberations, it is threatening to the fragile solidarity of a newborn team to acknowledge them. This is a good example of "pluralistic ignorance": everyone knows individually but assumes that no one else does. And as long as no one says it aloud, it might not even be true.

Thus, the members who feel out of the group cannot bring up their concerns because of the myth that everyone is in. People with less to contribute because they are less informed do not feel comfortable seeking help in getting more information to contribute more because of the myth that everyone has an equal chance to contribute. At the same time, the dominant participators might feel slightly guilty or uneasy about their absorption of a major share of the air time, so they decide to keep quiet for a meeting or two, thereby depriving the group of speedier motion toward solutions.

I have heard variants of all of these feelings expressed by members of participative groups. The more task-driven the group, the more they are muted in the urgency of the task, but they still exist where "participation" is assumed to mean rote equality, nondifferentiation. The task may get accomplished, but people harbor secret feelings that participation is not worth the emotional drain, or they may decide that involvement makes them feel *worse* rather than better. Or they simply stop coming to meetings.

Differentiation within a participative team is difficult, of course, in segmentalist corporate cultures that have not found a way to make people feel important or valued for their contributions unless they are in charge. Teams of peers, for example, might thus prefer to pretend that no one is any different from anyone else rather than have it appear that some are admitting to being less important on some issue. But again, where groups are task-driven they may manage to create some kinds of differentiation without too much trouble: individual assignments, choice of a leader or chair, nods to specialized competencies.

However, there are some kinds of distinctions among members that it is difficult for *any* team to make—some decisions that might be better handled by a hierarchy rather than by participation, as we see in the next dilemma.

"It is hard to fire your friend." If a team works, it often develops close bonds which mean that people cannot always be open and honest with one another for fear of hurting or because of norms developed in the group. Groups develop a variety of social and emo-

tional pressures resulting from friendship that make it difficult some-times for people to confront one another, rate one another accurately, or discipline one another. Thus, there are some issues for which managers need to step in and take responsibility. For example, if teams are asked to evaluate the performance of their members, they often resist singling out individuals either positively or negatively and want to give everybody the same rating. There are some issues on which it is a relief to have a higher-status authority figure who simply takes over and decides; it would be too difficult, or too emotionally pressuring, for the group itself.

DILEMMAS OF LINKING TEAMS TO THEIR ENVIRONMENT

"You had to be there": problems of turnover. Team spirit is ineffable; it does not reduce to a set of events that can easily be described to someone who did not share the experience—as anyone knows who has ever tried to tell a husband or wife about the "high" of everyone in the office up against a deadline, pulling together after hours to rush out a major report. Somehow, it cannot mean the same thing in the retelling. Newcomers, latecomers, and outsiders have not shared the group's experiences, cannot see what all the fuss is about, and may even be put off by the enthusiasm of team members. They are less likely to understand the previous problems that current practices were designed to solve, and therefore the practices.

For example, this dialogue took place in a work-unit team discussing its operations; two newcomers had just joined:

NEWCOMER 1: I feel we're supposed to be confessing something, and I don't know what I'm supposed to be confessing. . . .

NEWCOMER 2: I feel you're all dealing in abstracts all the time. I hope something comes out of this that makes us more efficient. I see you've got a "machine," but I don't know what it's supposed to make.

OLD-TIMER: Getting together like this is very helpful if everyone's desirous of improving. It's different in here in our attitudes, even our vocabulary. Inviting a newcomer does impose a problem for them. We have a responsibility to help them.

The newcomer situation is clearly less of a problem where the team is a clearly bounded work unit and new people simply do not enter. It is also less of a problem at innovating companies operating in integrative modes, like Chipco and Hewlett-Packard, where there is a shared culture in the organization, where newcomers are likely to understand "teams" because of having had similar experiences in another part of the system, or where there is an explicit education/socialization process for newcomers or late arrivals. But it is more of a problem for committees, task forces or "forums" that have looser boundaries and greater turnover of the population of attendees. At the Salem bearing plant, even production work teams suffered from too much turnover. The churning of team members because of movement between shifts to permit seven-day operation hindered the formation of strong relationships and the transfer of information. Supervisors, too, rotated every six months; just as they became familiar with the team, they were rendered "lame ducks."[24]

At the least, some team momentum is dissipated in the need to form new relationships and bring newcomers up to speed. Just as when there are long delays between meetings, the group may start spending more time catching up than advancing on the task. Moreover, the new people may all have new suggestions. There was so much variance in attendance for one fifteen-person task force at a high-tech company that five meetings all began by repeating the information provided at the first one. Regular attendees were so frustrated that soon *they* stopped coming, too.

Even more damaging situations can arise when the newcomer has managerial power and can undercut the group's work or take the team in new directions. By not understanding the team's history or the importance of what seem to a stranger like intangible and even "silly" aspects of team spirit, even a favorably disposed manager may unknowingly eliminate some factors responsible for the team's results: "Why do we need a conference room when space is so scarce?" One of the major reasons for the erosion of participation and autonomous work teams at the General Foods Topeka plant was a leadership vacuum after the managers involved in initiating the system left; later managers had too little understanding of it or vision of what it could be.[25]

Thus, turnover is a problem for teams not only because experienced people leave, but because new people enter. So that a team does not have to constantly repeat itself, revise early decisions, or find its work suddenly changed, continuity of people is clearly required, or at least the kind of socialization to a common culture more characteristic of innovating organizations.

The fixed-decision problem is another important continuity/stability issue. The process of "participation" implies the ability to be involved in a wider, and sometimes open-ended, set of decisions. But the requirements of efficient tasks make progress toward the group's goals dependent on "fixing" some decisions so that they are effectively removed from constant negotiation in the participation arena. Furthermore, it is frustrating both to team members and to those setting their own expectations based on team outputs to constantly have the ground shifted, to review and re-argue what appeared to have been settled. This angry dialogue occurred at a worker-management planning-council meeting at "Sound, Incorporated," discussing whether to reopen the question of a four-day week, which was low on the original priority list:

WORKER REP I [*Angrily*]: We keep going over the same thing.

WORKER REP II: Yeah, we've been going over the same thing, whether to issue something like this—

MANAGER I: I haven't attended all the meetings, but I thought we agreed this was down the list—

PROFESSIONAL: To me this is evidence of something we discussed before. If we made a decision, were we willing to stick by it? Now, we set up a list of priorities . . . we voted . . . okay, those three top priorities have been set up, and we have to stick to them.

REP I: You already—

PROFESSIONAL: Wait a minute—

REP I: You already changed the list.

WORKER REP III [*Sarcastically*]: No, we didn't.

REP I: How many times? My point is that those people want us to work on the four-day week, and no one seems to be willing to do anything about that.

MANAGER II [*Loudly*]: That is why we set priorities—so we could work at these things in order.

REP I: I thought we also decided that we could change anything we decided on at any time.[26]

Participation need not mean that everything needs to be created from scratch again as soon as new people are involved to give them a sense of "full" participation. This would be ridiculous and wasteful, as well as extremely frustrating for all the people who have already put in time. For example, once a team has invented a new performance-appraisal system, this issue has to be removed from the participation arena long enough for costs of the new system to be recovered. So some decisions are always fixed and operate as constraints, whether they are constraints from managers or simply past decisions or commitments to be honored. This needs to be made clear to each new team or each new participant.

But the fixed-decision requirement creates a paradox: participatory processes are established to involve organization members in decisions at one point, but then some things (parameters) get set, which later members have to live with, setting constraints on action. Thus, more "participation" is theoretically possible toward the beginning or in newer situations, when the ground rules are being established, or in situations where dramatic change is desired, than later in a system's life. This may be why standing committees on long-ago-decided subjects seem so boring and unappealing, while task forces inventing or creating a new system are so exciting: there is more participation and more power in the second instance. Thus, except in early stages or in revolutionary situations, participation will always be constrained by previous and, therefore, fixed decisions which place limits on debate and action possibilities. It is also likely that whoever the constraint setters are—managers or employee representatives or longer-term team members or simply the weight of organizational tradition—they will be the targets of some resentment from those who did not participate *then* and want to open an issue *now*.

At the same time, the reasons for participation include broader involvement and organizational renewal (i.e., better problem solving). Thus, continuity and fixed-decision requirements, which help the *team* do its work, need to be balanced by other organizational considerations: giving new people a chance, replacing older team members who have too much ownership of their points of view to consider new options in the light of external changes, broadening the team's perspective, suggesting revisions made possible by new technology, questioning what was "fixed" in order to search for the next generation of systems or programs. Especially in "entrepreneurial" and growing organizations, teams may never experience the degree of continuity and stability that they desire, and an air of constant negotiation will surround team deliberations.

The more that participation is localized and deals with issues under the team's control, the easier it is for the team to find enough stability to make progress on its tasks while permitting the organization to seek renewal by creating new problem-solving teams.

"Suboptimization:" too much team spirit. What if the team works together so well that it closes itself off from the rest of the organization, creating its own segmentation? A group can become *too much* involved in its own goals and activities and lose sight of the larger context in which it is operating. For example, the kinds of things that can help a group pull together—a retreat offsite to communicate better, a sense of specialness and unique purpose, private language and working arrangements—can also wall it off from every-

one else. At one innovative factory designed around self-regulated work teams with a great deal of power over their own activities, engaged in completing whole subassemblies of engine parts, the only serious problems the plant encountered involved cross-team competition and lack of interest in collaboration on issues affecting the whole plant.

This is what management theorists call "suboptimization": a group optimizing its own subgoals but losing sight of the larger goals to which they are supposedly contributing. This occurred in one major company that had established a series of employee-manager task forces to develop recommendations for new personnel-related programs. One of them was so carried away by its own momentum and enthusiasm for the task that it began to communicate directly with employees as though its recommended program were a *fait accompli*. None of them bothered to communicate with the others, even though the proposals they were developing had implications for the other groups' plans and, occasionally, overlapped them.

So while team spirit is a good thing for the team's operations, the group has to remember its relationship to the larger organization and has to be encouraged to remain open to the outside rather than closing itself off. The cultural-island problem that killed the Petrocorp Marketing Service Department's project can be generated by the group itself, especially if the next dilemma is not handled.

Stepping on toes and territories: the problem of power. The team-spirit dilemma leads us to the larger question of the link between a participative group and the rest of the organization: the power problems that arise because there are other constituencies who also feel that they have a stake in the problem with which the participative team is dealing. The team needs to be linked appropriately to these other parties, ideally from the beginning. Indeed, those whose legitimate territory involves the issues around which the participation is occurring should themselves be included in planning for it or carrying it out. An early task is to identify all the parties with a legitimate stake in the issue and decide how they will be involved and informed as the activities are carried out.

Power is a particular problem: managers do not want to give it up.[27] And why should they? In many segmentalist corporations, supervisors and middle managers, as I have shown elsewhere, feel sufficiently powerless anyway so that they may be even more resistant to schemes that take away what limited authority they feel they have and do not also give them something else to do to feel important and useful.[28] Sometimes they have reason to be fearful, as companies discover how participation in the form of team "self-management" cuts down the number of supervisors required. But more often they merely feel they are losing control (as in concerns by managers about flextime) without getting a meaningful new role to play, or that they are giving up rights to make decisions and will now be forced to bow to the will of the group.

Unions may also resist participatory programs out of concern about encroachments into their prerogatives if they cooperate with management. American unions are gradually coming to support various modes of worker involvement in planning, problem solving, or decision making, but there may still be skepticism about the potential for "co-optation": reducing union power by getting workers to identify with management.

"NIH" (not invented here): the problems of ownership and transfer. It is a familiar organizational phenomenon, especially in segmentalist cultures, that organization units want to do things their own way and are reluctant to adopt somebody else's solution. But the do-it-yourself mentality can conflict with building on the results of a participating team's efforts. This is less of a problem, of course, if the participants simply constitute a work unit that is solving its own problems with no implication that anyone else will need to do the same thing, and more of a problem if the team is set up to devise programs and procedures that could be potentially useful elsewhere: a task force, a model team, or a pilot project.

The NIH problem highlights another apparent contradiction of participation in large organizations: participation appears to suggest that by definition, everyone will have a voice in new systems and developments that fall within the arena for participation; but the realities of the division of labor and constraints of time mean that some people are going to be the recipients of programs designed by others. Some people are inevitably left out. The same ideas that some arrived at by high participation and that seem so intelligent and useful to them may be rejected by other parts of the organization simply because these creative team ideas are now imposed givens which they played no role in shaping. One division of a leading manufacturing organization, for example, set up a task force to review its pay program, within the limits of corporate policy. The new program was met with great success inside the division, and it received universal praise from top corporate executives. But task-force members seemed certain that no other division would adopt the same system, despite the team's willingness to turn over all its records and to make presentations about the program.

Simple jealousy may play a role. The fact that power goes to pioneers or plan developers rather than to administrators of someone else's plan can apply to teams or work units as well as individuals. Furthermore, a work unit may feel that it needs its *own* identity and flavor—and how can that program which works so well over there in that foreign territory possibly fit *us*?

There are always some NIH problems around diffusion of any organizational innovation, but somehow the problems seem worse when a participative process was involved in developing the innovation. Maybe managers and employees resent their peers' successes more than they resent having new systems pushed down as a result of top-management deliberations. Maybe the existence of participation has to imply decentralization: local units' putting their own stamp on things, coming to their own decisions, shaping their own programs—a diversity of expression within a common culture. But this latter assumption, which sounds reasonable on its face, has to be balanced against the need for an organization to learn from its own experience—meaning, in this case, the experience of its parts. Aside from those programs or systems which have to be organizationwide, for legal or efficiency reasons, there would be duplication and waste if each area invented everything from scratch—similar to the problems we saw earlier around the fixed-decision dilemma.

So transfer of results is important. In some kinds of organizations it would seem almost impossible if left voluntary. At one extreme, companies like "Southern Insurance" and "Meridian Telephone" operate in an atmosphere of enough scarcity and peer resentment that programs pushed down from the top have almost a better chance of being adopted. At the other extreme, companies like Chipco that reward people for having an entrepreneurial spirit and inventing their own jobs encourage managers to devise their own program rather than accepting someone else's—even if it conflicts with someone else's or reduces efficiency.

But there are also ways to set up problem-solving teams that are more likely to have their results and solutions diffused throughout the organization or repeated elsewhere. For example, representatives of other areas can routinely meet with the participating team. Or team membership can include some people from nonlocal areas, who will be the carriers of the idea back to their home territories (if *they* remember to keep their back-home constituents informed and involved).

"A time to live and a time to die." While it is easy for teams to take on a life of their own, participation needs constant renewal, for the sake of both team members and the organization. Even local work-unit teams seem to experience burnout after about eighteen months of intense activity; this is a common report from companies experienced with large numbers of quality teams or semiautonomous work teams. Periods of intensity need to alternate with periods of distance in order to give people the energy to continue, and so new teams form or rev up just as old teams begin to drop out.

In other forms of participation, such as task forces or councils, the question of life cycle is different. Bodies that are intended to be permanent may simply need to change their membership periodically to get new representatives and new ideas. But task forces or other ad hoc problem-solving devices may appropriately dissolve as their work is completed—even though some of them may want to continue the euphoria of organizational centrality and turn themselves into standing committees.

Organizations often do not think of how they will ensure the continuity of generations when they launch participation. But planning for the birth and death of teams is important for the organization's ability to reap the full benefits of participation. Those who are left out of one round of participative activity can know that they will be included in the future; those among the "elite" of participators know that they will be remerging with the organizational masses in their regular jobs. What former participants can leave behind is a legacy of their learning to be used to train and involve new participants, to smooth the passing for the dying and the birth for the neophyte. As these cycles become institutionalized, other linking issues become easier; participating teams and the ability to be effective team members are spread throughout the organization, cutting through it in many ways.

DILEMMAS OF EVALUATION

The "great expectations" trap: hoping for an organizational Utopia. It is important to remember that regardless of how well participation works, it will not solve *all* organizational problems. We have never created the perfect organization yet, and it is unlikely that we will in the future—if perfect means problem-free. However, by the use of participative techniques we can at least come closer to organizations that can involve their employees in staying ahead of change. Participatory systems are often painted by their proponents as organizational Utopias that will automatically improve everything. But it is not participation *per se* that has benefits for the company as much as other things associated with it.

With respect to productivity, experience shows that results can be mixed and can vary even in the same organization from department to department. Furthermore, the extent to which the organization allows for or facilitates participation may not be as large a contributor to productivity as formerly thought. Other factors, such as improved goal setting or training, account for increases in productivity in studies where participation has been offered as the major contributor. This was certainly true at both the Chestnut Ridge plant and the Petrocorp Marketing Services Department. Participation provided the *occasion* for broadening the skill base of employees and taking advantage of their talents to solve problems and invent needed programs.

If a system of participation does not automatically improve productivity, it also does not automatically counter the "alienation" and withdrawal of people in closely supervised, low-skill, repetitive jobs. The nature of the job itself—technology and job content—needs to be tackled first, and then interest in and favorable response to participation might follow.[29]

Increased *satisfaction* is more often a "guaranteed" result of participatory processes than increased *performance*, and satisfaction and performance are not always related. In one study, people with an "internal locus of control" (those who see outcomes as determined by themselves) performed better when they could plan their own assembly procedures, while those with an "external locus of control" (who see others controlling outcomes) performed better when a manager planned for them. (However, both groups were most *satisfied* in the self-planning condition.) But for some people, already "saturated" by involvement in *too many* decisions, even satisfaction may not result from participation; they would rather be *less* involved in meetings.[30]

Participation can have a few positive effects and no negative effects and still disappoint people. Disappointment with the results of participation, for both leaders and employees, is likely to be proportional to the amount that was initially promised. There is a tendency to try to sell participation by allowing expectations to float upward and hiding the problems in fine print. But organizations that appear to promise a great deal—that make it easy for "great expectations" to develop—can also disappoint people more. There is more frustration and cynicism if expectations that are aroused are not fulfilled. We can see some of this cynicism interwoven with the favorable reports in accounts of employee-owned-and-run companies like IGP (International Group Plans) in Washington.[31]

Part of the potential for disappointment in participative systems lies in the way in which most of us judge any new, change-oriented system. It is very easy for all of us to forget what the alternatives are when we evaluate our current situation. So new systems are often judged not in terms of how much better they are than others, but in terms of how far short of their promises they fall.[32]

It is just this problem of expectations that managers often mention as a reason (or an excuse) not to try new programs which could expand opportunity and power: "Let's not arouse expectations we can't meet and end up with employees who are more troubled than before." But this is a good excuse for inaction. It is possible to manage expectations by setting realistic and attainable goals. If people are given realistic information at the outset about exactly what they can expect, and are not promised everything, then they can calibrate their own personal goals accordingly. This means communicating clearly exactly what will and will not come out of this process for management and employees, exactly what benefits might or might not occur for employees' careers, and exactly how results will be measured.

Dashed hopes will not be totally eliminated just because there is better communication when expectations are set; the imagination/reality gap may be narrowed rather than closed. Some people may be initial skeptics about participation—e.g., managers doubting that there will be any results at all worth the time, workers doubting that management really means it—but others may be incurable romantics, imagining a more perfect end state than is realistic, and a less bumpy path to get there. And there are still others who may know better but still want a "quick fix"—the one-time program that can keep the employees happy and producing for all time so that managers will not have to worry about employee performance anymore.

Richard Walton, the Harvard Business School professor involved in many new-plant start-ups using participative teams, has pointed to two equally important errors in these situations: *management pessimism* and *management optimism*. Undervaluing the potential of employees and hemming them in with too many rules and restrictions can lead to as many problems as overestimating what employees will be able to do without guidance and periodic redirection. These two types of errors correspond to a paradox inherent in evolutionary processes: In the beginning of a new team system or organization being managed by participation, there is a "human-resource gap," in that people may have less skill or experience with a problem than the amount of opportunity for dealing with it. Later in an organization's or team's life, when many decisions have been made, there may be a "human-resource surplus," in that people now have more skill and experience than they can apply. People become more skilled just as the system becomes more routinized. Thus, managers have to be careful to let the reality of what the system lets people do correspond to their developmental stage: more attention to helping people get skills in the beginning, and not expecting too much of them; more attention to giving people a chance to use their skills once they have experience, and not suddenly reverting to lower expectations as the period of pioneering and invention dies down.[33]

I should mention another sense in which more participative

arrangements do not constitute organizational Utopias: they may guarantee more involvement *in general* without ensuring that all groups get equal access to the vehicles. A greater degree of organizational participation by itself is no guarantee that more equal treatment of women and minorities, for example, will automatically follow. "Participation" alone will not wipe out sexism and racism.[34] Although the more highly innovating organizations I examined also seemed to be more integrative in every sense, including equal opportunity, this is not necessarily an automatic result of greater employee participation; teams, as I showed earlier, can be exclusionary and politicized too. Finally, we know practically nothing about the effects of increased participation at work on relationships outside of work, such as family life. While past evidence would suggest positive "spillovers,"[35] several company human-resource managers are concerned that the increased absorption in work which accompanies intense participative problem solving might show up in family tensions.

"A little taste whets the appetite": what next? Let's assume that all goes well. The teams do their work effectively; the participative vehicles produce useful solutions and a satisfied sense of involvement. There is then one further dilemma of participation, but one that is still largely hypothetical rather than arising from experience: the question of whether participation is appetite-stimulating rather than appetite-satisfying. Once employees get a taste of decision making on a small scale, do they inevitably want to go on to decision making on a larger scale? Do changes in participation on the job lead to more political activities off it? Once more employees have a chance to develop their own work systems or personnel programs, do they want to go beyond this to play a role in other business activities? After tasting some participation, do they get hungrier for more—dissatisfied with the amount they now have, forgetting that they had even less before? Will "gain sharing," for example—a scheme for giving workers a piece of the action when productivity goes up, a system now being considered at places like Honeywell—have to become the standard practice?

There are no real answers to these questions yet, though they are being asked with increasing frequency. Some experiences with worker participation, such as the ones John Witte reported at "Sound, Incorporated," or my observations at Honeywell, Hewlett-Packard, and Chestnut Ridge, indicate, first, that there are sufficient problems with participation that no one wants to go whole hog very quickly; second, that there is a preference on everyone's part to keep some issues out of the participation arena; and third, that it is possible to keep participation bounded without negative effects on morale.

THE NEED FOR BALANCE

Participation would appear to work best when it is well managed. "Well-managed" systems have these elements: a clearly designed management structure and involvement of the appropriate line people; assignment of meaningful and manageable tasks with clear boundaries and parameters; a time frame, a set of accountability and reporting relationships, and standards that groups must meet; information and training for participants to help them make participation work effectively; a mechanism for involving all of those with a stake in the issue, to avoid the problems of power and to ensure for those who have input or interest a chance to get involved; a mechanism for providing visibility, recognition, and rewards for teams' efforts; and clearly understood processes for the formation of participative groups, their ending, and the transfer of the learning from them.

Note, of course, that most, if not all, of the means to resolve the dilemmas of participation are somewhat more readily attainable in integrative, as opposed to segmentalist, systems. At Honeywell's largest division, the management steering committee for the division's innovative projects has set up guidelines for handling each of those issues and a format for communicating them as part of the training for new team members.

Innovators needing to build teams to carry out innovative projects do better when they manage expectations from the start, neither promising nor expecting too much and allowing people to define for themselves the level of involvement they desire, opting out of participation in areas they do not care about. Political sensitivity is also important, finding a way to involve parties whose power might be at stake and giving them a reason to support the changes. When "more participation" *is* the change, this becomes even more critical.

But while encouraging participation, innovators still maintain leadership. "Leadership" consists in part of keeping everyone's mind on the shared vision, being explicit about "fixed" areas not up for discussion and the constraints on decisions, watching for uneven participation or group pressure, and keeping time bounded and managed. Then, as events move toward accomplishments, leaders can provide rewards and feedback, tangible signs that the participation mattered.

It is clear that managing participation is a balancing act: between management control and team opportunity; between getting the work done quickly and giving people a chance to learn; between seeking volunteers and pushing people into it; between too little team spirit and too much.

There are no rules or formulas for making participation work that substitute for the sensitive judgment of leaders about how to make the right trade-offs in a particular situation. A Chipco manufacturing manager, excited by the results from his team, cautioned his colleagues to use their own judgment when they heard about problems in a particular team: "Don't believe everything or every gripe you hear. A survey you run tomorrow may look radically different from the one you run today. People don't say what they mean when they're feeling down or frustrated. Subordinates may push you to 'act more like a boss,' but their interest is usually more in seeing someone else brought to heel than in getting bossed themselves."

Managing participation is a matter of balance—and patience. "Hang in there, baby, and don't give it up," an innovative manager encouraging the use of teams advised his reports. "Try not to 'revert' just because everything seems to go sour on a particular day." A vice-president of a progressive Midwestern communications company wrote me that she was discovering that her company's QWL programs were teaching her patience, a virtue that little in her previous career had taught her how to cultivate. But this is central to making participation work, to netting the greater gains that can come with team involvement. It takes longer to weld people into a team than to order them around; it takes longer to teach people a variety of jobs than to give them one simplified task. And it will take time for either the positive results of the participative system or any negative results of the authoritarian kind to reveal themselves. So wherever there is pressure for *quick* results, people will be unlikely to support participation.

But the long-term impact of well-managed participatory vehicles for energizing the grass roots and involving them in innovation should be a more adaptive organization, one that can more easily live with, and even stay ahead of, change. It is not so much that employee motivation will be improved as that the organization may be better able to tap and take advantage of employee ideas. Employees, in turn, may be more adaptable—more skilled and thus more flexible, more able to move with changes, and more favorably disposed toward management initiatives for change in which they know they can play a role.

The role of organization members in adaptive change, then, comes about through periodic integrative opportunities to step beyond their roles in the hierarchy of core jobs and get involved in team problem-solving efforts. These constitute episodes of drama and excitement, of almost communal solidarity, that punctuate an otherwise more routinized set of task activities which occur by necessity in the line structure.[36] These participative opportunities constitute the transition mechanisms by which the organization and its people see the need for changes and exercise control over the change process.

In short, participation is not a "program" or a "formula," and it may not necessarily be a permanent way of doing everything. Like the transition rituals of traditional societies that alternate with the more routine everyday structure, participatory processes should be seen as task-oriented, integrative rituals of high involvement and transformation—a way to engage many talents in the mastery of change.

CHAPTER

10

The Architecture
of Culture and Strategy
Change

> The present is a time of great entrepreneurial
> ferment, where old and staid institutions sud-
> denly have to become very limber.
>
> —Peter Drucker

IT IS HARD TO IMAGINE anything more frustrating to middle-level
corporate entrepreneurs and their teams than doing everything right
to develop an innovation, only to have it melt away because higher-
level executives fumble their part in the change process—by failing
to design and construct the new "platform" to support the innovation.

Corporate change—rebuilding, if you will—has parallels to the
most ambitious and perhaps most noble of the plastic arts, architecture. The skill of corporate leaders, the ultimate change masters, lies
in their ability to envision a new reality and aid in its translation into
concrete terms. Creative visions combine with the building up of
events, floor by floor, from foundation to completed construction.
How productive change occurs is part artistic design, part management of construction.

All the pieces can be right—new product prototypes already
test-marketed, new work methods measured and found effective,
new systems and structures piloted in local areas—and still an organization can fail to incorporate them into new responses to changing demands. As General Electric's Vice-President of Corporate
Research and Development remarked recently, it does not matter
what kinds of highly promising new ideas the company develops in
the laboratory: if the overall climate for innovation does not exist

throughout the whole organization—a readiness to readjust in response to the changes that use of the innovation will require—it is highly unlikely that even the best ideas will reach the economic mainstream.

The ultimate skill for change mastery works on just that larger context surrounding the innovation process. It consists of the ability to conceive, construct, and convert into behavior a new view of organizational reality.

I find it interesting that organizational theorists have produced much more work, and work of greater depth and intellectual sophistication, on the recalcitrance of organizations and their people—how and why they resist change—than on the change process. Maybe the first is easier, because "change" is an elusive concept. Not only is it notoriously hard to measure accurately—so how do we know when we have one?—but it can connote an abrupt disjunction, a separation of one set of organizational events and activities from others, in a way that does not match reality.

Many kinds of activities or tendencies are present in an organization at any one time. Some of these cohere and are called "the" structure or "the" strategy or "the" culture. But there may be other activities which contradict this core or begin to depart from it. Thus, at another time we could simply reconceptualize what the pattern is, emphasizing some activities in place of others *which may still linger*, and decide that the organization has "changed." Indeed, the act of making changes may involve merely reconceptualizing and repackaging coexisting organizational tendencies, as the balance tips from the dominance of one tendency to the dominance of another. The historian Barbara Tuchman once used the image of a kaleidoscope to describe this: when the cylinder is shaken, the same set of fragments form a new picture.

Acknowledging the elusiveness of "change," I use a modest definition of it here, one that stays close to the idea of innovation: Change involves the crystallization of new action possibilities (new policies, new behaviors, new patterns, new methodologies, new products, or new market ideas) based on reconceptualized patterns in the organization. The architecture of change involves the design and construction of new patterns, or the reconceptualization of old ones, to make new, and hopefully more productive, actions possible.[1]

It is important to remember that organizations change by a variety of methods, not all of them viewed as desirable by the people involved. The innovations implemented by entrepreneurial managers by participative methods or those designed and carried out by employee teams may reflect more *constructive* and *productive* methods of change, but they do not exhaust the possibilities, nor are they even typical in organizations with a high degree of segmentation and segmentalism. I catalogued earlier some of the authoritarian ways that segmented organizations introduce change: top-down announcements, rigidly controlled formal mechanisms, and the use of outsiders for whom some of the rules are suspended. Changes may also be brought about by internal political actions: for example, a *"coup d'état,"* in which officials plot to remove the chief executive; a rebellion, in which some members refuse to abide by the directives of the top and act according to their own rules; or a mass movement, in which grass-roots groups of activists mobilize to protest organizational policies or actions.[2]

The choice of methods—participative, authoritarian, or political —may be independent of the source of the pressure for change, although there is a strong likelihood that participative methods will be used when an organization's prime movers see the impetus for change as internally driven, based on choice and responsiveness, rather than externally imposed, based on coercion and resistance. In contrast, in a cascade-down effect, a change demand seen as imposed from without and not embraced by the organization's leaders may be handled in authoritarian or political fashion by the organization. (Both are segmentalist in nature.) To put it another way, organization leaders who are not sure they really want to change— whether in response to market pressures, government regulation, or the actions of competitors—may be more likely to restrict the chance for members to participate in shaping the kinds of changes that could occur, thus missing the opportunity to transfer potential threat into innovation.

The external world surrounding an organization and poking and prodding it in numerous ways is obviously important in stimulating change. But since the "environment" is itself made up of numerous organizations and groups—stakeholders and constituencies—pressing numerous claims, with varying degrees of power, and since a company is made up of numerous action possibilities not always expressed in official policy or strategy at any moment, any assumption of correspondence between what the environment "does" and what the company "does" has to be simpleminded and misleading, especially in the short run. If in the long run there appears to be adjustment by the company in predictable ways, that might be *mutual* adjustment—parties in the environment shifting in response to the organization's actions.[3]

But the fact that the environment is important—and its "discovery" represents one of the important developments for both real organizations and organization theory in the last two decades—does not mean that it "causes" change either automatically or directly.

This is a subtle point. Organizational change is stimulated not by *pressures* from the environment, resulting in a buildup of problems triggering an automatic response, but by the *perceptions* of that environment and those pressures held by key actors. Organizations may not respond to environments so much as "enact" them—create them by the choice to selectively define certain things as important. When an organization tries to "see" its environment, as Cornell social psychologist Karl Weick put it, what might it do to create the very displays it sees? And how might the environment change when it "knows" it is being "watched"? [4]

Clearly, decision makers, via their patterns of attention and inattention, intervene between a company and its environment. And this, of course, means that a company with a diverse group in the "dominant coalition" at the top—more fields and functions represented, more diversity in sex, race, and culture—is more likely to pick up on more external cues, as did the task forces at Honeywell or the management committees at J. C. Penney, than a company with a smaller, more homogeneous set of top decision makers or with a single function—whether finance, marketing, or any other—having disproportionate power to define the appropriate focuses of attention.

Furthermore, even if the "environment" looks objective and real in an industry, so that companies in it share *perceptions* of strategic issues, strategy does not automatically follow, and leaders may sometimes make strategic choices based on their own areas of competence and career payoff, rather than on what the best response might be to the anticipated character of the environment. A chief executive with a financial background might have more knowledge of balance sheets than operations and might get more credit in the press for acquiring and merging than for technical advances in manufacturing methods—and so long-term investment in productivity may be neglected. [5]

Innovation and change, I am suggesting, are bound up with the meanings attached to events and the action possibilities that flow from those meanings. But that very recognition—of the symbolic, conceptual, cultural side of change—makes it more difficult to see change as a mechanical process and extract the "formula" for producing it.

"Truth" is Impossible—
and Therein Lies a "Truth":
Accounts of Corporate Change

It is hard to tell the "truth" about organizational changes, and thus to learn what "really" makes them happen. I am not referring to something mundane and mechanical like the limits of participant perception and memory, but to rather more profound systematic forces built into the nature of organizational change itself. In understanding why change accounts are often distorted, we understand some important things about the architecture of change itself.

One limitation on the accuracy of models of change and even accounts of specific changes is shared by all historical analysis: the problem of when the clock starts running. In trying to reconstruct how a particular company got from state A to state B, we are also assuming there were a Time I and a Time II. But what is called Time I? Many current models of strategic planning or planned change begin at the point at which strategic decisions were made to seek an alternative course; recognizing a problem, leaders set out to mobilize the search for solutions or to move the organization in an envisioned direction. This, of course, reflects the rational-planning bias so nicely critiqued by business analyst James Brian Quinn on the basis of his examination of important strategic shifts at major companies like General Mills and Texas Instruments.[6] It also reflects a bias toward "official" history—the assumption that only leadership actions "count."

Generally, however, by the time high-level organizational odometers are set at zero to record change, a large number of other —perhaps less public—events have already occurred that set the stage for the "official" decision process, that indeed make it possible, like a successful experiment by a corporate entrepreneur. And still other events may have occurred that contradict the direction of change.

Thus, lack of awareness of this "prehistory" of change makes any conclusions about how a particular organization managed a change suspect, to say the least, and perhaps impossible to replicate elsewhere, a point to which I shall return. Most of us—and corporate actors are no exception—begin the recording of "history" at the moment at which we become conscious of our own strategic actions, neglecting the groundwork already laid before we became aware of it. What seems to us the "beginning" is, in another sense, a midpoint of a longer process, and not seeing this "prehistory" we may not understand the dynamism of the process already in motion, and we may be haunted later by some of its ghosts. Or we may try to repeat someone else's success based on his/her account of what he/she did "first," as I have seen many companies do, finding that little of it works, because the supposed "first step" was in reality preceded by a large number of other events that set the stage but go unreported because no "intelligent strategy" was involved.

In conceiving of a different future, change masters have to be historians as well. When innovators begin to define a project by reviewing the issues with people across areas, they are not only seeing what is possible, they may be learning more about the past; and one of the prime uses of the past is in the construction of a story that makes the future seem to grow naturally out of it in terms compatible with the organization's culture.

The architecture of change thus requires an *awareness of foundations*—the bases in "prehistory," perhaps below the surface, that make continued construction possible. And if the foundations will not support the weight of what is about to be built, then they must be shored up before any other actions can take place.

I see this repeatedly in the work of innovators and in my own involvement as consultant to change efforts. A new plant manager at Honeywell eager to move worker involvement into his plant, for example, tried to get the workers interested in quality circles or other forms of team participation, to no avail. When he stopped trying to "begin" and looked into the history of the plant, he found repeated instances of violations by management of promises to the workers to improve such comfort factors as the noise level and the intermittently broken air-conditioning system. Here the foundation was a negative one, and trust was low. So the manager had to reset his clock, give up on fast action, and go back; he fulfilled earlier promises, repaired the facilities, in order to build the trust to permit his innovation in worker involvement to take place.

But often the foundations enabling innovations to occur are positive ones: some experience with similar events that provides at least some skilled, knowledgeable people; a history of joint planning by the leaders who are going to have to act as a team to manage the innovation; preexisting relationships of cooperation and trust across segments involved in the change. Changes really "start" there. If those foundations do not exist, they have to be constructed first, and if they do exist, they may need to become part of the story that is told about the change, because foundations not only make change possible, they also provide security and stability—"grounding"—in the midst of it.

The complicated question of "beginning" is only one issue that can distort change accounts. The other is more subtle and complex: the *rewriting of corporate history* is often part of the innovation-and-change process itself.

The actual events in a change sequence (as seen by a detached on-the-spot observer) may seem very different from how they are rendered in retrospective public accounts, especially "official" ones.[7] The reconstruction itself serves important organizational purposes. For example, if the changes are ones that require many people's support to implement, then we are likely to see these well-intentioned "distortions" in official accounts:

Individuals disappear into collectives. What was initiated and pushed by one person may be redefined, because the person was successful in involving others and getting them to take ownership, as the will and the act of the group.

For example, one of my high-level corporate informants was concerned that I not make him too central in one of the accomplishments I recount or give him too much credit as an individual, because he pushed new concepts through the system by working with others to make them feel they had initiated and owned the change —"planting seeds" and then letting the harvest be reaped collectively. For organizational purposes, the whole group had done it, and to assert otherwise would be destructive of the new reality that had been built together. There was another striking example of this in the development of Data General's important new computer by several teams of engineers under the leadership of Thomas West. Though West had put the project together virtually singlehanded, one of his chief lieutenants predicted: "When this is all over there are gonna be thirty inventors of the Eagle machine. . . . Tom's letting them believe that they invented it. It's cheaper than money."[8]

In American companies, this transformation smooths power relations. The use of "power" is made possible partly by the power user's tacit agreement to keep his or her power invisible once others have agreed to participate. Others' participation may be contingent on a feeling that they are involved out of commitment or conviction —not because power is being exercised over them. Successful innovators know this, and so they often downplay their own role in an accomplishment in official organizational communications in favor of credit's going to the whole team. Or they spend time "convincing" their subordinates and giving them a piece of the action even if they could in fact apply visible power.

Early events and people disappear into the background as later events and people come forward. I saw this repeatedly in the course of managerial innovations. The account of a change at any particular moment has to feature most prominently those actors whose actions are most immediately connected to the foundations of the next necessary development. And so the earlier people and events may appear to be forgotten—not because memory fades with time, but because there is little to be gained in terms of continuing the momentum of the change by remembering them.

A Polaroid product-testing manager who worked for many months laying the groundwork for persuading the company to replace the product it was to display at a prominent site allowed the marketing group to take the credit and let his own role recede because he knew that they were the ones who would be taking what he had done further; the story now had to feature them and their work. Similarly, after West's computer design team at Data General managed the remarkable achievement of a superior machine in record time, their long months of intensive effort were turned into just a few occasional words in accounts of the development of the product; once it existed, other groups—manufacturing, sales—needed to take up more room in the story. And as far as I can tell, West seemed comfortable with this, understood it as part of the nature of innovation and change.

Conflicts disappear into consensuses. Just as in the treaties after a war, "losers" may disappear into allies. Pain, suffering, trauma, and resistance may disappear into "necessary evils." What was highly contentious at the time eventually gets worked out, and the price of the final agreement is to forget that conflict had existed, as in political systems where the final vote has to be unanimous despite the acrimony of the debate. The organizational memory, at least, cannot afford grudges, especially in integrative systems; segmentalist ones seem to nurse old wounds longer. One gets cooperation by agreeing to save the face of those who were critical or opposed and not embarrass them by reminding them of it. And the survivors of pain and trauma may, in their turn, agree to forget in exchange for some of the benefits of the change.

Equally plausible alternatives disappear into obvious choices. To get commitment and support for a course of action may require that it appear essential—not as one of a number of possibilities. By the time a decision is announced, it may need to be presented as the *only* choice, even if there are many people aware of how much debate went into it or how many other options looked just as good.

The announcers—the champions of the idea—have to look unwaveringly convinced of the rightness of their choice to get other people to accept the change. Unambivalent and unequivocal communication—once a variety of alternatives have been explored—provides security. One CEO of my acquaintance is not particularly good at this; he presents decisions about new procedures tentatively, expressing ambivalence about the favored option in the light of plausible alternatives—and others say to themselves, "If *he* isn't convinced, then why should we do anything about this?" The consequence is that change is stalled; no one ever does seem to get around to using the new procedures.

Accidents, uncertainties, and muddle-headed confusions disappear into clear-sighted strategies. There is a long philosophic tradition arguing that action precedes thought; a "reconstructed logic" helps us make sense out of events, and they always sound more strategic and less accidental or fortuitous later. Some analysts are even willing to go so far as to say that organizations formulate strategy *after* they implement it.[9]

But the importance of defining a clear direction, even if one is already almost at the destination, is to build commitment by reducing the plausibility of other directions, to reinforce the pride people take in the intelligence of the system, or to reward those leading the pack by crediting their vision, to remove any lingering doubts about what the direction is, and to signal to critics that the time for opposition is over. Thus, it may be organizationally important to present the image of strategy in the accounts that are constructed whether or not this rational model conveys the "truth." For the innovators to get the coalition to chip in with investments, for example, they have to feel that the entrepreneur knows what he or she is doing; strategic plans are one of those symbols which are highly reassuring to investors.[10]

Multiple events disappear into single thematic events. How a story about change is constructed also comes to reflect what the organization needs to symbolize, what images it wishes to create or preserve, what lessons it wants to draw to permit the changes to be reinforced or the next actions to be taken. Sometimes this reflects the preservation of power: e.g., creating the appearance that the leader did something that was really the result of a great deal of behind-the-scenes staff work, eliminating the messy events and focusing on the outstanding success, or telling about only the times when the leaders showed their commitment to quality of work life and ignoring the times they did not. Sometimes this simply reflects the human reality that too much complexity and detail cannot be grasped and remembered easily and thus interferes with a clear conception of what the situation now *is*. So a large number of things that might have occurred are reduced to just a few critical ones which tell a story that gives people a common image of what is now the right thing to do.

The fragility of changes (that exist alongside the residues of the old system) disappear into images of solidity and full actuality. Multiple organizational tendencies, including contradictory ones, often coexist, but these are ignored in favor of insistence that an innovation has taken hold simply because it exists at all. In some companies, this helps reward the innovators, builds commitment, and disarms the critics. But in others, where innovation is still threatening even when successful and productive, I suspect that this represents a kind of collective sigh of relief that "now we've done it; we don't have to think about change anymore"; some organizations are too ready to believe that all the hard work is over when *one* example is in place.

As if all these "distortions" were not enough, the organization's culture also influences the stories that must be constructed about change. Some prefer to submerge changes into continuity; others like to turn continuities into change. I have pointed out that segmented organizations may promote change-aversiveness and, thus, a preference for denying that major change has occurred. This makes it easy, of course, for the rest of the system to avoid adjusting in response; if the official story says that nothing much has happened any differently from what the system already knew, then no one has to move out of the safety of his or her segment.

On the other hand, change-embracing organizations like "Chipco" that pride themselves on entrepreneurship may do just the opposite: deny continuity and prefer to cast a large number of events—even those not particularly innovative—as "changes." At Chipco, credit went to originators, not to initiators; so people were pressed to claim "change" even when what they were doing was not greatly different, and people were consequently a little less able to build on one another's innovations than at other innovating companies.

There are tactical uses to these ways of constructing change, too. For example, announcing "change!" in one part of an organization's world to make clear the need for change in others is a stimulus to action in a way that stressing continuity is not. It is a power move, used well by leaders of social movements to rally the troops as well as by corporate staffs to influence the line organization. And younger people invoke change to show the older ones in power that their wisdom no longer fits, and they should step aside. In contrast, in "cultures of age," denial of change preserves the power of the establishment.

All of this tells us something important about the essence of the change process: *Organizational change consists in part of a series of emerging constructions of reality, including revision of the past, to correspond to the requisites of new players and new demands.* Organizational history *does* need to be rewritten to permit events to move on. (In a sense, change is partly the construction of such reconstructions.) To use a physical analogy: as each floor of a building is built, the supports need to be made invisible to permit focusing on the important current thing—the *use* of the space. Similarly, as an innovator "sells" each member of her coalition on the worth of her project, the influence process and perhaps even the origin of the idea may have to be "forgotten" and not revealed outside the group to permit attention to go to the people whose role is critical now. Official histories of changes, reports about projects, and even the way organization members tell one another about what happened to move the system from A to B always serve a present function.

The art and architecture of change, then, also involves *designing reports about the past to elicit the present actions required for the future*—to extract the elements necessary for current action, to continue to construct and reconstruct participants' understanding of events so that the next phase of activity is possible. "Power" may need to remain less than fully visible; "prime movers" may need to make sure others are equally credited; room at center stage may need to be given over to those people and activities that are now necessary to go on from here—e.g., the marketing people instead of the product developers becoming the "heroes" of the account.

Change masters should understand these phenomena and work with them; they should know how to create and use myths and stories.[11] But we should not confuse the results—an official, retrospective account of organizational actions—with lessons about guiding organizational change. We need to understand what goes on behind and beyond official accounts, to create models that are closer to events-as-they-happened than to events-as-they-are-retold.

In short, those who master change know that they can never tell the "truth," but they also know what the "truth" is. In their actions they exhibit knowledge at both levels, recapturing those aspects of a change process which have faded or disappeared in official accounts and rational models. Thus:

- Where groups or organizations appear to "act," there are often strong individuals persistently pushing.
- Where recent events seem the most important in really bringing the change about, a number of less obvious early events were probably highly important.
- Where there is apparent consensus, there was often controversy, dissent, and bargaining.
- Where the ultimate choice seems the only logical one, unfolding naturally and inevitably from what preceded it, there were often a number of equally plausible alternatives that might have fitted too.
- Where clear-sighted strategies are formulated, there was often a period of uncertainty and confusion, of experiment and reaching for anyone with an answer, and there may have been some unplanned events or "accidents" that helped the strategy to emerge.

- Where single leaders or single occurrences appear to be the "cause" of the change, there were usually many actors or many events.
- Where an innovation appears to have taken hold, there may be contradictory tendencies in the organization that can destroy or replace it, unless other things have occurred to solidify—institutionalize—the change.
- And where there appears to be only continuity, there was probably also change. Where there appears to be only change, there was probably also continuity.

These realizations constitute part of the "architecture" of change. But there are also a set of building blocks that together constitute the structure behind the process of change.

THE BUILDING BLOCKS OF CHANGE: FROM INNOVATION TO INSTITUTIONALIZATION

It is important to see how micro-innovations and macro-changes come to be joined together, how major change is constructed out of the actions of numerous entrepreneurs and innovators as well as top decision makers.

"Breakthrough" changes that help a company attain a higher level of performance are likely to reflect the interplay of a number of smaller changes that together provide the building blocks for the new construction. Even when attributed to a single dramatic event or a single dramatic decision, major changes in large organizations are more likely to represent the accumulation of accomplishments and tendencies built up slowly over time and implemented cautiously. "Logical incrementalism," to use Quinn's term, may be a better term for describing the way major corporations change their strategy:

The most effective strategies of major enterprises tend to emerge step-by-step from an iterative process in which the organization probes the future, experiments, and learns from a series of partial (incremental) commitments rather than through global formulations of total strategies. Good managers are aware of this process, and they consciously intervene in it. They use it to improve the information available for decisions and to build the psychological identification essential to successful strategies. . . . Such logical incrementalism is not "muddling," as most people understand that word. . . . [It] honors and utilizes the global analyses inherent in formal strategy formulation models [and] embraces the central tenets of the political or power-behavioral approaches to such decision making.[12]

An organization's "total" strategy is defined by the interaction of major subsystem strategies, each reflecting the unique needs, capacities, and power requirements of local units. Even when it is impossible to fully guide the organization from the top—i.e., predict in advance how these units will evolve—the right kinds of integrative mechanisms, including communication between areas, can ensure the coordination among these substrategies and micro-innovations that ultimately results in a company's strategic posture. In short, effective organizations benefit from integrative structures and cultures that promote innovation below the top and learn from them.

This kind of analysis of change provides a link between micro-level and macro-level innovation: the actions of numerous managerial entrepreneurs and problem-solving teams, on one hand, and on the other the overall shift of a company's direction to better meet current challenges. The buildup of experiences from successful small-scale innovations—or even the breakthrough idea that an innovator's work produces, in the case of new products or new technological processes—can then be embraced by those guiding the organization as part of an important new strategy.

In short, action first, thought later; experience first, making a "strategy" out of it second. Strategy may not so much drive structure as exist in an interdependent relationship with it.[13] In many cases new structural possibilities out of experiments by middle-level innovators make possible the formulation of a new strategy to meet a sudden external challenge of which even the middle-level innovators might have been unaware. Then the new strategy, in effect, elevates the innovators' experiments to the level of policy.

I see a combination of five major building blocks present in productive corporate changes, changes that increase the company's capacity to meet new challenges.

Force A. Departures from Tradition

First, activities occur, generally at the grass-roots level, that deviate from organizational expectations. Either these are driven by entrepreneurial innovators, or they "happen" to the organization in a more passive fashion.[14]

Some departures may be random or chance events reflecting "loose coupling" in the system—i.e., no one does everything entirely according to plan even if he or she intends to, and slight local variations on procedures may result in new ideas. They may be the result of "accidents"—i.e., events occur for which there is no contingency plan, or the organization's traditional sources are exhausted, so the company innovates by default, turning to a new idea or a new person just to fill a gap. Or a "hole" in the system may open up because another change is taking place: a changeover of bosses leaving a temporary gap, a new system being installed that does not yet work perfectly. All these constitute the "unplanned opportunities" that permit entrepreneurs to step forward even in highly segmented, noninnovating companies; they may work best at the periphery, in "zones of indifference" where no one else cares enough to prevent a little occasional deviance. The ideas or experiences resulting from deviant events then constitute "solutions looking for problems" [15]— models that can be applied elsewhere.

In innovating companies, in contrast to their less receptive counterparts, a high proportion of these departures from tradition are brought about through the actions of entrepreneurs who seek to move beyond the job-as-given. They may be stimulated by a plan, in the form of an assignment, but they may also be invented by the entrepreneur in response to cues he or she is getting that suggest there is a problem to be solved. As I have shown, there will be more such initiative-taking innovators in an organization with a large number of integrative mechanisms that provide incentives for initiative and make it easy to grab power. The presence of many integrative mechanisms also makes it more likely that the innovation will succeed, producing immediate payoff.

Departures from tradition provide the organization with a *foundation in experience* to use to solve new problems as they arise or to replace existing methods with more productive ones. This foundation in experience suggests the possibility of a new strategy—one that could not be developed as easily without the existence of organizational experiences. At the same time, those experiences condition the direction of any new strategies. In effect, it is hard to see where you want to go until you have a few options, but those options do not limit later choices.

One lesson is straightforward: an organization that wants to innovate to stay ahead of change should be just loosely enough controlled to promote local experiments, variations on a plan. It should make it easy for ambitious innovators to grab the power to experiment—within bounds, of course. It is those variations—sometimes more than the plan itself—which may be the keys to future successes. And there need to be enough experiments for organizational policymakers to have choices when it comes to reformulating strategies. This constitutes the internal equivalent of a "diversified portfolio" for turbulent times.

There is another important value that successful experiments or small-scale innovations have for the change process: they prove the organization's capacity to take productive action. An unfortunate number of change efforts seem to begin with the negative rather than the positive: a catalogue of problems, a litany of woes. But identification of potential, description of strengths, seems to be a better—and faster—way to begin. In my own experience helping corporations develop new modes of operating, I have found it valuable to look for the already existing innovations that signal ability to make the shift, and then to use these as the organization's own foundation for solving its problems and designing a better system. Exemplars—positive innovations—are better to highlight than trouble spots when one is trying to move a whole system.

But deviant events do not by themselves produce major change. Large systems are capable of containing many contradictions, many departures from tradition that do not necessarily affect the organization's central tendency. As in the case of the Petrocorp Marketing Services Department, innovations can easily disappear.

"Deviant" events result in overall change only under one or more of these circumstances: Perhaps enough similar instances of the event or idea accumulate slowly over time so that at some point, definition of the organization's central tendency changes in response to the new reality—a very slow process. (But these are much more common in integrative than segmentalist organizations.) Or, as a second possibility, the organization has mechanisms for the transmission of positive innovations to other sectors which might take advantage of them—e.g., informal or formal communication mechanisms for cross-fertilization. Or, finally, impending crisis or obvious problems that cannot be solved by traditional means lead the organization to search for a solution to grab, and so the deviant idea is pushed forward. The first circumstance is too slow and leaves too much to chance for today's competitive business environment. The second circumstance is more characteristic of high-innovation organizations than of noninnovating ones. But the crisis factor is central to major changes at both ends of the spectrum.

Force B. Crisis or Galvanizing Event

The second set of forces in the change process involves "external" ones, changes elsewhere that appear to require a response. By external-in-quotes I mean that they do not necessarily come from outside the organization—e.g., a lawsuit, an abrupt market downturn, the oil embargo, a competitor's new-product introduction—but may also be events within an organization's borders that are outside current operating frameworks—e.g., a new demand from a higher-level official, a change of technology, a recognition of change in the work force. The change-stimulating-change chain is one reason it is so hard to develop an orderly model of the change process; overlapping events intrude on one another. What is "external" to any change sequence we are trying to describe—or *manage*—may be the A force (tradition departure) or C force (strategic decision) in some other change sequence.

The critical point for the people involved is that the event or crisis has a demand quality and seems to require a response. If the crisis is defined as insoluble by traditional means, or if traditional solutions quickly exhaust their value, or if the external parties pushing indicate that they will not be satisfied by the same old response —then a nontraditional solution may be pushed forward. One of the grass-roots experiments or local innovations may be grabbed.

I propose that organizations with segmentalist approaches to problems will be less "externally" responsive. A tendency to isolate problems—more accurately, pieces of problems—in segmented subunits, and a reluctance on the part of each subunit to admit to being unable to handle its piece adequately, will result in less ability to perceive earlier crises before they add up to full-blown disasters. The "seen it all before" syndrome in a culture of age may result in few things being seen as "crises." Danger signals may simply not be attended to as events requiring response. And even if one unit—perhaps assigned to scan a particular part of the environment—sees the signs, there may be few mechanisms for transmitting this information to other units, or for getting others to cooperate in a response. Thus, perception may be restricted, and so may action.

On the other hand, integrative approaches may mean that an organization "sees" more galvanizing events in general and "sees" them earlier. A tendency to tie problems to larger wholes is one aspect of this. Rather than writing off potential external problems, an organization characterized by this approach may instead see small crises as symptomatic of larger dangers and prepare earlier preventive responses. Since information flows more freely across integrative structures, and since the culture encourages identification with larger units and issues rather than smaller units and specialties, it is easier for the signals of "external" change seen by one part of the system to be added to those seen by others, helping to define a "crisis" demanding response. In addition, the culture of change we saw in innovating companies may promote in people the desire to define events as "crises" that can be used to mobilize others around the search for an innovative response.

While I have generally refrained in this book from making comparisons to Japanese management, there is one generalization worth mentioning. MIT sociologist Eleanor Westney has found that large-scale Japanese organizations are prone to identify crises with what seems to the Western observer to be almost hysterical rapidity. Despite a decade of rising circulation for the national press, for example, Japanese newspapers have been finding "indicators" in demographic developments of an impending "newspaper crisis" in Japan, and using it to galvanize marketing strategies.[16]

At the same time, effective response to crisis may depend on the tradition-departure factor. That is, the organization already has in hand the possibility for a response with which it has experience, and thus it can move much faster to make the changes the crisis seems to demand. Thus, random departures from tradition that had occurred almost by accident, it seemed, helped General Motors respond quickly and innovatively to a number of crises facing its industry in the 1970s, as we shall see in the next chapter.

But neither deviance nor crisis alone guarantees changes without the next two conditions in place: leadership for making strategic decisions in favor of change and creating an orderly plan, and individuals with enough power to act as "prime movers" pushing for the implementation of changes once the decision has been made.

Force C. Strategic Decisions

At last we get to the point in the process familiar in most of the "change management" or "strategic planning" literature. This is the point at which leaders enter, and strategies are developed that use Force A to solve the problems inherent in Force B. A new definition of the situation is formulated, a new set of plans, that lifts the experiments of innovators from the periphery to center stage, that reconceptualizes them *as* the emergent tradition rather than as departures from it.

The notion at Chipco that leaders *select* strategy in part from among solutions developed from grass-roots efforts, rather than defining it in total in advance and thus constraining innovation, is only a more explicit and consciously designed example of the way much positive change seems to take place. It is just that the prior existence of Force A and Force B often goes unacknowledged in official accounts of change or in technical models of planning.

While "strategic" is clearly an overused word, and many companies are dropping it as an automatic modifier to "planning," it does express an important idea for this part of the change process: deliberate and conscious articulation of a direction. Strong leaders articulate direction and save the organization from change via "drift." They create a vision of a possible future that allows themselves and others to see more clearly the steps to take, *building on present capacities and strengths, on the results of Force A and Force B,* to get there.[17]

If one can never get the leaders together to do that kind of strategy formulation, to build on a set of innovations, then it is likely that the innovations will drift away. Or that so many kinds of innovations will float by that none of them will even gain the momentum and force to take hold. But what makes the leaders ready to engage in this formulation is the experience the organization has already had.

It may be more accurate to speak of a series of smaller decisions made over time than a single dramatic strategic decision, but I am stressing the symbolic aspects of strategy after events are already in motion. Thus, there may be key meetings at which a critical piece of what later became "the" strategy was formulated, a plan or mission statement generated that articulated a commitment, or an important "go-ahead" directive issued. For example, as we shall see in the next chapter, though it is hard to identify a *single* strategic-decision point in the case of General Motors' decision to downsize its fleet of cars, there were clear turning-point events. One was the executive-committee decision in December 1973 to speed downsizing of the fleet, at both the high and low ends. At Chipco's "Chestnut Ridge" plant and in various Honeywell divisions with change programs, the key event was formulation of explicit strategy by the executive steering committee, thus legitimizing earlier actions and permitting new ones to take place at greater speed. Thus, leaders' articulation of what may have been only embryonic up to that point represents "change" to the organization. Such leader action is important to crystallize change potential once departures from tradition have given the organization some experience with the new way.

Not surprisingly, more integrative systems have an advantage here too. More entrepreneurs, pushing more innovations, create pressure to do something with them. More overlaps and communication channels and team mechanisms keep more ideas circulating. And the existence of teams at the top, drawing together many areas and exchanging ideas among them—as contrasted with segmented officials running fiefdoms—are in a better position to engage in forward planning to tie together external circumstances and grass-roots experience. The preexistence of coalitions and cooperative traditions makes it easier to get moving; precious time does not have to be expended forming the coalitions that will make the strategic decisions.

Of course, as we saw earlier in the contrast between Chipco and General Electric Medical Systems, innovating organizations can sometimes try to keep so many possibilities alive that they avoid strategic decisions, avoid strong leadership that imposes focusing mechanisms to promote some actions over others. The key is to allow a continual creative tension between grass-roots innovation in a free-wheeling environment and periodic strategic decisions by strong central leaders.

So now new strategies are defined that build new methods, products, structures, into official plans. The crystallized plans serve many purposes other than the obvious, as Karl Weick has pointed out:

> Plans are important in organizations, but not for the reasons people think. . . . Plans are symbols, advertisements, games, and excuses for interactions. They are *symbols* in the sense that when an organization does not know how it is doing or knows that it is failing, it can signal a different message to observers. . . . Plans are *advertisements* in the sense that they are often used to attract investors to the firm. . . . Plans are *games* because they often are used to test how serious people are about the programs they advocate. . . . Finally, plans become *excuses for interaction* in the sense that they induce conversations among diverse populations about projects that may have been low priority items.[18]

In short, strategic decisions help set into motion the next two major forces in "change."

Force D. Individual "Prime Movers"

Any new strategy, no matter how brilliant or responsive, no matter how much agreement the formulators have about it, will stand a good chance of not being implemented fully—or sometimes, at all—without someone with power pushing it. We have all had the experience of going to a meeting where many excellent ideas are developed, everyone agrees to a plan, and then no one takes any responsibility for doing anything about it, and again the change opportunity drifts away. Even assigning accountabilities does not always guarantee implementation if there is not a powerful figure concerned about pushing the accountable party to live up to it. Hence the importance of the corporate entrepreneur who remains steadfast in his or her vision and keeps up the momentum of the action team even when its effort wanes, or of a powerful sponsor or "idea champion" for innovations that require a major push beyond the actions of the innovating team. Empowering champions is one way leaders solidify commitment to a new strategy.[19]

Prime movers push in part by repetition, by mentioning the new idea or the new practice on every possible occasion, in every speech, at every meeting. Perhaps there are catchphrases that become "slogans" for the new efforts; John De Butts, the former AT&T chairman, used to repeat "the system is the solution" and built it into Bell's advertising, and Tom Jones, the CEO of Northrup, liked to reiterate, "everybody at Northrup is in marketing." [20] At Honeywell, then-President (now Vice-Chairman) James Renier instituted the "Winning Edge Program," a remarkably long-lasting corporatewide motivation program, and the slogan "We are the Winning Edge" has found its way onto everything from memo pads to coffee mugs. It is currently fashionable to draft "corporate philosophies" or compile lists of seven or eight "management principles" which stimulate the executive team and sometimes employee groups, to discuss the key phrases that represent a shared vision, phrases that can then be tacked on every wall and woven into every speech.

What is important about such communications is certainly not that they rest on pat phrases but that they are part of unequivocal messages about the firm commitment of the prime movers to the changes. It is easy for the people in the company to make fun of the slogans if they are unrelated to other actions or not taken seriously by the leaders themselves. Prime movers pushing a new strategy have to make clear that they *believe* in it, that it is oriented toward getting something that they want, because it is good for the organization. They might, for example, visit local units, ask questions about implementation, praise efforts consistent with the thrust. The personal tour by a top executive is an important tool of prime movers.

This is especially important for changes that begin with pressures in the environment and were not sought by the corporation—changes in response to regulatory pressures, shifts to counter a competitor's strategy. The drive for change must become internalized even if it originated externally, or prime movers cannot push with conviction, and the people around them can avoid wholehearted implementation. I have seen numerous instances of this around affirmative action, for example; those companies where prime movers found a way to see the changes as meeting *organizational* needs and convey an unwavering commitment have a better track record with respect to women and minorities than places with weak or equivocating leadership.

People in organizations are constantly trying to figure out what their leaders *really* mean—which statements or plans can be easily ignored and which have command value. Leaders say too many things, suggesting too many courses of action, for people to act on all of them. Thus, prime movers have to communicate strategic decisions forcefully enough, often enough, to make their intentions clear, or they can run into the problems that arise when zealous subordinates, trying to interpret vague statements from the top, take strong action in the *wrong* direction. I call this the *"Murder in the Cathedral* problem," after T. S. Eliot's play. The drama, set in 1170, describes the events that followed when King Henry II of England said idly, not clearly meaning it, that he wished someone could get rid of that "pesky priest." His aides promptly slaughtered Thomas à Becket in Canterbury Cathedral, and Henry spent the rest of his life doing penance. How many other unintended or even wrongful acts have occurred in organizations because leaders were unclear about what they really wanted—or didn't want?

A few clear signals, consistently supported, are what it takes to change an organization's culture and direction: signposts in the morass of organizational messages. The job of prime movers is not only to "talk up" the new strategy but also to manipulate those symbols which indicate commitment to it. The devices which can be used to signal that organizational attention is redirected include such mundane tools as: the kinds of reports required, what gets on the agenda at staff meetings, the places at which key events are held, or the reporting level of people responsible for the new initiatives.[21] At one high-tech company the CEO shook his executives out of their customary modes of thought by staging an elaborate hoax. The annual executive conference began at the usual place with a rather long-winded recital of standard facts and figures by the CEO himself. His soporific speech was interrupted after an hour by the arrival of helicopters, which flew the whole group to another, more remote place where the "real" meeting began—still punctuated by surprises like elephants on the beach—with the theme of creativity and change.

A few events can have symbolic value far beyond their statistical importance; General Motors' commitment to quality of work life in cooperation with the United Auto Workers, for example, was signaled by such events as the promotion of Alfred Warren, long associated with programs in furtherance of that commitment, to the vice-presidency of industrial relations and the presence of Irving Bluestone, a UAW leader, as a speaker at a major executive conference. At Honeywell, a set of embryonic innovation-producing practices such as employee problem-solving task teams picked up speed when the general manager, Richard Boyle, who had decided to personally chair the steering committee, rearranged a crowded calendar to be present at every meeting and hooked up a speakerphone so that an operations head at a remote level could also "attend" the meetings.

Prime movers push—but to complete the process they need ways to embody the change in action.

Force E. Action Vehicles

The last critical force for guiding productive change involves making sure there are mechanisms that allow the new action possibilities to be expressed. The actions implied by the changes cannot reside on the level of ideas, as abstractions, but must be concretized in actual procedures or structures or communication channels or appraisal measures or work methods or rewards.

"The map is not the territory," philosophers warn us—but we cannot even begin to find our way around the territory to the people with whom we must interact *without* a map. This is not quite an obvious point—or at least, not obvious to all managers. I have seen too many ideas adopted by organizations as matters of policy while members at lower levels scratch their heads wondering what this means they should *do*. Quality of work life and participative management themselves often suffer from this syndrome: official endorsement and prime movers pushing, but no new vehicles to support change. Sophisticated proponents of QWL often decry the reduction of the concept to its manifestations in particular programs ("QWL is more than just a quality circle, it is a principle"), but without the specific vehicles it may be impossible to realize the principle. Organizations always have to steer a course between the need for an identity to the change expressed in concrete actions and the danger of falling into faddism or isolation of the practice as "just another campaign that will pass in time."

The problem is not the association of an idea with a program, but rather the existence of *too few* programs expressing the idea. Changes take hold when they are reflected in multiple concrete manifestations throughout the organization. After all, people's behavior in organizations is shaped by their place in structures and by the patterns those structures imply. It is when the structures surrounding a change also change to support it that we say that a change is "institutionalized"—that it is now part of legitimate and ongoing practice, infused with value and supported by other aspects of the system.

"Institutionalization" requires other changes to support the central innovation, and thus it must touch, must be integrated with, other aspects of the organization.[22] If innovations are isolated, in segmentalist fashion, and not allowed to touch other parts of the organization's structure and culture, then it is likely that the innovation will never take hold, fade into disuse, or produce a lower level of benefits than it potentially could.

The first step, of course, is that something has to work; premature diffusion of an innovation is a mistake some companies make—trying to solidify something before it has proved its value. The new practices implied by an innovation need to produce results and a success experience for the people using them. Then the new practices can become defined and known to people. They take on an identity and perhaps a name, balancing the dangers of faddism with the need for identity. People can see their presence, recognize their absence, attribute results to them, and evaluate their use.

A number of integrative actions can help weave the innovation into the fabric of the organization's expected operations. Changes in training and communication are important. People need to learn how to use or incorporate the new structure or method or opportunity. This is aided by training for any new skills required, help provided for people to make the transition—why companies with successful employee-involvement programs invest so much in consultants and training, as Chipco did at Chestnut Ridge. Then communication vehicles (e.g., conferences, networks, informal visits) spread information about them, help transfer experiences from earlier users to newer ones. At Honeywell a variety of conferences, floating resources (staff available from a Center), training tapes, brochures, and traveling road shows are spreading the idea of participative management.

Furthermore, it is important that rewards change to support the new practices. Successes in using them get publicity and recognition or maybe even formal rewards, like the way parallel-organization participation was written into job descriptions and appraisals at Chipco. This can mean the development of measures of their use and accountability for doing so—e.g., in performance appraisals. (At Nashua Corporation, for example, the CEO made participation in a quality program the criterion for 20 percent of the bonus in 1980 and 50 percent in 1981.) Leaders or prime movers have to demonstrate that they want the changes and continue to push for them even when it looks as if things might slide back. In successful change efforts there is a continuing series of reinforcing messages from leaders, both explicit and symbolic. And individuals find that using the new practices clearly creates benefits for them: more of something they have always wished they could have or do.[23]

Other structures and patterns also need to change to support the new practices: the flow of information, the division of responsibilities, what regular meetings are held and who comes to them, the composition of teams, and so forth. Furthermore, incorporation of innovations is further aided when people are encouraged to look for broader applications, so that the new practices move from being confined to a few "experimental" areas off to the side to being broadly relevant to tasks of all sorts.

All of these ways of embodying change in the structure create *momentum* and critical mass: more and more people use the new practices, their importance is repeated frequently and on multiple occasions. It becomes embarrassing, "out-of-sync," not to use them.

It is also possible to go even further to build strength into the change that has been constructed. For example, new practices can become "contractual": a written or implied guarantee to customers, a written or implied condition of work in the organization, etc. They can become a basis for the selection of people for work in the organization: Can they use the new practices? Do they fit with the new posture? And there can be mechanisms for educating new people who enter the organization in the practices. By this time, the practices are no longer "new" but, rather, simply "the way we do things around here."

The "failure" of many organizational change efforts has more to do with the lack of these kinds of integrating, institutionalizing mechanisms than with inherent problems in an innovation itself.[24] We have seen that some very positive innovations, as in the case of the Petrocorp Marketing Services Department project, are never taken advantage of because of oversegmentation and hence, over-isolation of the innovation. Some new product or technological process ideas are never developed, as in the small steel company that sold an ultimately important invention to the Japanese rather than exploiting it—to its later regret. And even in innovations embraced by the organization as important strategies, like the QWL effort at General Motors described in the next chapter, neighboring systems may not change to support them, such as selection systems, reward systems, and extension to other kinds of tasks (salaried workers or middle managers).

In short, innovations are built into the structure of the organization when they are made to touch—and change—a variety of supporting systems. But the action vehicles also need to be derived from a good theory, in order to avoid the "roast pig" problem.

The "Roast Pig" Problem

Pervading the time of institutionalizing innovations, when leaders want to ensure that their benefits can be derived repeatedly, is the nagging question of defining accurately the practice or method or cluster of attributes that is desired. Out of all the events and elements making up an innovation, what is the core that needs to be preserved? What *is* the essence of the innovation? This is a problem of theory, an intellectual problem of understanding exactly *why* something works.

I call this the "Roast Pig" problem after Charles Lamb's classic 1822 essay "A Dissertation on Roast Pig," a satirical account of how the art of roasting was discovered in a Chinese village that did not cook its food. A mischievous child accidentally set fire to a house with a pig inside, and the villagers poking around in the embers discovered a new delicacy. This eventually led to a rash of house fires. The moral of the story is: when you do not understand how the pig gets cooked, you have to burn a whole house down every time you want a roast-pork dinner.

The "Roast Pig" problem can plague any kind of organization that lacks a solid understanding of itself. One striking example comes from a high-technology firm I'll call "Precision Scientific Corporation." Precision grew steadily and rapidly from its founding to a position of industry preeminence. To the founders, many of whom still manage it, this success is due to a strongly entrepreneurial environment and an equally strong aversion to formal bureaucratic structures. But recently, growth has slackened, margins are down, competition is up, and Precision is even beginning to contemplate cutting back the work force. Increasingly Precision's leaders have the feeling that something needs to be done, but cannot agree on what it should be.

One obvious issue at Precision is waste and duplication. For example, there are a dozen nearly identical model shops on the same small site, neighboring operating units have their own systems for labeling and categorizing parts, and purchases tend to be haphazard and uncoordinated. But although this is well known, and although a number of middle-level "entrepreneurs" surface from time to time with systems innovations to solve the problems, the leaders express considerable reluctance to change anything. So each wave of good ideas for operational improvements that washes up from the middle goes out again with the tide.

The reason is simple: Roast Pig. Precision senior executives have been part of a very successful history, but they do not seem to fully understand that history. They have no theory to guide them. They do not act as though they knew exactly which aspects of the culture and structure they have built are critical, and which could profitably and safely be modified. They are afraid that changing *anything* would begin to unravel *everything*, like a loosely knit sweater. In the absence of a strong theory, they feel compelled to keep burning down the houses, even though house costs are rising and other villages are reputed to have learned new and less expensive cooking methods.

In many companies, management practices are much more vulnerable to the Roast Pig problem than products, because the depth of understanding of technology and markets sometimes far exceeds the understanding of organizational behavior and organizational systems. So among a dozen failures to diffuse successful work innovations were a number that did not spread because of uncertainty or confusion about what the "it" was that was to be used elsewhere.[25] Or I see "superstitious behavior," the mindless repetition of unessential pieces of a new practice in the false belief that it will not work without them—e.g., in the case of quality circles which companies often burden with excessive and unnecessary formulas for their operation from which people become afraid to depart.

Beliefs may indeed help something work, but beliefs can also be modified by information and theory. The consequences of failing to perform this intellectual task are twofold: first, as one innovation gets locked rigidly into place, further experimentation may be discouraged—house burning may become so ritualistic that the search for other cooking methods is stifled; and perhaps more important, the company may waste an awful lot of houses.

The other extreme also poses problems, of course: reductionism, or the stripping down to apparent "essentials," thus missing some critical piece out of the cluster of elements that makes the innovation work. This fallacy of understanding also needs to be corrected by theory and analysis. As usual, the issue is balance between inclusion of unnecessary rituals and the elimination of key supports.

Thus, the task of conceptualization is as important at the "end" of a change sequence, when the time comes to institutionalize an innovation, as it was at the beginning. Behind every institutionalized practice is a theory about why things work as they do; the success and efficiency of the organization's use of the practice depend on the strength of that theory.

THE VISIONS AND BLUEPRINTS OF CHANGE MASTERS

It has become fashionable among organizational-behavior theorists to apply the word "art" to management practice, setting it up in opposition to the idea of "technique." Thus, Richard Pascale and Anthony Athos speak of the "art" of Japanese management, covering a range of human sensibilities and sensitivities quite different from the analytic skills taught in business schools. "Great companies make meaning," they said, showing us that leadership deals with values and superordinate goals and not merely with technical matters. Warren Bennis has made a similar point in his discussions of the "artform" of leadership; leadership involves creating larger visions and engaging people's imaginations in pursuit of them. Even discussions of corporate strategic planning are beginning to stress the intuitive side, the artful crafting of an image of possibilities out of the materials provided by organizational subunits—dealing in symbols, creating coalitions with shared understandings, building comfort levels. Indeed, among the most popular models of the policymaking process today are those which focus on "accident" of circumstance more than rationality or even intelligent intuition; decision makers are shown as "muddling through" instead of rationally calculating.[26]

Such discussions add an important dimension to our understanding of how to guide organizations to achieve higher levels of success. They clearly do not replace the need for rational, analytic techniques such as budgets; reporting and control systems; objective setting and reviews; financial, production, and related quantitative measures; market analyses and environmental scans; and other tools in the modern manager's bag. But the role of such tools has to be seen in perspective. They are part of the management of ongoing operations, keeping the organization on course. They may even suggest areas where changes or improvements are necessary, and they provide the data to back up the argument and get a change effort moving. But overused to guide organizations, they may also stifle innovation, reduce creativity, and prevent organizations from benefiting from the departures from plan that produce entirely new strategies.

The art and architecture of change works through a different medium than the management of the ongoing, routinized side of an organization's affairs. Most of the rational, analytic tools *measure what already is* (or make forecasts as a logical extrapolation from data on what is). But change efforts have to *mobilize people around what is not yet known*, not yet experienced. They require a leap of imagination that cannot be replaced by reference to all the "architect's sketches," "planner's blueprints" or examples of similar buildings that can be mustered. They require a leap of faith that cannot be eliminated by presentation of all the forecasts, figures, and advance guarantees that can be accumulated.

"Do the right thing," Chipco tells its people; but who ever really knows in advance what that is, if change is involved? A comment by Charles L. Brown, Chairman of the Board of AT&T, a company now facing a dramatically new environment, is revealing in this respect: "I think we can do the internal job [of changing] without fear of failure, once we're given some decent understanding of what is expected of us. But the complexity of trying to change ourselves, when we don't know what the future rules are going to be, injects a degree of uncertainty that creates a lot of anxiety."[27]

Blueprints and forecasts are important tools and should be provided as much and as frequently as possible. But they are only approximations, and they may be modified dramatically as events unfold. And they are fundamentally different from the emotional appeal—the appeal to human imagination, human faith, and sometimes human greed—that needs to be made to get people on board. And of course, to the extent that change efforts raise concerns about loss and displacement—the negatives that people can easily imagine—architects of change also have to take these issues into account. They have to manage the politics and the anxiety with inclusive visions that give everyone a sense of both the direction of action and their piece of it.[28]

Thus, it is not surprising that I find myself concentrating on the symbolic or conceptual aspects of the change process—on new understandings, on the communication of those, and then on the inevitable reformulations as events move forward. The architects of change have to operate on a symbolic as well as a practical level, choosing, out of all possible "truths" about what is happening, those "truths" needed at the moment to allow the next step to be taken. They have to operate integratively, bringing other people in, bridging multiple realities, and reconceptualizing activities to take account of this new, shared reality. I know exactly how managers feel who come into a meeting with an excellent plan reflecting long hours of toil, only to find it reshaped by their colleagues in small respects even when there are no major flaws; I have been disquieted when my "perfect" proposal for a participative team has been revised by that very team. Here's the paradox: there needs to be a plan, and the plan has to acknowledge that it will be departed from.

In short, the tools of change masters are creative and interactive; they have an intellectual, a conceptual, and a cultural aspect. Change masters deal in symbols and visions and shared understandings as well as the techniques and trappings of their own specialties.

Thus, those of us interested in promoting change should be wary of excessively logical "how-to" approaches, whether in the form of strategic-planning models or that of other one-two-three guides. Those kinds of models can be extremely useful as a discipline and a structure for discussions resulting in plans—I say this not as a throwaway line but out of my own experience running top-executive strategy sessions. But they fit only one piece of the change process and, by themselves, provide no guarantee that action will *fit* the plans. Many companies, even very sophisticated ones, are much better at generating impressive plans on paper than they are at getting "ownership" of the plans so that they actually guide operational activities. Instead of a formal model of change, then, or a step-by-step rational guide, an outline of patterns is more appropriate and realistic, a set of guiding principles that can help people understand not how it *should* be done but how to understand what might fit the situation they are in.

Perhaps a key to the use of others' experience with change—indeed, a key to the process of innovation itself—is to learn to ask questions rather than assume there are preexisting answers, to trust the process of operating in the realm of faith and hope and embryonic possibility. Repeating the past works fine for routine events in a static environment, but it runs counter to the ability to change. What an innovating organization does is open up action possibilities rather than restrict them and thus trusts to faith as well as formal plans. A well-managed innovating organization clearly has plans—mission, strategies, structure, central thrust, a preference for some activities/products/markets over others—but it also has a willingness to reconceptualize the details and even sometimes the overarching frameworks on the basis of a continual accumulation of new ideas—innovations—produced by its people, both as individuals and as members of participating teams.

Change masters are—literally—the right people in the right place at the right time. The *right people* are the ones with the ideas that move beyond the organization's established practice, ideas they can form into visions. The *right places* are the integrative environments that support innovation, encourage the building of coalitions and teams to support and implement visions. The *right times* are those moments in the flow of organizational history when it is possible to reconstruct reality on the basis of accumulated innovations to shape a more productive and successful future.

The concepts and visions that drive change must be both inspiring and realistic, based on an assessment of that particular corporation's strengths and traditions. Clearly there is no "organizational alchemy" capable of transmuting an auto company into an electronics firm; there is only the hard work of searching for those innovations which fit the life stage and thrust of each company. But all companies can create more of the internal conditions that empower their people to carry out the search for those appropriate innovations. And in that search might lie the hope of the American economic future.

PART V

CAN AMERICA DO IT? REALIZING A CORPORATE RENAISSANCE

CHAPTER

11

Trying to Turn Around
an American Archetype:
The General Motors Story

Will we be smart enough, fast enough, innovative enough to take full advantage of these
opportunities? ... Yes ... we can give the
status quo the good, hard, swift kick in the
pants that it deserves. Nobody in automotive
history has had a better shot at the status quo
than we do, and if we blow it, we may never
have such an opportunity again.

—Speech by Elliot M. Estes,
then President of General Motors

To my mind, the important point is to recognize that we have a serious problem and there
is nothing to be gained by searching for
scapegoats. There's no sense in castigating
the government, the unions, our employees,
or the foreign manufacturers—or even the
management of the auto companies. All that
is academic. Let's not worry about how we got
here but rather how we are going to get our
country, our industry, and our company back
on the fast track.

—Speech by Roger B. Smith,
Chairman, General Motors

Until 1970, nothing had changed here since
Sloan.

—Frequent comment
from long-service
General Motors managers

CAN AMERICA DO IT? Can a range of American companies, in a range of industries, rise to the challenge of the *new* environment and increase their innovative potential? I think that the answer is yes, and that the organizational strategies and lessons I have outlined are applicable beyond high tech, beyond younger companies in newer fields—as many of my examples show. But to further answer this question of whether and how traditional firms in traditional industries can transform themselves—and what might impede their progress—what better case could we turn to than the most dramatic reflector of the *old* environment: General Motors?

General Motors is an American archetype. It has represented the furthest extension of industrial environment that, until recently, dominated the United States. For a long time it was the biggest and most powerful industrial giant, a true impersonal corporation (unlike Ford's status as a "family" company), a creature of the American boom years. It has symbolized the best of America's successes and the worst of its unsolved problems. Its products have been a part of the romance of the American dream. Its pioneering organizational structure, invented by Alfred Sloan in the 1920s ("decentralization" into more than thirty divisions divided into Groups and dominated by five competing car lines, held together by financial controls and a committee structure culminating in the Executive Committee of about five top officers), has been studied avidly in business schools and copied by other companies. While Ford pioneered mass production, GM was the innovator in marketing and consumer finance.

General Motors has also been peculiarly American in its regionalism and its familiar middle-class folksiness, a company whose dominant strategies emerged in the 1920s and 1930s and continued to partake of the flavor of that era. The General Motors headquarters building in Detroit, begun in 1919, is a monument to its Art Deco origins and was once the largest office building in the world in terms of floor space but is now not particularly awesome; its fifteen stories look like any other graying office building in the downtowns of Northern industrial cities, and it is possible to greet the president coming out of the lobby barbershop. Even the executive floor at the top is not formidable; banks do a better job of impressing visitors. There are security guards in glass barriers in the lobby, but at the top a very cheerful, warm receptionist—who is also exceedingly professional—banters lightly with executives as they pass her desk and makes visitors like me feel at home.

In short, there are few images of power conveyed in the formal trappings that surround General Motors managers and executives. It is only in the deployment of people that signs of power appear. At a dinner I attended with GM directors, several aides in three-piece suits, M.B.A.s from the New York treasurer's office, were stationed at the door to run errands and to keep strays from the dinner. This is fitting. Like others of its vintage, General Motors is a labor-intensive company in at least this sense, and labor has been part of its wealth: adding more staff to do more things.

Also old-fashioned American were the company's work values. John de Lorean, the maverick executive who left to start his own company (and was later accused of drug dealing to save it), ultimately did not fit in—not so much for his designer suits, glamorous lifestyle, and iconoclasm, but rather—insiders claim—because he was too often absent from the job and did not always do what they thought was his share of the work. General Motors is composed not of high-tech geniuses whose personal expressions of nonconformity must be tolerated to maximize their creative breakthroughs, like the atmosphere at "Chipco" or other young computer companies, but of solid Midwestern citizens whose work lives oddly mirror that of their products: productivity as a simple mechanical matter of putting in more time and effort to turn out more product.

General Motors' vast work force has reflected regional America more than cosmopolitan America, starting with Detroit. Until the 1970s, less than 30 percent of the salaried work force and less than 10 percent of the foremen—the first line of manufacturing supervision—had college degrees, and college backgrounds were more likely to reflect smaller, regional centers than national ones anyway.[1] A high proportion of the company's managerial talent has come through the General Motors Institute in Flint, the degree-granting technical college operated by the company until it was made an independent engineering school in 1982; in the past, about 80 to 90 percent of GMI graduates went to work for General Motors, and through 1972 a great many of those enrolling were friends or relatives of GM employees.

General Motors was archetypically American regional in its product/market strategy, too. Its international operations were fragmented and kept at a distance, managed out of New York until 1978. America was the market, and Americans could be persuaded to upgrade their car purchases just as they upgraded their lifestyles with a move to the suburbs. The assumption of continuing expansion in the national standard of living set the tone for American business strategy in the era in which General Motors prospered.

The wealth and success of General Motors—the number one industrial corporation on *Fortune*'s list until 1974—represented the wealth and success of American business itself, and the company's products were easily understood and close at hand. Thus, it is not surprising that General Motors became an easy target for critics of corporate power or waste or unresponsiveness as social consciousness rose in the United States in the 1960s. General Motors became a key symbol to the public of the smug conviction of American business of its own superiority and its unwillingness to admit to the need for change. "What's good for General Motors is good for America" entered the national vocabulary as a slogan to be condemned, parodied, and turned into book titles—even though that was *not* what Charles Wilson said in his confirmation hearings before the Senate when Eisenhower appointed him Secretary of Defense.

The company's size, and the size of the industrial army that composed it, made its factories the obvious sites for reformers describing the problems of the industrial era. Automotive workplaces came to symbolize everything alien and alienating about the traditional twentieth-century factory: mindless assembly lines; the roar of an engine test room; the oppressive heat of a foundry; armies of form fillers in the office; occasional drug dealing, sex, and violence in the parking lot; and proportionately few opportunities for promotion at the bottom.

The pressures of the transforming era of 1960–1980 hit General Motors particularly hard, as befits an archetype: what all corporations faced to *some* extent, the older corporate giants faced to a *large* extent. And especially General Motors.

There are two General Motors Corporations.

The first General Motors is the ailing auto giant. The industry's financial woes and declining sales have been the subject of a constant barrage of headlines over the last several years; in 1981 the Commerce Department estimated that more than 1 million people were jobless because of the auto slump.[2] To many Americans a symbol of power grown fat, lazy, and complacent, General Motors is accused by portions of the American public of refusing to make changes when it was obvious that the country wanted and needed smaller, higher-quality cars. (Never mind that market data—high sales and profits on big cars—may not ever have made this clear to the company; the public wants instant response.)

This General Motors is the object of suburban dinner-table scorn. Cars are a common topic of conversation, and everyone I meet seems to have an anecdote explaining why American car manufacturers are losing ground to the Japanese. In all these stories, cost is not the issue so much as a perception of quality difference. Meanwhile, auto-industry analysts in 1982 were calling GM's "J" cars "another Edsel." And *Time* magazine made much of the fact that when GM offered every buyer of a Detroit-area home it was selling for transferred employees a free new car, it sold forty-six houses— but had no takers for the cars.[3] (Of course, the cash equivalent of the sticker price was more desirable to Detroiters, who could easily find discounted cars.)

That's one General Motors. There is another General Motors that has also been in the public eye, but perhaps not quite so visibly or dramatically as the first.

The second General Motors is a pioneering employer acclaimed by professionals and headlining every conference on new work systems. This second General Motors is seen as a well-run company that quickly adapts to change and stays ahead of its competition by recognizing the need for new practices. A *New Yorker* article heralded General Motors' speedy downsizing decision—first in the domestic auto industry. *Fortune* titled a congratulatory article, "How General Motors Stays Ahead." The more than $700 million the company lost in 1980 was the first net loss since 1921, the year GM almost collapsed—the company made money through the 1930s even though many of its plants had to be temporarily declared "surplus" and removed from the balance sheet. In the spring of 1982, GM's sales pickup was far ahead of domestic competitors', in part because its financial strength allowed it to offer a lower 12.8-percent interest rate, and in 1981 it was the only U.S. auto firm to make a profit—albeit not on operations. In 1981, and until I went to press in early 1983, it was the only U.S. auto firm to pay a dividend.

Because of General Motors' 1973 agreement with the United Auto Workers to form the first private-sector national joint Quality of Work Life Committee, and because of a variety of "quality of work life" programs in its plants that have increased employee involvement, General Motors is considered by human-resource professionals, in the survey I ran to identify "progressive companies," to be the most innovative company in the United States in new designs for workplace organization. Its showcase QWL factories—which include its *oldest* as well as newer plants—are mandatory stops on the professional touring circuit. Indeed, these plants are so accustomed to the presence of visitors who are there to learn "how General Motors does it" that even shop-floor workers have a planned patter. One can easily imagine the tour buses pulling up at the plant doors and the tour guides launching into their pitch.

Which is the real General Motors—unresponsive giant or pioneering well-managed company? Or is it really some of both?

"What's Wrong with General Motors Is Wrong with America"

The pressures of the 1960s to 1980s hit General Motors harder and earlier than other American companies, and thus in the GM experience we see in microcosm how these external change pressures affected the inner life of an American corporation.

For example, consumerism and the environmental movement found rallying points around General Motors and the auto industry on issues of pollution and safety; Ralph Nader's career was "made" by the GM Corvair issue. General Motors was a prime target of regulatory agencies such as the Equal Employment Opportunity Commission and the Environmental Protection Agency, and of legislation such as the 1970 amendments to the Clean Air Act. One of the worst conflicts of the civil rights movement exploded in 1967 in Detroit, on GM home territory, inevitably contributing to Detroit's temporary slide downward as a one-industry town. The strike at GM's Lordstown, Ohio, Vega plant in 1972 became a national *cause célèbre*, coming to symbolize the assumed discontent of a new breed of younger workers and resistance to the "modernized" higher-speed factory.

With rising and continuing inflation in the 1970s, an earlier industry strategy that could be characterized as "buying" worker compliance with unusually high wages and generous time off (rather than job improvements) came back to haunt the auto companies in the marketplace. It was difficult to compete with the Japanese in cost because of a wage-plus-benefits differential of about $8, and it was difficult to compete in quality because of an apparent neglect of plant modernization and of the worker-as-person for so long.

Then a shift in world markets toward growing internationalism also hit the whole industry hard. Every automobile company in the world recognized that real success lay in penetrating the U.S. market —the world's largest and most lucrative—and foreign producers, already skilled in making small fuel-efficient cars, had a major lead on GM, whose high margins came from larger cars. And General Motors, as the largest U.S. manufacturer, also had the most to lose. Imports rose from a trivial 1-percent market share in 1955 and 8 percent in 1958 to almost 16 percent in 1974. Then the two oil crises of the 1970s zapped the auto companies more than anyone else (except consumers, of course), and imports gained a formidable 23-percent market share in 1979,[4] rising to as high as 32 percent thereafter.

External pressures were not the company's only worry in this era. During the last half of the 1960s, a series of operational measures showed systematic and serious decline in effectiveness and performance, accompanied by increases in labor-related problems. For example, net income as a percentage of sales for the corporation fell from 10.3 percent in 1965 to 7.0 percent in 1969. (It fell even further, to 3.2 percent, in 1970, although that was something of a temporary aberration due to the impact of a long UAW strike in the fall, as it recovered to roughly 7 percent through 1972, then declined again.) Along with this systematic decline in performance were a number of extremely significant changes in labor measures. For example, over the same 1965–1969 period, absenteeism rose roughly 50 percent, turnover rates were up more than 70 percent, grievances rose 38 percent, and disciplinary layoffs rose more than 40 percent.[5]

Thus, the extremity of the pressures on General Motors to innovate and change shows us how the challenge facing American business in this transforming era becomes living reality for some companies, an inescapable set of problems. Through the experiences of General Motors as both pioneering industry leader and suffering giant, we can see in microcosm how the problems of the 1960s and beyond provoked the need for innovations in the 1970s and 1980s.

In this archetypal company, external events were the triggers that allowed corporate entrepreneurs the room to build on departures from tradition and translate problems into innovations in strategy and systems. The external environment combined with the actions of internal leaders, the corporate entrepreneurs and "prime movers" of innovations, to produce change. The history of General Motors in the 1970s illuminates the integrative approaches that make responsive change possible, along with the residues of segmentalism that can hold it back.

It is hard for some people who read today's financial news to realize just how innovative General Motors has been, even before the 1970s. It produced a number of "firsts" in product and procedure that paid off financially and socially: development of the catalytic converter; downsizing its fleet of cars two years ahead of the domestic competition; developing the first national joint labor-management committee on quality of work life with the United Auto Workers, as well as a number of cooperative projects netting remarkable turnarounds in factory productivity. The company also organized two committees of the Board of Directors that were models for other major companies—a public-policy committee and a nominating committee of only outside directors. It reorganized to build an international strategy. Its new personnel system, called a "human resource management system," was one of the first to give line managers accountability for people decisions under a set of uniform and equitable policies.

But General Motors executives do not claim that the company is innovative; they are as aware as anyone of its slow response to some of the pressures of our times. What they do say is: "When the trends are clear, we play an awfully good game of catch-up."

Both downsizing and the move to an international strategy represent important changes in focus and product for General Motors. But while essential, such aspects of strategy do not necessarily help an organization *continue* to produce innovation, continue to respond to the demands of change. So the other kind of accomplishment—involving changes in organization structure, management philosophy, and the treatment of people—are in a certain sense more fundamental, and they deserve more attention in the consideration of adaptation and innovation—why I have emphasized them so strongly here. Changes in an organization's human systems are its *innovation-enabling innovations*.

Obviously, one kind of programmed change—the annual model changeover—had long been a part of the GM routine. There are also constant consolidations, reorganizations, and facility openings and closings, as in any large company, and especially in a seasonal, cyclical business. And the company's mobilization for war production during World War II was fast and successful, one of a number of triumphant achievements as America's premier manufacturer.* But transforming changes in underlying philosophy, such as those which have occurred in the past dozen or more years, seem novel.

More than once I was told, "Until 1970, nothing had changed here since Sloan"—the 1930s. That is what makes the accounts that follow both noteworthy and heartening. Even older corporate giants are capable of moving toward the structures and cultures that can produce more innovation and mastery of change. In September 1982, GM initiated the most extensive review of its organizational structure since Sloan's work of the 1920s.

STRATEGIC CHANGES:
WINDOWS ON THE WORLD

An important foundation for an innovative organization is the crossing of boundaries, the ability to seek and tie together diverse sources of data and experience. Thus, the recomposition of the Board of Directors in the 1960s and the infusion of outside experts in management changed GM from "one of the most insular and inner directed companies around" in the 1960s to becoming more "outer directed and strategic."[6]

Until the 1970s, GM's Board and key managers were largely insiders. But by the time Chairman Frederic Donner retired in 1967, half of the Board was composed of outsiders, and GM began to seek

* In confining the account that follows to the 1970s as well as to matters where General Motors' experience can be representative of many other industrial corporations, I am obviously leaving out a large number of important events in GM's history —and stopping artificially in 1982.

some outside executives, too, to head staff where GM had little internal expertise or where issues were changing fast. Stephen Fuller and Ernest Starkman were recruited from Harvard and Berkeley, respectively, to become Vice-President for Personnel Administration and Development and Vice-President of Environmental Activities. When Starkman died, David Potter returned from a stint in Washington to take over. Later, Betsy Ancker-Johnson replaced Potter, and Marina Whitman joined General Motors as Chief Economist —both outsiders, both Ph.D.s, and both women.

As outsiders came in, insiders reached out. Said one executive, "Under Donner, the feeling was 'Let's stay out of the newspaper'; the more data the public has, the more rocks they will be able to throw. This changed under Roche and increasingly so through the '70s." The greater public contact of high-level corporate executives reached its zenith in the Murphy-Estes era, the last half of the 1970s. Chairman Thomas Murphy was "Mr. Outside" and President Elliot "Pete" Estes "Mr. Inside," but both met the public. Murphy made more than ninety speeches one year, about five hundred in total over the span of his chairmanship (1974–1980). Murphy commented, "I would have said it was not in my makeup, in my thinking. But people in business have to change, move, adapt to the times." Estes made a "mere" thirty or so speeches a year, choosing the "best" three invitations per month.

Indeed, Murphy takes particular pride, he told me at his New York apartment a few months after his retirement, in having been Chairman when the corporation was "responsive and responsible." Except for Wilson, who left GM to become Secretary of Defense, "no Chairman spent as much time in Washington as I did." Of course, he hastened to add a footnote disclaiming personal credit, a style I found characteristic of GM executives: "most of what I became involved with started with Jim [Roche] or Dick [Gerstenberg]."

The presence of outsiders on the board, along with continued public pressure on the company, led to further innovations. After the 1970 annual meeting, the Board of Directors created a Public Policy Committee, to "give matters of broad national concern a permanent place at the highest level of management," the first committee of its kind established in a major corporation. Primary emphasis was on the business–public interface rather than day-to-day operational aspects of the business—e.g., affirmative action, automotive safety and emissions, operations in South Africa, corporate planning, public transportation. I spoke to the committee in 1979 about organizational structure and career success for women and minorities over an elegant dinner at the Hotel Pierre in New York the night before a GM Board meeting, my first encounter with several of GM's elite. Both women directors served on the committee then, and Catherine Cleary, a banker, chaired it.[7]

In March 1972, General Motors again set American corporate precedent by establishing a Board nominating committee of non-employee directors only to choose Board members and conduct continuing studies on the size and composition of the Board.

But among the most outer-directed moves the company could make in the 1970s was to learn to compete in the international marketplace and to improve its citizenship of the countries where it had operations.[8]

Never as reliant on international markets as Ford, General Motors had entered the international arena in 1911. Until the mid-1920s, the corporation merely exported American cars overseas. But when American cars became too large to successfully compete there, GM decided to build cars and trucks abroad; the company acquired Vauxhall Motors, Ltd., in England in 1925, and Adam Opel AG in Germany in 1929. For the next fifty years, almost all of GM's business outside the United States and Canada was conducted by the GM Overseas Operations Division, run by a corporate vice-president/division general manager who ranked as a peer of the vice-presidents running car divisions. GMOOD was based in New York City, and it served as the administrative umbrella for the offshore vehicles, components, and sales companies which were organized as wholly owned subsidiaries of General Motors Corporation.

Then in 1978, in response to the changed international car market, GM reorganized and fully integrated its overseas and North American operations. It took responsibility for the European components plants away from GMOOD and gave it to the U.S. divisions with similar product lines. Then GMOOD was abolished as a division, its general manager elevated to the status of a vice-president/group executive, and the managing directors of the principal offshore vehicle companies became corporate vice-presidents reporting to the new VP/group executive. Thus, the managing director of Adam Opel, its largest manufacturing complex on one site, became a vice-president ranking with the vice-president/general manager of all U.S. auto divisions. At the same time, the headquarters functions of GMOOD were moved from New York to Detroit and merged into their counterpart central office staffs. GM also created new corporate structures—such as the Overseas Executive Committee and the overseas advisory councils in Europe and Australia—to develop strategy and exchange information across the world.[9] The "world car" concept was promoted, and in 1979, GM announced plans for the most expensive expansion of foreign operations in its history, committing more than 20 percent of its long-range investments outside North America, where the company now recognized its growth opportunities lay.

One of the important moves to support an integrative international focus was the development, in the late 1970s, of new personnel policies governing overseas assignments.

Under the old segmented system, a key personnel executive told me, if an international operation needed someone, its Personnel Department would call up a friend back here, who had little incentive to recommend his or her best managers. So there was a common assumption, perhaps a wrong one, that the *less* effective managers were sent overseas—perhaps one factor in General Motors' poorer results than Ford's outside North America. Those sent overseas were essentially "retired," with no right to return to their divisions—like being sent into lifetime exile.

The new policy limits foreign assignments to three years, renewable for another term only with the consent of the home division, which has to keep open a space for the manager. It is important now, as General Motors sees itself as a worldwide company, to send the best people overseas to get experience, a philosophy increasingly accepted. And future promotions to high executive positions in GM will be more likely for those who have had work experience outside the United States. Indeed, several people speculated to me that Estes' international experience as head of overseas operations put him on the inside track for the GM presidency. And in 1982, Robert Stempel was appointed vice-president and general manager of Chevrolet, GM's biggest division, after eighteen months as managing director of Adam Opel—a tour overseas between heading Pontiac and heading Chevrolet.[10]

General Motors' corporate windows on the world had local reflections down in the factories, too, as I saw on some of my visits. At about the time of the international reorganization, I picked up a copy of the employee newspaper at the Detroit Diesel Allison Division and was struck by the new flavor reflected there: an image of a company trying to be responsive to a changing domestic and international economic environment, more socially conscious and responsible. Page one noted proudly that DDAD had received an award from the Defense Logistics Agency to Indianapolis operations for contractor excellence in assisting minority business enterprises —one of nine firms in the United States so honored. (This was particularly noteworthy because DDAD's Indianapolis operations had been the target of one of the most extensive class-action civil rights cases ever brought against a GM unit.) Just below was the announcement that DDAD's Brazilian operations, as part of the international restructuring, were to be a separate subsidiary of the corporation. (DDAD no longer has manufacturing operations in Brazil, but it is now involved in joint ventures in Taiwan and Mexico.)

Downsizing the Fleet

General Motors' moves to open its windows on the world ever wider helped lay the foundation for a major change in product strategy: rapid downsizing of its fleet, about two years ahead of its domestic competitors.

In terms of the products themselves, the turnaround was remarkable. At the end of 1973, GM had the worst average gas mileage of the U.S. auto companies—12 miles per gallon—but by the 1977 model year, its average of 17.8 mpg was the industry's best, and its big cars alone averaged 3 mpg better than its entire 1974 fleet.[11] But significant as the changes in product lines were, the organizational changes behind them are perhaps even more significant; GM's rapid response to a changed environment was made possible by new integrative organizational devices.

The 1977 appearance in showrooms of General Motors' downsized automobile lines was not the result of any single pressure or policy decision but, rather, fits my change model: a mix of departure from tradition, environmental pressures, strategic decisions, and a key "prime mover" in the Chairman, Richard Gerstenberg. General Motors delayed for a long time meeting the challenges of small-car competitors, but when it decided to act, it acted fast.

Cheap domestic gasoline had allowed the American public to have its love affair with large, gas-guzzling cars throughout the 1950s and 1960s. Europeans, paying three or four times as much as U.S. consumers for fuel, produced and drove small, fuel-efficient cars. Some of those cars, particularly Volkswagen, found a small but well-defined niche in the American car market. But GM felt that *its* customers wanted big cars. Estes told me, recalling the 1960s:

> In contrast with Japan and Europe, the bigger we made the car the better it sold. However, in late 1959 and early 1960, although at Pontiac we were selling in record numbers and had moved into third place in the industry, we decided that our cars were too big and heavy for good fuel economy and too low in height for maximum passenger comfort. As a result, the 1961 regular-size GM cars were made narrower and higher for improved passenger comfort and lighter by two hundred pounds for improved fuel economy. The 1961 models were not well accepted by our customers and 1961 was a bad year for sales, even though we introduced an additional new line of interme-

diate cars with small fuel-efficient engines. So, in 1962 through 1965 models, we redesigned the regular cars to *look* longer and wider. The four-cylinder, small V-6 and small aluminum V-8 engines were discontinued in the intermediate-size cars. The four-cylinder engine was transferred to GM de México, the V-6 to Willis and the small aluminum V-8 to British Leyland. By 1968 we were building only big V-8 engines. We were responding to the customer, and the result was record sales. The customers' desire and preference, as a result of low-cost fuel (supported by government control), was responsible, at least in part, for what later became the industry problem.

Besides the lost sales in 1961, General Motors had had other bad experiences with small cars. The Corvair was a national scandal, thanks to Ralph Nader, and the Vega was plagued by labor troubles and disappointing performance. Furthermore, the goal of saving energy seemed opposed to that of meeting government pollution and safety regulations; it took the development of the catalytic converter to solve this apparent contradiction.

Repeated warnings from energy and environmental experts in the early 1970s led GM to appoint an in-house Energy Task Force in 1972. In March 1973 the Task Force concluded, in its chair, David Collier's, words: "First, that there was an energy problem. Second, the government had no particular plan to deal with it. Third, the energy problem would have a profound effect upon our business." [12] In early October 1973 the Energy Task Force presented updated findings to the GM Board of Directors. But in the midst of this deliberate planning process, the Middle East war erupted in mid-October. The oil embargo imposed by the OPEC nations quickly affected the American consumer. By Thanksgiving, gas lines formed across the nation, and the price of gas jumped upward. The American public's buying preferences in cars quickly reflected the scarcity and higher price of gas. GM, ill-prepared to meet the public's sudden demand for fuel-efficient cars, lost its usual share of the car market in the first quarter of 1974. However, when the oil embargo ended in the spring of 1974, the American public returned to large, gas-guzzling cars, creating record profits for GM in 1976.

But despite the return of the market to big cars, and despite some voices inside the decision-making councils that did not want to invest in what they saw as a "short-lived" trend to smaller cars, the Executive Committee, led by Chairman Richard Gerstenberg, decided to speed up downsizing plans. Estes, then a member of that Executive Committee, recalled:

In the spring of 1973, we [the Product Policy Committee] made a decision in the next [1977 model] redesign to reduce the size, weight, and cost of the regular-size car. The target was a weight reduction of four hundred pounds. In the fall of that year, the Mideast oil embargo and resultant fuel shortage and price increase changed the demand almost overnight from a high proportion of regular and large cars to sixty-five percent small and compact cars. We reassessed our decision to reduce weight four hundred pounds and changed the target to a thousand-pound reduction and established a goal of five-mpg improvement in fuel economy. Several members of the Executive Committee observed that the "move to small cars won't last" and that "we're spending too much for the redesign."

Other insiders made clear that Gerstenberg's foresight was the driving force, *not* market data or customer surveys, "contrary to what a lot of people want to hear."

During the first oil embargo, GM made the crucial decision to downsize its entire fleet so that all its dealers would be competitive. In a series of meetings during December 1973 and January 1974, the corporation decided to reduce its fleet by 1,000 pounds, instead of the planned 400, and also, more critically, to rush the introduction of a new small car for the bottom of its line, the Chevette, while speeding up the already planned development of a luxury car at the top of its line, the Cadillac Seville.

The Chevette was to go from planning to market in eighteen months; the company had never developed a new model so fast before. New methods were called for, and GM seized on earlier departures from tradition in solving the current problem. Two new integrative tactics were used to quickly implement its decision: first, the establishment of a centrally coordinated "project center" with engineers drawn from all divisions—an organizational innovation for the auto industry; and second, the deliberate use of international experience—GM expertise in building small cars abroad.

General Motors had earlier hit upon the project-center idea in developing the catalytic converter in response to the Clean Air Act amendments of 1970; it took advantage of the nontraditional experience of a unit on the GM periphery, Delco Electronics, a contractor to NASA, which used project teams in the U.S. space program, pulling together a temporary group of experts from various functions and divisions to work on a specific, large-scale project. Finding this innovation useful in the success of the new engine design, GM adopted it to respond to the energy crisis. Thus, when GM decided to downsize its fleet, it used the integrative concept of project centers, a form of matrix organization, to do this quickly and economically. Temporary groups of talented engineers from each division were lodged in the corporate engineering staff, with project managers reporting administratively to a director of current product engineering (who in turn reported to the vice-president over the engineering staff), and also accountable for the substance of their work to the chief engineers of the car divisions.

The project center was clearly integrative, especially in the co-operation among divisions that ensued. Now all divisions were in on the act simultaneously, centrally coordinated and monitored, kept on course without segmentalist divisional rivalries' intruding. Corporate staff monitored cost and fuel economy carefully, using a system set up by the financial staff. Meanwhile, project-center engineers were themselves learning the new style of management-by-persuasion required to make an integrative innovation work—especially in a matrix situation. "We became masters of diplomacy," a former project-center head commented. "A center can't force a common part on a division." [13]

A second, earlier departure from tradition toward the new international focus also proved valuable in responding with a new strategy to the energy crisis. At the time of the downsizing decision, "Pete" Estes had just transferred back to Detroit as executive vice-president after a stint as head of Overseas Operations, where General Motors *did* have success in building small cars. Estes' overseas assignment was then unusual for General Motors—perhaps a "lucky break" for the company—and he was dubious at first about its value. As Estes told me, sitting behind his massive desk in his last week before retirement, glancing at the humorous model cars given him at retirement parties, "It hadn't been usual before to go the Europe track before becoming President. I'm not sure that's what they had in mind when they sent me overseas. My first reaction was 'Gosh, I don't know what that means.' After being at Chevy, overseas was just a trouble spot." But he soon found important expertise that could be used: small-car experience in the "T" car, the model for the Chevette.

The "T" car was truly international—designed in Germany; produced, in part, in Brazil; and redesigned in Britain. The project-center engineers integrated it into their designs, though not without reluctance. ("The engineers wanted to start from scratch," Estes recalled.) But this time—unlike unfortunate experiences with the Corvair and Vega—"we went to the experts in the United States and abroad who were waiting in the wings, and they gave up their NIH [not invented here]. Imagine what the Chevy engineers said about being sent to Europe!" In the spring of 1974, GM flew a team to Germany to pick up the drawings. Estes later commented: "The job had to be done in a hell of a hurry. . . . We even stayed with the metric system to simplify our problems in debugging the car over here." [14] As is typical of many innovations, GM's experience at its periphery—designing small cars overseas—helped it meet a quick production-date deadline for what was now a central corporate thrust.

GM's downsized cars came to the market in 1977. Since the American public had renewed its romance with large cars with the disappearance of the gas lines in 1974, GM found itself downplaying advance press comments about its downsized fleet. But then, in January 1979, the Shah of Iran was deposed. From April through September of 1979 gasoline was once again in short supply, and once again the car market was shaken. And General Motors, with its downsized cars, more than held its own during this second oil crisis.

MODERNIZING PERSONNEL SYSTEMS TO COPE WITH CHANGE

Occasional "wins" with a product strategy are no guarantee that a company will continue to be innovative or responsive. So another kind of strategy is important: the development and use of people.

To talk of *a* General Motors personnel system at all before the 1970s is misleading; there were at least as many systems as there were divisions, and perhaps as many as the number of managers, who exercised almost total control over hiring and promotions outside the unionized ranks. While no one could give me figures—the absence of a coherent system also means the absence of measures—there were enough tales of favoritism in selection (even a former GM president remembered a few) to make me conclude that (a) treatment of people was highly idiosyncratic before the 1970s, and (b) this was not a corporate priority anyway. Preparation then for a management job too often consisted of pulling someone off the line and telling him—it was practically always "him": "Wear a white shirt tomorrow; you're a supervisor."

The winds of change blowing across this kind of practice—labor unrest, affirmative-action scrutiny, and a competitive economic situation requiring a more skilled, adaptive management—would find it a fragile structure indeed, easy to knock down.

So the first innovation was the establishment of a more solid, and integrated, structure for making personnel decisions. That structure was brought in by a reorganized corporate staff, the Personnel Administration and Development unit, under a new vice-president, Stephen Fuller, which created a brand-new, state-of-the-art human-resource-management (HRM) system. Established in 1972, it provided all organizational units with a vehicle for handling personnel decisions in integrative rather than segmentalist fashion—i.e., to see them in light of the needs of the whole corporation rather than as reflections of narrow parochial interests. An HRM Committee was established in each unit, comprising the chief operating executive and his staff—a sign of the importance people issues now had. The committee was to provide the necessary direction and support in the introduction of human-resource-management concepts as well as serving as a permanent human-resource function in each organization, acting on every personnel transaction. The committees were supported by staff work, especially by the Personnel Department, and were to meet at least monthly.[15]

The HRM system was approved by the Executive Committee in late 1973. In the spring of 1974 implementation gradually began, with full implementation by mid-1976. A formal performance-appraisal system was introduced in 1976, with training eventually for more than twenty thousand supervisors. A central HRM database inventory was designed to track every appraisal across the corporation. Whenever a position was open, the HRM Committee was required to do an internal search, handled by the HRM staff in line with specified standards or self-identification of interested candidates; cross-organizational searches were required in special cases, as when the division did not have adequate women or minority candidates of its own. Somewhat later, a career-development system and inventory and a series of developmental programs were added to the HRM system.

The HRM system was a brilliant and innovative move. It not only developed coherence, equity, and an integrative element to personnel service, but provided a vehicle in each facility that gave line managers responsibility and accountability for personnel decisions. Furthermore, there was now a mechanism for monitoring these actions and for helping people move across as well as within divisions and levels.

Reforms in selection and training were another major objective of the early 1970s because of a clear need to improve supervision in the light of appalling statistics from the factories—and also to prepare for a future, enlightened breed of upper-level manager. Strides were made in both areas. In 1979, mandatory companywide training for all new supervisors was introduced, about 80 percent of it oriented toward interpersonal skills—what was referred to as "blue-blooding." But this program went out with the budget cuts that followed the company's financial losses. However, upgrading of recruitment was accomplished more readily. By the end of the decade, college recruitment was way up, and the percentage of GMI students with links to the company was substantially down, signifying a broader recruitment base. GMI had its first woman valedictorian, and she was headed to Harvard Business School for an M.B.A. and a doctorate.

The innovations envisioned by the new personnel-administration-and-development staff were helped along by three factors: external pressures, participative design processes involving a coalition from across the corporation, and the backing of forceful leaders. In the early 1970s the tide was running in favor of the HRM system. The corporate staff masterminding its development assigned the design work to a task force of carefully picked divisional salaried personnel directors who had both technical skills and the power to influence their division general managers. Housed in a hotel across the street from the Detroit GM building, the task force worked intensively for a month.

Once their product was pulled together, the process of working on approval began. As with the innovations discussed in Chapter 8, objections were overcome through persuasion—but Estes helped too. Then executive vice-president in charge of the corporate personnel staff along with other central-office nonfinancial staffs, he simply "got mad and said we're going to *do* this." He helped the plan by "blessing" it in other ways. When it was first presented to the Executive Committee, they tabled it because the presenters had not consulted the division general managers, and the Executive Committee wanted to see division support. At the next Executive Committee presentation, the division general managers were present, but Estes set the tone by making clear his own support in an opening speech. A personnel staffer answered the first question, and then Estes answered all the rest. The matter was settled.

Details of the HRM system were influenced by a lawsuit in Atlanta. This was one of a number of external events that drove GM's personnel innovations; government regulatory agencies, for example, had singled out GM as a "Track I" company whose compliance with affirmative-action regulations would be challenged. The Atlanta action involved a minority plant worker, Jake Rowe, who complained of having been passed over unfairly for a foreman's job; General Motors lost. The judge, pointing out that decisions about promotions were in the hands of only the next-level manager, ruled that thenceforth in selecting supervisors GM should demonstrate that at least two levels of management were involved. Personnel staffers agreed with the principle—but should a judge be writing personnel procedures for the company? In another case, a judge provided detailed and, in the staff's view, naive descriptions of how to fill jobs in a warehouse. It was becoming clear that if General Motors was to control its own personnel destiny, there was need for a logical, objective cross-company plan, and the HRM system was "the only game in town." Thus, there was another example of a departure from tradition being taken advantage of to solve an external crisis, and built into a new strategic decision, which was then supported strongly by prime movers at the top.

High-level pushing of the company's new personnel stance continued. At one point the chairman had even publicly declared that there was no room in management at General Motors for anyone who could not support affirmative action; his speech was shown on videotape all over the company. Overt acceptance of the new system was common; a staff member recalled, "When the HRM system was put in, we expected backlash. There was not much, but rather a degree of resignation. What the hell. People do the job. They're good soldiers."

But now for GM's personnel-policy innovations of the 1970s to be translated into action vehicles, living realities for the organization, a corporate entrepreneur was required who could develop the power and the professionalism of the personnel function itself while increasing the divisions' "ownership" of the new personnel procedures.

The designated chief architect of these changes was Stephen Fuller, the first vice-president of the separate personnel-administration-and-development staff. Like most innovators who build a strong team, Fuller himself is clear about the large number of key people who deserve credit for GM's personnel innovations, from the Executive Committees under whom he served to the divisions themselves. There are also mixed reactions from top executives and other staffs about whether credit is deserved by *anyone*, let alone an individual who came from outside GM and then returned to academia. But it was clear to me, as an observer and recorder, that despite the rewriting of history to give collective credit, someone like Fuller had been necessary to pull together the disparate threads only casually connecting personnel activities before he came, and act as the corporate entrepreneur gathering power capital to invest in innovative actions. During Fuller's ten years at General Motors (November 1971–March 1982) he was the leader, the focus, the instrument through which others acted.

As one of his staff explained it to me, "Steve Fuller was a symbol at a time when the corporation was in the agonies of change. He became a central figure in a play for which the script was already written, and the audience was ready to applaud."

In retrospect, GM's actions in splitting off the PAD staff from Industrial Relations in 1971 (they had been part of a combined central personnel staff), and then recruiting Fuller from the Harvard Business School as its head, seems almost as much historical accident as the result of a careful plan, like other departures from tradition that lay the groundwork for innovation. A series of totally unpredictable events—early retirements, heart attacks, and other problems—took all the obvious candidates out of the running. Meanwhile, an early organization-development project that social psychologists at the University of Michigan were carrying out with General Motors had been producing substantial financial payoff. "Therefore," an insider recalls, "they said we need a professor who is a social psychologist and knows about quality of work life. They only knew two of them. Steve Fuller was one." But instead of confining himself to the quality of work life, Fuller took on the whole personnel function.

When Fuller arrived on the scene, the personnel function at GM was weak, though better than at many older manufacturing firms. It was also the victim of segmentalist traditions. Before Fuller, and until 1971, the corporate salaried-personnel administration staff had been in a combined department with industrial relations, and it was clear that industrial relations had the clout; indeed, practically all divisional personnel directors had come up through labor relations, not salaried personnel. Furthermore, the action in personnel was in the divisions. Despite many bits and pieces of policies and programs covering the salaried work force (the *Salaried Policy Manual* alone was about two and one-half inches thick), "it is fair to question how effective and integrative these were," William MacKinnon, then a key staff member and now Fuller's successor, recalled. But the corporate staff, which was oriented upward toward the Executive Committee, was out of touch with the divisions and even with the division personnel directors, who reported directly to the division general manager.

The isolation of corporate personnel from the divisions was reinforced by the isolation of many personnel directors within the divisions; without a strong corporate function behind them, they were often shunted aside. In some cases this was symbolized by offices located in the factory rather than with the other division executives. The weak and segmented nature of the personnel function contrasted with the more integrated and more powerful financial function designed by Alfred Sloan decades earlier. Division controllers, for example, reported on a "dotted line" to the corporate financial staff (and directly to the division managers),[16] and generally it was the controllers, not the personnel directors, who sat in on bonus and "P&S" (progression-and-succession) reviews—in short, making personnel decisions. Personnel staff at that time wondered if they were given short shrift and treated as "happiness boys," "giveaway artists," or "blabbermouths"; they worried whether the company's tendency to label many things confidential was just to keep the information away from personnel—including sales data and occasionally data on executive salaries. "In the beginning was the financial staff," I was told. "It was created on the first day. Personnel was not even created on the seventh."

The new PAD staff thus set about to establish a more appropriate balance. Fuller and his new staff had to counter segmentalism with an integrative strategy that both pulled the personnel function together as a whole and strengthened its local influence.

First, the PAD staff tried to achieve strong, intimate, constantly maintained relationships with the divisions, to respond to the neglect field staff had felt and to make sure that all programs out of corporate reflected division concerns.

While reaching out to the divisions through frequent visits and almost daily calls to each one from a corporate liaison and through regional conferences,[17] the staff also involved them in developing a plan to integrate the personnel function. One of Fuller's first acts in office was to form a committee of division personnel directors, headed by the personnel director of Buick, to consider the selection and development of personnel professionals. About nine hand-picked personnel directors from major divisions sat on the committee with two of Fuller's staff to monitor and advise. The Weiser committee issued a report about a year later establishing a progression sequence for personnel directors, running from smaller to larger divisions and across a range of experiences. This system gave the corporate staff, *de facto*, a much larger voice in the selection of personnel directors, since division general managers relied on them for candidates outside their own divisions.[18]

The PAD staff and colleagues in the divisions also sought to upgrade the staff by rotating line managers through relevant corporate personnel jobs, such as directing the quality-of-work-life efforts. Fuller attracted talented people from other parts of the company, including William MacKinnon, then a Yale and Harvard Business School graduate on the financial staff.

Then the next step was to gain responsibility for those functions theoretically "belonging" to the personnel area but currently being executed by someone else—especially bonus, salary, and succession reviews for the top five thousand executives, "the people running the place." It was a delicate matter to make the switch, but opening wedges were provided by the retirements of local controllers—suggesting that this was a good time to turn the function over to the personnel director—or by the entry of a new personnel director, if the general manager liked him. (This succeeded in all but two or three divisions where the personnel director still has no responsibility for top-management appraisals.) In the case of the progression-and-succession reviews, the personnel staffs simply filled a vacuum, searching for replacements inside and outside the division and preparing lists in anticipation of known retirements; this alone served as a strong organizational integrator and increased the influence of the personnel function.

Finally, the corporate personnel staff not only "reclaimed" territory but also found "new lands where no one has sent out any explorers, so that you can plant your national flag. One was organizational structure. For example, should the Executive Committee be five or eight? Should there be three or four candidates for the presidency?" The staff managed to get a stake in these issues—though not every top executive was willing to cede personnel a role—and on one of my last visits to GM headquarters, I learned that a PAD staff member was doing the legwork for a special project to examine span of control in response to the need to reduce overhead costs.

Overall, success bred success; the personnel function gained reputational power from the success of its new programs and policy changes.

Fuller was also "blessed," he recalled to me in his Detroit office one wintry day, about a year before he retired, with having worked under chairmen and presidents who made themselves accessible to him, asked questions, and supported changes in personnel policies. Gerstenberg, Murphy, and Estes were particularly helpful. For the first seven-plus years of his term at General Motors, Fuller reported directly to a member of the Executive Committee, and after six months into the job, that person was Estes. He got into the habit of talking through his recommendations with Estes first (and later with Terrell, the vice-chairman), sending copies to the president so that the personnel staff could retain its identity as a source of ideas, rather than having its ideas transmitted through an intermediate executive. When Estes became president, the good relationship he and Fuller had established gave Fuller's staff voice in executive selection for general managers and above. This inner-circle membership brought reflected power back into the personnel function.

Staff report varying successes in implementing aspects of the HRM program. The HRM committees, for example, can sometimes be more concerned with raises and promotions than with transfers, and so some decisions sneak by them that staff consider nonmeritocratic, smacking of the old parochial and nepotistic ways. But there seems to be a consensus that because of the HRM system a much higher-quality person—and a more people-oriented style of manager—is moving to higher levels, and HRM-style decision making occurs in the upper executive ranks. The Executive Committee still spends a week a year reviewing the top six thousand people in the company, using input from the Group Executives (a tradition established well before the HRM system), "instead of going into a closet, and later a new chief engineer appears in a puff of smoke," a personnel official said. But there are also new elements. A "team audit"—a shared appraisal process—was applied to bonus-eligible managers in 1980, and in 1982 it was decided to extend this to another ten thousand. Furthermore, high-level managers can now see their appraisals for the first time.

The transformation of personnel practices and the increased stature of the function was brought home to me as I sat in Fuller's office while we talked. Just across a narrow hallway on the executive floor was the office of the new vice-president of industrial relations, and he dropped in on Fuller while I was there, evidently a several-times-daily occurrence. Alfred S. Warren, Jr., appointed in 1980, was a former personnel director for Fisher Body, regarded by some as a Fuller protégé and strongly identified with quality of work life; he had headed the corporate staff concerned with QWL matters in the mid-1970s after a successful career as a GM manufacturing executive.

No longer was personnel a poor stepchild of labor relations. On the contrary, it was given some of the credit for General Motors' other major achievement of the 1970s: the development of a set of joint activities with the United Auto Workers to improve the quality of work life. Many of the same players active in personnel reforms for management were also involved in the foundations of QWL.

Quality of Work Life: Union-Management Cooperation

The modern American corporate movement to improve employee "quality of work life" via cooperative efforts involving the union as well as management was nurtured, if not born, at General Motors. Irving Bluestone, head of the General Motors Department of the United Auto Workers, gave this concept its name and persuaded the company to establish the first national joint labor-management committee on quality of work life among major corporations. But General Motors' own experiments in what was then called "organization development"—improvements through participative problem solving and employee involvement—made the climate right for QWL.

Thus, the same company whose Lordstown plant had become the national symbol for worker discontent and labor conflict was the one to usher in what *Business Week* termed a "new era" in American industrial relations. And though an important part of GM's agreement with the union involved treating QWL as an end in itself rather than an adjunct to productivity, the benefits to the company for changing its treatment of workers have been clear: higher quality, lower absenteeism, higher productivity, fewer grievances—and positive publicity about the dedicated workers in QWL-oriented plants.

QWL was a direct counter to the segmentalism dividing workers and management as well as segments of the work process itself—a segmentalism that had shown up in hostility, grievances, absenteeism, and poor product quality.

The history of QWL at General Motors is more complicated than that of the other major achievements of the 1970s for reasons bound up with the nature of the phenomenon itself: it represents a major change in philosophy and culture, it involves voluntary grass-roots participation at many locations more than policy or strategy set at the top, and it requires mending a large number of conflictful relationships. Furthermore, General Motors was plowing new ground and had to invent many of the tools, first convincing skeptical observers—on both sides, labor as well as management—that the efforts were serious and meaningful. But my micro-to-macro change model still holds: fortuitous departures from tradition, external crisis, strategic decisions, prime movers, and action vehicles wrought a transformation.

When asked about the beginnings of QWL at General Motors, some people go back to 1947, when Alfred Sloan established GM's employee-research section, thereby leading to a long history of surveying employee attitudes.[19] Attitude surveys in the 1960s made it clear that something was wrong; in a study of grievances in a Pontiac assembly plant, the employee-research staff found—not surprisingly —that supervisors' behavior made a difference in worker attitudes and in plant results.

As the idea of seeing differences in plant performance as a function of working environment became more sharply focused, Edward Cole, then president, made a personal visit to all 130 major GM facilities, with an increasing sense of astonishment, insiders recall, at the different characters displayed by the plants. So Cole was receptive to new ideas from Rensis Likert, professor at the Institute for Social Research (ISR) at the University of Michigan and a dinner-club partner of Cole's. ISR was then working to build better communication for other major companies, but General Motors was not using this resource in its own backyard. Cole began to meet regularly with Likert and his colleagues, often on the company plane as Cole flew around the country.

Ultimately, Cole asked Likert if ISR could develop a project to demonstrate the value of participative management in practice, and to help reduce the chances for an almost certain strike in 1970 when the next UAW contract was to be negotiated. With Delmar Landen as project manager (head of the organization research-and-development staff, part of the personnel-development group under Fuller) and a joint team of GM and ISR people, the project was launched in 1968. "So many people were involved," Landen recalled, "that every time we held a meeting we practically had to hire a convention hall."

This project was a watershed innovation which later could be used to address growing labor problems.

The GM/ISR project involved work in two pairs of matched operations, in Atlanta (Lakewood and Doraville GMAD assembly plants) and in Saginaw, Michigan (central foundry and Chevrolet Division metal castings). In each pair, one was a relatively high performer and the other was a relatively low performer. Lakewood, in particular, was a very poor plant with the project transformed into a "sweetheart operation." [20]

A number of things with an integrative flavor were critical in the success at Lakewood. Its plant manager, Frank Schotters, the corporate entrepreneur in this case, brought his experience at Doraville, which he had previously turned into a high-performing plant over an eight-year period. Secondly, training became a way of life for the entire work force; hourly employees, for example, received twenty thousand hours of classroom training in the first year alone, and salaried employees were also trained. Operating information was made available so that people could make judgments about their own effectiveness compared with others', and great stress was placed on the notion that people had to work as teams behind and in conjunction with their supervisors.

Lakewood was a dramatic turnaround, especially against the background of steadily deteriorating labor relations and profitability across the corporation. Although I was asked not to make exact figures public, it is clear that General Motors felt the substantial investment to ISR and support of the program through both salary and out-of-pocket expenses was eminently worthwhile; the gains to GM involved substantial amounts. By the end of the program, General Motors had the Lakewood innovation as a foundation in experience and a symbol of the success that could be obtained from these new approaches.

Several top executives wanted that kind of success to be repeated. And one in particular, noting the emerging external crises, was willing to be the GM "prime mover." So in November of 1971, Richard Terrell, then executive vice-president of North American automotive operations, called a meeting of about seventy top managers: group executives, general managers, division personnel directors, manufacturing managers, and Fuller, who had just taken over as vice-president of the personnel administration and development staffs. Terrell was concerned about the causes of GM's current labor problems and their implications for profit margins, return on invest-

ment, and other bottom-line measures, issues Landen and his staff
had been examining. Thus, this meeting would be used to try to
develop a program of action.

After a round of speeches and reports, Terrell announced they
would meet again in six months, appointed a division general manager as the chairman, and said that between this meeting and the
next one he wanted every division to start on some sort of program.

To the second "Terrell meeting," six months later, in Atlanta—
a site choice perhaps highlighting the success at Lakewood—managers brought their reports. Some managers merely dusted off a traditional idea and relabeled it—e.g., a suggestion system—but
nevertheless, the assignment heightened consciousness about the
issues. As Terrell insisted on a continuing series of meetings, old-
ideas-repackaged gave way to new and increasingly innovative approaches as divisions competed to improve performance through
new people programs.

Meanwhile, from an entirely different direction, the name QWL
had itself become established, and the UAW was playing a significant role. An innovator at the union helped lay the groundwork there
for eventually weaving the threads of innovation into a new labor
and organizational strategy.[21]

Irving Bluestone, author of the QWL name, became a critical
figure in the GM labor-management work-improvement efforts
when he became the UAW director for the General Motors department in 1970, and then a UAW vice-president in 1972. When I first
met him in 1980 at an executive seminar we were jointly conducting
for an AT&T operating company, I could see why he had long been
regarded as an unusually intellectual figure within the American
labor movement; he swapped social-science research references and
professor stories with me with ease—and joined the faculty of
Wayne State University after retiring from the UAW.

Bluestone began gearing up for QWL in 1968 by reading: "I got
wind of what GM was doing in what was then being called 'organizational development,' so I started reading about job improvement
myself." Naturally, Bluestone and his colleagues were vitally interested in GM's activities in any area that could potentially impact on
the union's membership, but his position was quite clear: "Unless
the union is involved as an equal partner, these programs amount to
paternalism. Whatever management gives, management can take
away."[22] Bluestone's first suggestion of a joint union-management
committee on QWL in 1970 was viewed with some suspicion by the
union as well as management. (Bluestone recalled that in 1970 he
had no other QWL supporters on the twenty-six-member UAW
Board; by 1980 he had a clear majority.)

But increasingly, the skeptics on both sides began to lose

ground. Within GM itself—not only at Lakewood but in a new project at Tarrytown—it was becoming increasingly clear that these new approaches not only were *acceptable* to the rank-and-file and local union leadership, but under the right circumstances could be enthusiastically embraced. Moreover, from the company's point of view, it became clear that, at least in this instance, morality and pragmatism appeared to run together.[23] In both Lakewood and a second showcase QWL innovation, very significant savings were realized, along with the potential for more harmonious labor relations—in striking contrast to the nationally publicized discontent at the Lordstown, Ohio, Vega plant, an event pressuring GM.

The combination of innovations netting a positive foundation in experience and clear external crises set the stage for a major strategic decision.

The breakthrough came in 1973, as a part of the negotiations between the UAW and GM. Bluestone again asked for a joint commitment to QWL. Now he had growing support from the union, and GM's experience with successful workplace innovations made it more receptive.[24] The result was a letter of understanding in the 1973 GM-UAW contract and formation of a joint committee to improve the quality of work life, composed of four people—Fuller and George Morris (VP for labor relations and chief negotiator, as a conservative anchor), and two from the UAW—Bluestone and Don Ephlin, an administrative assistant to UAW President Woodcock who later became vice-president heading the UAW's Ford Department. Even though it took a few years before the momentum began to build, this agreement, more than any other single thing, was the critical event. This put QWL on the industrial map for the first time, not only at GM but elsewhere. And now the union was a clear partner in GM's QWL efforts.

"In the development of the QWL concept, several guidelines were generally adopted which are basic to the introduction and acceptance of the process," Bluestone told me. "There will be no 'speedup' by reason of the QWL process; no one will be laid off by reason of the QWL process; participation by the employees is voluntary; all contract provisions, in both the national and local agreements, remain inviolate; and the union and management are coequal in the planning, designing, and implementation of the QWL process and receiving feedback information concerning it."

Once the strategy was clear, several plants became models of the "action vehicles" through which the innovative, integrative QWL strategy could be realized. An early model was the Tarrytown, New York, car-assembly plant.

If Lakewood was "pivotal" in moving GM management more in a QWL direction, then Tarrytown was the activity that brought all the threads together and became a corporate, union, and national symbol of the potential value of QWL. The Tarrytown project began in April 1971, developed a more intensive and sophisticated character about three years later, and continued with strong support for about four years thereafter.

In 1971 Tarrytown was an embarrassment both to the union and to management—to the union because of an internal political split (according to Bluestone, an avowed Communist leading second-shift dissidents), and to management because it was very ineffective and unproductive—it had come very close to being shut.

When the Tarrytown plant manager approached officers of the union in April of 1971, he suggested creating a new management philosophy which could be helpful to both sides but which would require support and action by both union and management. Joint problem solving began with involving the workers in decisions about layout, as two departments were to be moved.[25] Then in April of 1974, after the UAW/GM letter of agreement, three-day training programs for both union and management personnel were developed, on QWL philosophy, effective teamwork, and problem solving. Eventually, by 1978, more than 3,300 employees had participated, largely from the union side.

Several crises indicated the importance of this continuing thrust. Toward the end of 1974, at the height of the OPEC crisis, when the entire second shift was shut down and 2,000 workers laid off, the newly developed relationship between the union and management meant that, in the union president's view, "Everyone got a decent transfer, and there were surprisingly few grievances. We didn't get behind. We didn't have to catch up on a huge backlog."[26]

By 1979, Tarrytown was unrecognizable in comparison with its past. According to the plant production manager, the $1.6-million QWL program (counting only out-of-pocket expenses) "has been absolutely positive," in terms of cost and efficiency, "and we can't begin to measure the savings that have taken place because of the hundreds of small problems that were solved from the shop floor, before they accumulated into big problems." The union agreed. It saw Tarrytown going from one of the poorest plants in quality to one of the best, with absenteeism dropping from more than 7 percent to less than 3 percent and, by the end of 1978, with a total of 32 grievances listed in comparison with 2,000 seven years earlier.[27]

Other turnarounds were just as remarkable. General Motors' Fleetwood plant, visited in June 1981, was an aging factory in Detroit, externally run down. Roughly 3,600 people were working at Fleetwood in the hourly ranks, along with about 400 supervisory and professional staff, producing Cadillac C bodies. But the plant was quite remarkably clean and uncluttered, especially in the body shop and neighboring areas. Things were well organized; there was no sense of great rush. There was a friendly atmosphere on the floor, and there seemed to be plenty of space in the aisles and good facilities for the people—for example, there were a couple of large banks of a dozen pay phones with, typically, two or three people talking on them. Parts were neatly piled up, people seemed comfortable, and there was no sign of pressure. Workers spoke easily to the supervisors and the foremen. Generally, it gave the appearance of a plant that had its act together. The production line was running at a speed of 55.2 units per hour out of a maximum design capability of 60. That provided reasonable output while giving people enough time to do their work without the sense of being steadily pressured.

In 1976, the plant had had about the highest cost level in the corporation, and the lowest quality. The grievance load was about 5,000, with roughly 150 grievances per month coming in, and there were crisis negotiations two or three times a year. Fleetwood was the last to settle the 1979 union contract—it didn't settle until 1980; indeed, in previous labor negotiations, there had never been any local settlement at all because the union and the management could never agree on anything. The tradition at Fleetwood, for both management and union, had been constant animosity. As a manager put it, "Our job was to screw the union; their job was to screw management. It was a way of life." It was one of the worst plants in General Motors with respect to labor—full of sabotage, very high absenteeism—and neither the union nor management wanted much to do with it.

Then, in 1978, a new manager was brought on board. Known as a "people manager," Les Richards clearly was installed to do something about Fleetwood—the kind of assignment on which corporate entrepreneurs thrive.

Richards first sought education for the staff—a total of about 190 hours of training on MBO and team communications for every manager or professional at the plant—before anything was said to the union at all. The evolution, according to the plant personnel manager, involved a lot of time called "working on people's attitudes"— e.g., helping supervisors learn to be less punitive, to show they really were concerned with people; stressing good relationships with the people as well as good housekeeping so that the plant would be well organized. Finally, when managers were on board— voluntarily, because, like other innovators, the plant manager worked in a participative fashion and did not use his authority to *order* involvement—an offsite meeting was held to which the union was invited. This was a year and a half after Richards had come to

Fleetwood. Two union leaders volunteered, and about half of the rest were finally persuaded to come. This meeting convinced the union the effort was worthwhile, and gradually union support grew.

The next step was the development of a joint program, centering around "employee participation groups" (EPGs). These provide "structure to the QWL process by bringing members of a work group together to become more involved in making decisions which affect their work life." There are a number of integrative mechanisms making these groups effective: two coordinators of this program reporting to the production manager; union coordinators providing overall motivation and monitoring; a team of six facilitators available to assist; a joint union-management QWL advisory group—about a dozen people from all levels acting as a sounding board for the program; and a joint union-management policy group of high-level leaders, including the local president and the plant manager. EPG membership is voluntary. The group members work on paid company time, and they choose their own leaders.[28]

Unmistakably, management has indicated its seriousness about these efforts. Management put in a good deal of money and energy, in terms of both the use of staff and worker time and real out-of-pocket cash. This was, by all accounts, the real clincher as to whether management "meant it" or not. Like other corporate entrepreneurs, Les Richards evidently did some things that went beyond the norm and took some risks in the process. He brought 100 people back in soon after a layoff, though they couldn't entirely be justified, but he spread them across the plant in order to provide a little flexibility so that more people could participate in the QWL activities. That fact—bringing back people from a layoff, despite the fact that the company as a whole was not doing it and there was no real push from the union—was seen as a major sign of commitment.

The results for the company: quality ratings that shot up, absenteeism and grievances that shot down.

"This bad plant, Fleetwood, that we thought we might sell four or five years ago is now our number one model," Charles Katko, GM vice-president and general manager of Fisher Body, told a Yale audience in November 1981. "And there's total support from top management."

Les Richards remained as plant manager for two years and was responsible for turning the plant around, with the help of Al Warren, then Fisher's personnel director, who helped sell the QWL concept to plant management. Current plant management seems just as committed. And so do the workers.

Pride and team identity at Fleetwood are revealed in small ways. In addition to a plantwide newsletter reporting company news and featuring reminders about the QWL training and the suggestion system ("in the month of May alone 37 Fleetwood employees were paid a total amount of $26,441.03 for their ideas"), sections have their own papers. Department 314, the body shop, produces the "Fleetwood Express," highlighting team activities on and off the job: EPG groups and sports events. (The word "team" appeared about two dozen times in two pages.) "Presence" rates (percentage of a supervisor's work force in attendance) were also recorded. A typical item:

PULLING TOGETHER
by Joe Wilson

Joe Wilson would like to take this time to thank everyone in his group for their time and effort in improving both the housekeeping and quality. This is the kind of teamwork we need to make Fleetwood Number 1 all the way. Thanks guys!

"Of course, it's no Utopia," the personnel director commented. The plant continues to have problems. On the morning of our visit there was an unpleasant incident. A foreman and a general foreman saw a picnic table with a checker set painted on it. They barked at the men sitting there, but nothing happened, so they broke the table —just smashed it up—instead of talking to the union committeemen. The issue was that checkers was illegal in the work spaces, and these managers had evidently said on several occasions that they were going to do something about it. Their patience was finally exhausted. The union promptly filed a grievance. Later in the day, we discovered that the resolution was quite straightforward. The supervisors involved apologized; the company agreed to rebuild the table in its original form, presumably with a reprimand to the supervisors; and the union undertook to see that the table was moved out of the illegal space and put back in an area where it would have been acceptable. This seemed like a reasonable way to handle what in earlier times would have been an extremely inflammatory incident.

Some parts of Fleetwood still have "problems" with the QWL concept (especially staff services), not all supervisors are happy, and there are concerns about how to sustain momentum and continuity. But overall, it is hard to doubt that this plant has become a markedly different place from what it was several years ago. And after Richards was reassigned to GMAD's new Lake Orion assembly plant, his successor, Thomas Clifford, a retired black Air Force major general, continued the QWL efforts and broke new ground.

QWL as Strategy for the Future

Into the 1980s the QWL activities at GM both broadened and deepened.

Other corporate—and union—entrepreneurs drove local change efforts, like Albert Christner, president of UAW Local 599 for Buick, when he ruled out of order a motion that "there will be no QWL programs at Buick" and helped Buick transform a foundry into a model team-based production facility for torque converters, Factory 81. And action vehicles—the fifth force in my change model —were increasingly visible; people had models of exactly *what* to do, of what QWL meant in practice.

Results continued to be impressive, like a drop in grievances at GM's large Assembly Division from about 40,000 to 10,000 in one year. By 1979, there were more than forty joint Quality of Work Life committees in place within General Motors plants. They cut across virtually all of the corporation, and many had developed elaborate programs.[29] All of GM's new plants, since about 1975, had been designed with QWL principles in mind, and with representatives of union and management as well as of the organizational research-and-development staff.

Line management of General Motors continued to reinforce the importance of QWL and its relevance to performance and the reward system. For example, the organization research-and-development department launched an annual companywide QWL survey in 1976, with union approval, designed to provide a continually revised data base on the attitudes of GM employees about their work and their supervision. In 1979, F. James McDonald, executive vice-president for North American automotive operations, who became GM's president in 1981, shared with Estes the "prime mover" role and "strongly encouraged" every GM plant to launch a QWL program. As Fuller commented, "In all cases specific approaches to QWL are optional, but as of 1979, quality-of-work-life improvement is mandatory." More evidence yet came in 1980, when GM's Second Executive Conference on QWL, involving more than 350 executives from all over the world, included Bluestone as one of its main speakers. The sight of a union officer sharing the platform with GM's top officers made a clearer point than almost anything else that could have been done by GM. Then Warren's appointment as vice-president of industrial relations added a further signal to the system of the importance of QWL as a GM strategy for the future.

A Legacy of Segmentalism: Can General Motors Stay Ahead?

Change is both amazingly fast and painfully slow at a company the size of General Motors.

New policies and programs can be designed and conveyed almost instantaneously, because of a large corporate staff apparatus with fast turnaround time and a tight vertical chain of command through which it is possible to pour information quickly. In the mid-1970s, GM sent a negotiating team to Indianapolis and Washington, for example, to negotiate a single affirmative-action compliance agreement with the government, and within a matter of weeks the entire corporation was blanketed with videotapes, seminars, and materials making the company's stand clear.

Implementation, however, is highly variable. Some policies lend themselves to rather immediate use, but others, by their very nature, must take time (e.g., twenty thousand supervisors cannot be trained simultaneously), and still others require the cooperation or the reeducation of numerous layers of managers in order to be meaningful beyond the letter of compliance with a formal system.

So there are, not surprisingly, conflicting views throughout the company about the extent to which General Motors has, or uses, a wide array of progressive practices and whether the innovations I identified shape actual operations. A number of employees to whom I spoke felt that *nothing* had changed, and several who were in plants with QWL programs, for example, did not know that these programs existed.[30]

In my view, General Motors' most significant accomplishments of the last decades were made possible by its departures from segmentalism, by integrative activities building bridges across organizational segments and between the company and its environment. They involved looking at problems in their larger contexts and bringing multiple sources of information to bear on solutions, accepting a variety of related organizational changes as natural fallouts of defining problems in broader ways.

But countering these integrative tendencies may be a heritage of segmented structure and segmentalist culture that has made it difficult at times to sustain or build on these accomplishments. A few innovations in strategy and practice that are themselves treated segmentally—as isolated decisions kept from touching other decisions—are not enough to help American companies become more innovative and master the challenges of change.

Here is some of what was happening in the middle of 1982 as I wrote this. GM's 1970s stance of open contact with the public was undercut by such widely reported moves as an attempt to reduce the time for shareholder questions at the Annual Meeting and even to prevent shareholders from attending the main event; even if the newspapers were wrong, public relations were harmed. The "J" cars, plagued with problems of cold-weather starting and slow acceleration, were labeled "potential Edsels" by industry experts, and

GM announced it would buy 200,000 subcompacts from Japan under an agreement with Isuzu, perhaps replacing its own small-car manufacturing with imports. Furthermore, the company was reported to have returned to its old system of basing layoffs on seniority rather than on performance-appraisal data, thereby appearing to undercut a critical chunk of the human-resource management system.[31] And relations with the UAW were shaky.

Were GM's problems a sign of segmentalism's returning to displace the cooperation and integration required for innovation? I could only wonder whether the ghosts of segmentalism would rise to haunt General Motors, as they might haunt all American companies that manage to innovate in some areas but leave other basic premises unexamined and untouched.

Organizational "Ghosts"?

The innovation-limiting impact of both organizational segmentation and a segmentalist approach to problems has surfaced throughout this book. Many of the innovations of the 1970s that helped General Motors solve problems in its environment and with its work force were largely cross-divisional or cross-functional, involving coherent overall direction and coordinating mechanisms for exchange or expertise across divisions or areas. Even QWL, which operates locally at the interest and initiative of local labor and management, rested on a corporationwide arrangement, the GM-UAW joint committee, and was pushed initially by corporate leadership.

At the same time, the forces that may undercut or limit in utility the achievements of the 1970s involve in part the separateness of the divisions and, concomitantly, the vulnerability of "philosophy" or "style" changes, the problems of application and implementation of excellent corporatewide programs, and the snail's-pace spread of good ideas—five years for a valuable change in the sales department of one division to be picked up by a second. "We don't dictate to the divisions," a top executive said (obviously an untrue statement, because financial controls and objectives represent a clear form of "dictation").

Recognizing the need to spread ideas faster in the light of the company's financial woes, a line executive on special assignment to the president's office in 1981 began daily lunch-hour meetings where the group executives could exchange ideas. But perhaps this mechanism was "too little, too high." The meeting leader recalled that as general manager he could push his ideas and strategies as hard as he liked, but as group executive he had to work through independent-minded division general managers, with no guarantee of success. Still, the use of more integrative devices for cross-fertilization can be helpful, if divisional competition does not interfere.

The project centers, so useful for downsizing and later for the "X" cars, the latest emission system, and passive restraints, are one sign of more cross-fertilization at GM, of a more open marketplace of ideas, of a slight weakening of separatist divisional identity in favor of a wider organizational view. Another is a greater concern in recent years with transfer of more managers across functions and divisions to create a broader perspective and weaken "segmented" parochial identification with just one division or just one function. (All executive candidates will now have experience in at least two functions.) Still another is the recently initiated monthly meetings of group executives to talk about people: overseas assignments and transfers across divisions. (Some top executives told me that the corporation had always done this as a matter of course, but personnel staffers felt that there were only isolated instances until very recently.) In 1982 the Executive Committee, on the recommendation of the PAD staff, adopted a formal "management development strategy" which stressed the importance of cross-functional and cross-divisional (including overhead) experience in developing potential senior executives. Emphasis was also placed on university executive-development programs. Much of this was not new (James McDonald had been pushing cross-divisional moves strongly while still an executive vice-president, and vice-chairman Howard Kehrl was interested in the Japanese development system), but its articulation even in the midst of a terrible recession was significant: the strategic decision that brings grass-roots innovations to the forefront and gives prime movers the impetus to push for action.

There are also attempts to create more of the structural arrangements I earlier identified as supports for innovation: younger managers on freewheeling assignments as troubleshooters, product managers in a matrixlike situation pulling together teams across engineering, production, and finance.

But will these kinds of integrative tendencies take hold? The project-center concept worked well and quickly to respond to a particular and immediate need, stimulated by external pressures and driven by the top. But since 1981 observers have questioned whether this structure would be used as often as in the 1970s. And how will this type of integrating device fare against years of divisional competition, now newly emphasized in company strategy, and the "lead division" concept for R&D (each division specializing in specific improvements)—both of which are notably slower?[32]

For all the applause greeting current Chairman Roger Smith's announcement of reduced bureaucracy (e.g., elimination of some ninety-two formerly standard reports) and the rewards to be had by divisions if they made their profit objectives (Smith's "3 Rs" in an offsite speech to executives early in his chairmanship were "risk, responsibility, and reward"), there were also corporate executives arguing that General Motors' problems required *more*, not less, centralization—or, at least, integrative central coordination for what is essentially a single-product-line company.

Will the staffs themselves overcome a segmentalist legacy? "Staff" departments are theoretically the problem solvers and integrators, providing perspective on issues that cut across line functions. Indeed, the PAD staff served as corporatewide change agents in the 1970s. But the ability of the staffs at General Motors to function in this way has been limited in the past by two factors: the reputed "power" of the financial staff, and occasional divisive territorial behavior on the part of the other staffs.

Chief-executive succession is but one of many "signs" of the "power" attributed to the financial function—e.g., the treasurer's office serves as staff to the Board of Directors. As one of the few "integrated" functions with dual reporting responsibilities, and a long history of cross-divisional career moves, the financial staff has developed a cohesiveness that, when joined with its role as "corporate scorekeeper," as Thomas Murphy put it, gives it the appearance of great power.

One myth about General Motors is a supposed "tradition" that the chairman (and CEO) is a financial person and the president (and COO) a production or operations person, and the chairman as financial voice has with the power to choose his successor.[33] Recent events reinforce what might only be myth. Chairman Murphy was due to retire December 31, 1980; President Estes, January 31, 1981. In January 1980, Murphy appointed a subcommittee of the Board to recommend candidates for chairman and president. In order to get this important committee headed by the former chairman, the committee was convened before Gerstenberg's scheduled retirement from the Board in May. Then, however, it surprised no one that Roger Smith was the favored candidate, because Smith had worked under both Murphy and Gerstenberg. Thus, the influence of the financial staff and of the present and former chairmen was strong. There was evidently some dissension from operating executives with a strong product and people orientation because of their concern that the financial side had too long dominated the company, and that "the corporation uses figures too much." But the "financial votes" prevailed.

In contrast, other staffs have been assumed to have lower power and were known in the past to occasionally indulge in internal bickering—a problem widely shared in American industry. One sign of the relative status of the staffs was that for years industry–government relations reported to finance (its first vice-president was a for-mer assistant treasurer) and not to the legal staff, because the issue was considered so important. (Now it reports to an external-affairs group.)

Many of the rifts have been mended, as in the efforts of person-nel staffs to increase their ability to work with the financial staffs as equals. But even through the innovative 1970s there were, reput-edly, continuing conflicts between staffs that have pieces of related problems: e.g., marketing and customer satisfaction; labor relations and personnel; research and environmental activities. Some issues were carved up and assigned to different staffs, with the result that cooperative proactive problem solving was very difficult. (For ex-ample, different pieces of affirmative action and EEO were handled by legal, labor relations, and personnel staffs, a segmentation of the issue that provoked in past years what one executive called "petty territorial battles.") If the very parts of an organization supposedly responsible for adaptability—the staffs—occasionally bog down in empire building and infighting, segmentalist tendencies are rein-forced.

In addition to the problems of structural segmentation, there are also potential problems stemming from segmentalist—compart-mentalized—approaches to organization changes themselves—e.g., QWL. QWL at General Motors was innovative a decade ago and is still impressive; the company was clearly a pioneer. But if QWL is not integrated with other organizational and business decisions, it can be a fragile, vulnerable, and still limited achievement. A number of questions remain.

For example, how far will management let it go beyond the first stage of opening and improving communication? Will it compart-mentalize QWL and cast it in narrow terms? As one GM personnel executive commented, "QWL is just asking people what they want. How many times can you do that?" A high-level line manager told me QWL was "just treating people well—why do we need an elab-orate program for *that*?"

And how well does management understand QWL? In one case an order came down to a plant manager who sat on his plant's QWL steering committee: have ten more problem-solving groups by next week. The plant manager had to explain to his management that he was only one vote out of fourteen—a clash with authoritarian man-agement wanting the results of QWL without understanding how they come about. There are also signs that involved managers do not always know *why* what they are doing is working—the "Roast Pig" problem—and thus, how to repeat it elsewhere. At Fleetwood, for example, I was told that the key variable was "changed attitudes." But what I noted was a number of changed organizational *struc-*

tures: vehicles for increasing workers' power and options, new ways
to contribute, channels for circulating ideas.

And what about the "missing middle," the vast middle ranks of
the company? Salaried jobs seem barely touched by QWL, even as
a philosophy, let alone as a program, unless on an *ad hoc* basis a
particular local manager decides to act in a participative fashion.
Obviously, I could not be everywhere in GM to see this for myself,
but this observation from extensive interviews at more than eight
sites was confirmed by numerous staff people and later by the hints
of unionizing interest in the salaried work force. Many corporate
staff seemed to consider QWL a "joke," in that it was not a management philosophy with any bearing on *their* jobs. Furthermore,
first-line supervisors and middle managers were under pressure to
operate according to QWL principles without getting similar support
or attention from *their* management.

Even in the factories, there was nearly universal agreement that
the second and third levels of supervision (general foremen and
superintendents) were the keys to whether any program or style
worked, and yet they were rarely the target of change efforts—or of
sympathetic attempts to improve *their* work situation so that change
could cascade downward. And in a company used to tight controls
over the use of time, bosses, especially of salaried and professional
workers, could easily object to QWL-type activities.[34]

How far has QWL really spread throughout the company? It is
difficult to get a fix on how much of the work force and how many
sites are affected by QWL, although, as I reported, there appear to
have been about 75 to 100 operational joint labor-management committees at the end of 1981. It also appears that QWL has caught hold
more strongly at two extremes: in new plants (so-called "greenfield
sites") where team processes could be designed in from the beginning, and in plants where crisis precipitated cooperation and the
search for tools to ensure survival. Some enlightened top managers
have pushed QWL in their divisions, but others—especially those
making money and with other good indicators without it—could
avoid wholehearted support.

QWL seems to have been put in a compartment labeled "relations with the union," and thus it became difficult to take other steps
to tie it into the company's way of operating—a devolution of what
began as a generally applicable organizational improvement technique into a program of union–management relations aimed at the
hourly work force. It will never last in the long run, a key executive
commented, unless supported by a different kind of selection system
and an increase in training and education—let alone different policies for handling business fluctuations and other critical matters that
affect the treatment and contribution of employees.

Furthermore, there have been other signs of segmentalism even within the QWL-as-labor-relations box.[35] One example of treating QWL as a segmented activity, isolated from other business decisions, was a plant closing in California, early in 1982, at GMAD-Southgate. Despite years of QWL at General Motors, and despite a statement on plant closing in the GM Public Interest Report that promises advance notice where possible, many workers first learned about this closing on the radio. Reports of great hostility ensued, and one worker published an angry op-ed column in *The New York Times*.[36] Then, in April, about a month after the UAW agreed to a new contract with General Motors involving an estimated $2.5 billion in wage and benefit concessions to the company under a principle that management and labor would share equally in financial sacrifices, GM announced a new bonus program for executives which made them eligible for bonuses at lower levels of profits. Douglas Fraser, UAW president, commented that the timing of the new bonus plan was particularly irritating: "I've never seen a situation where workers were so upset. They felt they were double-crossed." Related union-member discontent occurred at GM's announcement that it would buy 200,000 subcompacts from Isuzu in Japan, perhaps "exporting" U.S. jobs.[37]

In segmentalist fashion, one set of agreements appears to remain isolated from other decisions that should instead take those related factors into account. But perhaps this is connected with another problem older organizations face in trying to change: a culture that has looked backward more than forward.

"Looking Backward":
A Culture of Organizational Immortality

General Motors is only one of a number of major corporations accused by financial analysts and the public of being shortsighted and short-term in its corporate thinking.

Ironically, GM executives *do* express a long view, I found—but this is likely to be a long view backward into company history. Alfred Sloan was invoked ritualistically and modern management thought rarely mentioned in my interviews. Roger Smith told me he was trying to "return this company to the way Sloan intended it to be managed." And apparently, there have been until recently, few mechanisms or incentives for making decisions with long-range payoffs,[38] other than "forward-model programs" concerned with products.

There are signs of change here, too. Smith is personally identified with a long-range computer model of the company. In 1981, the

348 CAN AMERICA DO IT?

new president, F. James McDonald, asked each division for five-year objectives and detailed one-year objectives. As another executive put it, "We are engaging in more long-range planning because of external events. You take cognizance of them if they hit you in your income stream: the cost of fuel, intervention of the government, cost problems, Japanese competition." One symbolic touch was the plaque former president Estes put in his office in 1965, when he was general manager of Chevrolet, with a slogan he used in almost every speech: "If something has been done a particular way for 15 or 20 years, it's a pretty good sign, in these changing times, that it is being done the wrong way." In successive moves, he crossed out "15 or 20" and put in first "5" and later "1."

But there is also an air of what I can only call "organizational immortality" surrounding the words of top executives. The "organizational immortality" theme consists of these kinds of ideas, echoed throughout my interviews: *The organization is larger and more important than any one person. Nothing is without precedent. History is a guide*: and therefore, as a corollary, *nothing has been unanticipated. Nothing is forced upon the corporation that it did not already want. Good, modern, progressive ideas and practices exist already in the corporation, certainly in policy and probably somewhere in practice. Thus, continuity is more important than change. The company can outlast anything.*

It is not hard to see where these kinds of ideas come from. With GM's vast size, it is indeed likely that someone out there somewhere is already doing whatever it is that critics think the company should do more of; of course, the serious question is whether that reflects the corporate thrust or just random deviance. But it does give the corporation an advantage when it seeks solutions to problems: top management feels that there is little need to go outside the organization because someone can be located with the relevant expertise. And indeed, in only a very few cases—and those largely staff rather than line—has the corporation looked outside.

If the lessons drawn are large enough, if the sense of historical continuity is strong enough, then precedents can be found. And because the corporation is, after all, composed of reasonably intelligent and informed people, and because issues—even the energy crisis—rarely spring full-blown with no signs that they are coming, it is possible to say that responses have been anticipated at least in discussion. William Safire called this the "classic State Department position, which always holds that no disagreement exists and nothing ever represents a change from the previous statement."[39]

But whether those responses were part of a strategy, were dominant before required, is another matter. And precedent can be an excuse for inaction. Former President Estes ruefully commented that Charles Kettering, in the 1930s, had predicted that only a ten-year supply of petroleum was left and the company should make engines to save fuel. But then every ten years there was ten years' more, and ten after that. "We've always had ten years of fuel." So the company stopped paying attention.

I was struck by both the comforting and the troubling aspects of organizational immortality in all of my top-executive interviews. Stability and continuity are important in managing a vast corporation, and a sense of history can be an important element in providing a culture of security and pride. But denial of the realities of change can make it difficult to hear or admit new ideas. To insist that no unprecedented or unanticipated actions occur is to engage in a distortion of history that makes it hard to credit innovators or stimulate others to seek new and better ways. To say that "we've always wanted it that way anyway" or that "our policies espouse this" when practice decidedly departs from that is both to weaken faith in official pronouncements and to make it difficult for anyone to question the effectiveness of implementation.

This describes a culture of age rather than one of youth. The wisdom of age is to move slowly because fads, trends, and pressures have come and gone. Similar crises have appeared before and been weathered, so why get aroused over this one? In contrast, the naïveté and enthusiasms of youth involve jumping to respond to every change, involve deliberately seeking the new for its own sake. Having *not* seen it all before, the young are likely to exaggerate the magnitude of crises or pressure and thus be more likely to act.

Perhaps not only the youth of the forms but the youth of the work force gives high-technology industries an advantage over autos in responsiveness to change, in willingness to embrace change, above and beyond any differences in market conditions. Companies dominated by and respecting the young may be more activist, less cautious, in responding to new developments, internally and externally. It is not the chronological age of the individuals alone, of course; GM has always appointed *some* relatively young people to senior positions.[40] But *Fortune* once pointed to General Motors' "Council of Elders" (former top executives) as a strength; where is General Motors' "Council of Youth" to keep it in touch with new ideas or encourage innovation?

Thomas Murphy voiced the immortality theme directly: "We're in business for all time. Any span as chief executive is only a breath. We must see that it provides continuity to relate well to our customers."

With these words, he began our interview, echoing these ideas throughout the conversation. I noted how often examples came from the distant past, and how often the "we've already done it" or "we've always done it" themes were invoked in response to my questions. I admired his dignity, his personal commitment to a number of civil rights and feminist causes, and his stellar performance at the pinnacle of General Motors during a period of great responsiveness, but I also felt faintly troubled when he seemed to dismiss the profoundly transforming nature of the period from 1960 to 1980 for American business by saying: "The seventies—yes, they were an interruption, but there was never once a question about Americans' needing automobiles and we'd be making them. We had some problems getting our product in step. *That was all.*"

FROM GENERAL MOTORS TO AMERICA

As an archetype, emblematic of the corporate America we have been, General Motors also tells us about the corporate America we must become. Both the past and the future are reflected in contrasting organizational cultures, structures, and problem-solving styles. The remaining legacy of segmentation and segmentalism found in many American corporations is strikingly different from the integrative character of the firms that produce continual innovation and succeed in staying ahead of change.

The "Roast Pig" problem that I identified earlier illuminates the necessity for intellectual understanding of the essential elements in a change: "If you don't understand why the pig caught in a burning house gets cooked, you're doomed to waste lots of houses just to eat roast pork." Let me now suggest a second precept equally appropriate to American companies struggling to change in these transforming times, one stressing the necessity for full commitment: *"If you want to continue to eat roast pig, you'd better go whole hog."*

Piecemeal—segmentalist—change is not enough to help companies develop the innovations they need to survive and prosper. Innovation derives from the whole structure, culture, and approach to problems characterizing an organization. The General Motors experience shows how far American companies—even the fattest giant of them all—can go to transform themselves, even if the transformation is not yet complete or secure.

A GM executive put the issues for the company's future—and in a sense, that of American business in general—in terms of the need for a new strategy, one stressing innovation and, thus, a better use of people:

Everyone who reads is aware of the impact of Japanese competition and the cost differentials on the US auto industry. The labor differential is almost two to one. So it is totally unrealistic to think that however many new contracts we write we will close the gap. Therefore North American car manufacturers will have a cost disadvantage in perpetuity. But companies can succeed on bases other than cost. Maybe cost control and volume —our traditional strategy—needs to be replaced. We need a strategy of survival, relying on something other than our traditional strengths. In everything the company does, we must have a high rate of successful innovation. And that brings us back to the people, to the company's human resources. So the problem is not in Tokyo, it is in Detroit.

The challenge for American corporations is whether human resources across every slice of the company—division and corporate staffs; personnel and finance; labor and management; hourly work force and salaried work force; even black and white, male and female—can so unite and integrate, eradicating a historical commitment to segmentation, that a new strategy of success by innovation is made possible.

CHAPTER 12

Reawakening
the Spirit of Enterprise:
Toward an American
Corporate Renaissance

> The Renaissance, which bridged medieval
> and modern times, is not considered a unique
> blooming, but part of a process of develop-
> ment. Just the same, something happened; it
> happened by way of contact of cultures, it
> happened through the reacquisition of classi-
> cal learning, it happened as the age of explo-
> ration doubled the extent of the known world,
> it happened as new classes of people thrust
> their way into the political process, and as cu-
> rious natural philosophers began to ask ques-
> tions.
>
> —Elizabeth Janeway,
> *Powers of the Weak*

RIP VAN WINKLE went to sleep for twenty years. When he awakened,
the American landscape had changed dramatically, and his bones
creaked with age. The world around him was no longer one he
understood or in which he could function well.

Will history someday see that classic story as a parable for much
of corporate America today—falling asleep in the 1960s and waking
up too old, too tired in the 1980s? Or can more organizations learn
to operate in the integrative and highly people-centered ways of the
innovating companies I have identified?

If American organizations use this opportunity to arouse the
potential entrepreneurs in their midst—the people at all levels with

new ideas to contribute—then, unlike Rip van Winkle, they could be renewed, refreshed, and readied for a changed world. The spirit of enterprise could thus be reborn, heralding a kind of Renaissance for corporate America.

The first question about a "corporate Renaissance" is its feasibility, especially for older, troubled industries. Doubters wonder whether the existence of more corporate entrepreneurs would really make much of a difference for them. They question whether the success of high-technology firms is really due to better management rather than to growing markets that will absorb virtually anything. In short, are practices characteristic of companies like Hewlett-Packard, "Chipco," and high-tech sectors of GE really transferrable to matured corporate giants?

At extremes of market conditions, of course, the quality of the organization probably does not matter much. Having the right product in a world hungry for it masks a large number of organizational sins: the company can get away with being poorly managed and still do well financially. At the other extreme, where the product is clearly the wrong thing at the wrong time, then no amount of organizational change by itself is likely to guarantee success.

We would search in vain for the organizational alchemy to transform a smokestack industry into a high-technology firm; and attempts to effect a "transmutation" via divestitures and acquisitions deflect important managerial attention from improving the quality of internal operations. Companies take on their shape, and in part become locked into it, from their industry and their history. More recently founded companies can adapt practices from a more modern era and can support more change while their basic systems are still being developed, making it more likely that computer and electronics companies will indeed have more innovation-stimulating practices, while industries with ample time to experience hardening-of-the-organizational-arteries, such as autos and insurance, will have fewer.

Yet despite these broad tendencies, individual firms even within the same industry still vary, and there is a range of differences with respect to investment in people and the encouragement of employees' participation in innovative problem solving. In my comparative study, those firms with early and progressive human-resource practices, when compared with a similar company in the *same industry,* had been significantly more profitable over the last twenty years.

Similarly, it is not only high-technology firms that are characterized by the kinds of practices that keep innovation alive for long periods of time. Procter and Gamble operates within one of the oldest industries, but has continued to break new ground in product introductions as well as workplace practices. (Indeed, the company considers information about its team-oriented manufacturing facilities "proprietary.") Cummins Engine, well known for its quality-of-work-life programs, occupies a spot in one of the so-called declining industries, but has performed consistently better than other diesel-engine manufacturers. And an even more striking illustration of the potential for transformation of older industries is the speed with which many banks are trying to reorient and develop new products and services as the rules governing financial institutions change. If banks, long considered among the most traditionbound of American corporations, can do it, then so can other industries.

Constraints and environmental conditions count; they make it more or less difficult to carry out organizational objectives, and they present leaders with problems as well as opportunities. It would be naive to fail to recognize this. But within those boundaries, it is the capacity to engage and use human energies effectively that sorts successes from failures, in the long run if not always in the short. Innovations such as new products or market applications, as well as effective organizational problem solving at all levels, depend on people—and the need for these competitive advantages grows with environmental turbulence and a less expansive economy.

Today more than ever, because of profound transformations in the economic and social environment for American business, it should be a national priority to release and support the skills of men and women who can envision and push innovations. This requires, in turn, corporations that operate integratively, that help individuals make the connections and get the tools to move beyond preestablished limits and break new ground, working through coalitions and teams. Making the power available to people at all levels of organizations to take action to introduce or experiment with new strategies and practices, often seen as a luxury of rich times, is in fact a *necessity* for survival and success in difficult times.

"Corporate entrepreneurs" are often the authors not of the grand gesture but of the quiet innovation. They are the ones who translate strategy—set at the top—into actual practice, and by doing so, shape what strategy turns out to mean. Top leaders' general directives to open a new market, improve quality, or cut costs mean nothing without the ability of relevant managers and professionals to design the systems to carry them out or redirect their staffs' activities accordingly. So the meaning of change and the extent to which it can significantly affect an organization can be, in many cases, determined almost exclusively by the initiative and enterprise of people at middle levels and below, who themselves design new ways of carrying out their routine operations that may quickly or eventually add up to an altered state for the organization.

Indeed, without sufficient flexibility to permit random creativity in unexpected—and nonpreferred—places in the organization, many companies would not have developed new programs, new products, or new systems that were eventually adopted as organizationwide initiatives, to the great benefit of the whole. Because innovators have their finger on the pulse of operations, they can see, suggest, and set in motion new possibilities that top strategists may not have considered—until a crisis or galvanizing event makes them search for a successful departure from tradition in the company's own experience.

In short, individuals do not have to be doing "big things" in order to have their cumulative accomplishments eventually result in big performance for the company.

It is in this sense that individuals in the right circumstances are the keys to innovation. They are only rarely the inventors of the "breakthrough" system. They are only rarely doing something that is totally unique or that no one, in any organization, ever thought of before. Instead, they are often applying ideas that have proved themselves elsewhere, or they are rearranging parts to create a better result, or they are noting a potential problem before it turns into a catastrophe and mobilizing the actions to anticipate and solve it.

By being able to get the power to act, individuals are helping the organization stay ahead of change.

The environment—e.g., industry conditions—and history— e.g., a company's past investments—constrain the *arenas* in which corporate entrepreneurs can maneuver, but they do not eliminate the *possibility* for productive innovations in *some* arena.

If the physical or product side of the organization cannot be quickly modified, for example, owing to capital investment and other heavy sunk costs, there is still room for innovation in the organization itself—in modifying production methods, in changing product details to be more responsive to customers, in identifying novel uses, in improving quality, or in taking advantage of the knowledge of the organization's people to respond better and more quickly to the environment—to improve service, to improve relationships, to reduce labor conflict or other forms of costly friction in the company. Repeated studies of innovation in American companies make clear that it need not decline as the company ages or as a product matures but that the domain shifts away from rapid technical changes to changes in method or form of organization.

Even so, one should not rule out the possibility of a dramatic breakthrough in product form or use itself. It is when a company stops believing that it can always do better, regardless of the domain in which excellence is sought, that innovation is stifled.

Thus, the innovating organization needs the people ready to see and act on these possibilities. It needs the people capable of adapting as circumstances change because they already have a broader base of skills, of organizational knowledge, and of relationships in advance of any demand for change. They are flexible and deployable, with restraining time reduced.

In short, a company can make itself adaptable by removing more of the barriers to major changes in *advance* of external crisis or threat. It can have people familiar with problem solving and with working through others in teams, who will be ready to step forward as innovators when they see a need and a possibility. These skills are less necessary in routine operations, of course, where segmentalism can work—units separated from each other, levels working independently, individuals pursuing defined-in-advance jobs without looking for or thinking about improvements other than incremental ones. But under circumstances demanding change, demanding new responses, then an adaptable, cooperative, and even entrepreneurial work force is required.

It is in this sense that modifications in organization design and improvements in human-resource systems constitute innovation-promoting innovations. They make the organization ready to both stimulate and take advantage of unprogrammed innovations that come from participating teams at the bottom and entrepreneurial managers in the middle and higher.

What is important to note about the "failures" of older companies with respect to innovation is not the utter impossibility of change, but rather, how easily it *could* have been done. In both the smaller example of the "Petrocorp" Marketing Services Department and the larger example of General Motors, the pieces are there for transforming changes—toward a more integrative form—with the potential to solve major problems for the company. But if the organization's leaders do not recognize this, and do not put the pieces together, then the potential for innovation-enhancing innovations is lost. If there is any "failure" in these accounts, it is not in the impossibility of transforming change but in the lack of ability on the part of those guiding the systems to understand the change process sufficiently well to take advantage of the innovation in their midst. Maybe these isolated instances of microchanges would not, one by one, "save" the company—but they might change the odds; they might tip the balance.

The true "tragedy" for most declining American companies struggling to keep afloat in this environment is not how far they are from the potential for transformation but rather how close they might come and not know it. How many quiet corporate entrepreneurs have put forward the seeds of ideas for major improvements? How many of them have silently improved their own operations, or slightly shifted production methods, or raised quality standards, or involved their workers in problem-solving teams, and gained in the process an important kind of learning for the company, if leaders had only known? How many quality-of-work-life programs have shown so much potential and yet been limited in their application and spread? How many people *know* that they could contribute more but feel that their company does not care about their efforts?

All any company has to do to explore its own potential to become a more innovating organization is to see what happens when employees and managers are brought together and given a significant problem to tackle, along with the power tools—the resources, information, and support—to help them meet the challenge. The energy and excitement that are unleashed, the ideas that come bubbling forth, the zest for work that suddenly appears in employees who had looked like "deadwood" are just a few of the indications that there is more "entrepreneurial spirit" to be tapped in most organizations, more willingness to cooperate in solving problems when the roadblocks stemming from segmentalism are removed.

Many executives who are otherwise hard-boiled and tough-minded evaluate the impact of their innovation-stimulating organizational change efforts in just such qualitative terms. When I push them for measures, they counter with intuitive yardsticks. For example: "All I have to do is walk through the factory and I know it feels different around here. People say hello in a different way. They smile. They are tackling their work with energy. They push back. I hear from them. There are ideas bubbling up from the bottom. . . . I used to avoid walking through the plant if I could help it, but now it's a pleasure."

The executive who spoke those words was clear that there would be a payoff in both productivity and innovation to his support for a more integrative, participative organization, and it would clearly show up in financial results. But he did not feel that he had to justify the creation of an energetic organization merely on the basis of bottom-line results. After all, that would defeat one of the purposes of his new style of organization: to begin to treat people as contributing individuals rather than an anonymous mass whose primary purpose was fitting into the slots the company had made available.

Thus, if there is a realm where economic and human interests coincide, it is here, in the creation of innovating organizations. Reporting the impact only in terms of numbers is to deny a very important part of the reality of these kinds of changes. After all, as people, we live out our work lives not only through abstractions like numbers but rather through the numerous daily encounters that give us opportunities for contributing—or not—as the organization's structure and culture allow.

How to Begin

Obviously, not everybody in an organization should be involved in innovation and change all, or even much, of the time. Even while considering change, companies have to manage a wide range of ongoing operations where efficiencies may require repetitive and routinized tasks, tightly bounded jobs, and clearly defined authority. Specialization of organization segments and limited contact between them can be an excellent strategy where no change is required, where repeating what is known *works* because both the demands and the activities to meet them are predictable. So innovation is not the *only* task of a successful company; the structure must also allow for maintenance of ongoing routines.

When trying to visualize the kind of organization that has *both* an array of routine jobs and opportunities for innovation, I am reminded of a common magician's trick using a set of large "magic rings." To set the stage, the magician hands five separate rings perhaps 8 to 10 inches in diameter to a volunteer from the audience for inspection. The rings each appear perfectly smooth and unbroken, and try as he or she might, the volunteer cannot get the rings to connect. Then the magician takes the rings back and tosses them into the air, and they immediately interlock. For the next few minutes, the magician dazzles the audience with displays of all the possible configurations of interlocking rings.

This provides an intriguing metaphor for the innovating organization. For a large part of its ongoing operations, an innovating organization may look on the surface just like a segmented one. It has a clear structure; its organization charts may show a differentiation into departments or functional units, there may be stated reporting relationships, and people may occupy specific jobs with specific job descriptions and bounded responsibilities. Just like the magic rings, the parts can be separated and, for routine purposes, dealt with separately. But with the toss of a problem, the additional connections between and across segments become clear: executive teams considering decisions together; "dotted line" reporting relationships to another area or more; multidisciplinary project teams; regular meetings of councils representing several areas; crosscutting task forces; territories shared by more than one function; teams of employees pulling together to improve performance; networks of peers who exchange information and support each other's projects.

It is the possibility that the separate rings can indeed be easily connected, when the need arises, that gives the organization its potential for innovation.

Integrative, participative mechanisms do not *replace* the differentiation of definable segments that carry out clear and limited tasks; they supplement it. They prevent the existence of segments from turning into segmentalism. This is the idea behind a "parallel" organization presented earlier, a second organization that links the separate rings of the maintenance-oriented organization in flexible and shifting ways to solve problems and guide changes. There is a clear structure for routine operations overlaid with vehicles for participation. There is a predictable routine punctuated by episodes of high involvement in change efforts.

Top executives, whose mandate is to define the organization's structure, are the appropriate "magicians" in this case. They are the ones who can allow the tossing of the rings to connect people in new ways, across segment boundaries, so that they can participate in solving problems.

The idea behind having a second, or parallel, organization alongside routine operations only makes explicit what is already implicit in an integrative, innovating company: the capacity to work together cooperatively regardless of field or level to tackle the unknown, the uncertain. In a formal, explicit parallel organization, this is not left to happenstance but is guided—managed—to get the best results for both the company and the people involved.

Note that the parallel organization is not itself another specific "program" to temporarily solve an immediate local problem (isolating bits and pieces of problems from each other in segmentalist fashion and not allowing solutions to affect the whole system) but rather a means for managing innovation, participation, and change to ensure a *continuing* adaptive organization and an adaptive population within it. But under the leadership of the parallel organization may be any number of specific "programs," improvements, R&D efforts, and problem-solving groups. Its steering committee may manage an array of integrative vehicles—from standing committees linking parts of the organization on issues of major policy concern to temporary problem-solving groups or teams to innovative projects initiated by corporate entrepreneurs.

As the issues and problems change, so does the configuration of the "rings" making up the parallel organization. The steering committee, as manager of this flexible and responsive system, makes the links among the rings, creating new links in new circumstances—and connects the parallel organization to ongoing operations. This connection is important, in both directions. It helps make the problem-solving efforts responsive to the needs of the rest of the system, and it helps ensure that they in turn can take full advantage of the innovations derived from parallel-organization efforts.

There is a parallel organization guiding change at the divisions making up Honeywell's Defense and Marine Systems Group, for example. There the steering committee is chaired by each division general manager and consists of all his direct reports plus other key operations heads. At different times, the steering committee is managing different projects; in mid-1982 they included an advisory committee monitoring the implementation of a new performance communication system it had designed; a standing committee on community relations; nine task teams on major employee concerns; a study group considering how to involve the union on the steering committee; and a number of individuals to whom divisionwide tasks had been assigned. At the same time, the steering committee continued to take a broad view of policy and to examine long-range goals.

What is striking about Honeywell's participative activities is their careful *management*. The steering committee, with its staff, was watching over budgets, writing guidelines for proposals, establishing accountabilities, communicating to the rest of the organization, and generally handling all the logistics needed to make the parallel organization function effectively. In short, the parallel organization at Honeywell is a coherent vehicle for ensuring that the conditions supporting innovation and productive change are present in the division. (People still do their regular jobs in the routine hierarchy, but they also have a second way to contribute, through teams and task forces, above and beyond the limits of their job.) It is not surprising that Richard Boyle, the division general manager leading this activity, is known as an "entrepreneur" around Honeywell. For him the division's outstanding financial results, well above plan, and its future projections are inextricably connected to the commitment to participative management; besides, "it saves us [the executives] time when we can get more of the organization involved in helping us be ready for the future."

Thus, an appropriate place to begin replacing segmentalism with the integrative approaches that support innovation is with the creation of a second structure, a structure for change, parallel to and connected with the company's ongoing structure for doing business as it has already learned how to do. Building a steering committee to guide this structure for change is itself an integrative step: the team at the top working cooperatively across functional lines to view their territory as a whole and combine data about needed changes from many perspectives—problem seeking as well as problem solving. This group can look for and reward innovations that already exist—the departures from tradition that suggest new options—as well as stimulate and encourage other innovations. It can set broad guidelines that give direction to action, channeling the entrepreneurial instincts of innovators in productive directions. And it can decide whether and how to change the way ongoing activities are handled.

Top management has many options for stimulating more innovation. It can assure that a portion of each job definition is loose, that roles and interests overlap enough to force people to work together across disciplinary or hierarchical boundaries, through multiple connections fostering cross-segment initiatives and teamwork. It can support and coordinate the actions of innovators, providing legitimacy, information, and resources to potential corporate entrepreneurs. By seeing the connection of decisions to one another, and encouraging and supporting coalitions, it can avoid segmentalism and make changes that will help support valuable initiative so that it does not slide away. It can make sure that all kinds of people and all kinds of levels in the organization feel included in an integrated whole, with the chance to participate in making a difference for the organization.

And then, the new approach can cascade downward. Teamwork to guide a parallel organization at the top can be matched by similar teams at the head of each major operation, serving integrative functions on more local levels, and moving downward to create integrative mechanisms for middle management, across level, across function, across barriers of race, sex, or employment category. When this occurs, the opportunity for lower echelons to participate in improvement-oriented teams (such as quality circles or committees or task forces or simply staff meetings) can be related appropriately to a consistent organizational culture and coherent strategic directions —not detached pieces handled segmentally but integrated groups connected to an organizational style that begins at the top and that supports local flexibility and initiative.

With my findings about innovating organizations in mind, it is not hard to imagine an action program to remove roadblocks to innovation at a "Southern Insurance," a "Meridian Telephone," or a Petrocorp. These would be among the important elements to be managed by the executive team or its designated steering committee:

Encouragement of a culture of pride. Highlight the achievements of the company's own people, through visible awards, through applying an innovation from one area to the problems of another—and letting the experienced innovators serve as "consultants."

Enlarged access to power tools for innovative problem solving. Provide vehicles (a council? an R&D committee? direct access to the steering committee?) for supporting proposals for experiments and innovations—especially those involving teams or collaborators across areas.

Improvement of lateral communication. Bring departments together. Encourage cross-fertilization through exchange of people, mobility across areas. Create cross-functional links, and perhaps even overlaps. Bring together teams of people from different areas who share responsibility for some aspect of the same end product.

Reduction of unnecessary layers of hierarchy. Eliminate barriers to resource access. Make it possible for people to go directly after what they need. Push decisional authority downward. Create "diagonal" slices cutting across the hierarchy to share information, provide quick intelligence about external and internal affairs.

Increased—and earlier—information about company plans. Where possible, reduce secretiveness. Avoid surprises. Increase security by making future plans known in advance, making it possible, in turn, for those below to make their plans. Give people at lower levels a chance to contribute to the shape of change before decisions are made at the top. Empower and involve them at an earlier point —e.g., through task forces and problem-solving groups or through more open-ended, change-oriented assignments, with more room left for the *person* to define the approach.

Before these kinds of organizational changes can be made, of course, corporate leaders must make a personal commitment to do what is needed to support innovation. They must believe that times are different, understand that the transforming nature of our era requires a different set of responses. They need a sense of sufficient power themselves that they can be expansive about sharing it. They need a commitment to longer-term objectives and longer-term measures. And they as individuals must think in integrative rather than segmentalist ways, making connections between problems, pulling together ideas across disciplines, viewing issues from many perspectives. In short, top executives need at least some of the qualities of corporate entrepreneurs in order to support this capacity at lower levels in the organization.

If there is any domain over which top executives have control, it is organizational culture and structure, the setting of the context for others around and below them. Even if ideas bubble up, as I showed at Chipco and other innovative organizations, organizational style bubbles down. Even the most effective of corporate entrepreneurs soon reach the limit of their own ability to push innovative improvement when the environment set at the top does not support their activities for the use of their results. Indeed, until corporate leaders see the nature of this environment in its full-blown implications, they are doomed to make segmental, and therefore ultimately less effective, responses.

Instead of continuing to think that they can run the organization from the top, effective leaders will be those who know how to take advantage of the capacity of those below. They will be those who appreciate the fundamental transformation in the way organizations and the people in them must work to fit the economic and social challenges of our time. And thus, they can help contribute to a resurgence of the entrepreneurial spirit even within large organizations, a virtual Renaissance for corporate America.

THE AMERICAN WAY TO A CORPORATE RENAISSANCE

The models for the innovating organization are not particularly new, although they have received greater public attention and legitimacy in just the last few years, when the news reached us that certain of our successful foreign competitors might be beating us because of more people-conscious, commitment-producing workplaces. During the last twenty years, a large number of tools have been made available to American companies to stimulate the highest performance from their employees and managers, from more meritocratic performance-appraisal systems that reward individual achievement to cross-level problem-solving teams.

The tools are there—if we care to use them.

To me, that is the central issue: not inventing still another fancy new management system with its own acronym or alphabet label, but using what we already have. The issue is to create the conditions that enable companies to take advantage of the good ideas which already exist, by taking better advantage of the talents of their people. By encouraging innovation and entrepreneurship at all levels, by building an environment in which more people feel included, involved, and empowered to take initiative, companies as well as individuals can be the masters of change instead of its victims.

New ideas will "save the American economy." New ideas will provide our competitive advantage.

The source of new ideas is people. That's why an organization's way of educating and involving people, distributing them among assignments, and rewarding their efforts are so critical in its ability to innovate. Selecting "good" people, certainly. But there are not enough creative geniuses to go around. And there are too many problems in most American companies in this era for them to be able to afford to have only a handful of people thinking about solutions.

Individuals make a difference. That's the positive side of "American individualism"—entrepreneurs not afraid to break the mold in seeking to break a record or competing to win a game. In organizations, this initiative is best expressed through teamwork, and thus we saw that managerial entrepreneurs with innovative accomplishments were most likely to have participative/collaborative styles, to involve a team of others to bring their idea to fruition. Innovation and participation are linked. Strong individuals, along with a tradition of teamwork, bring productive accomplishments into being.

It is hard to mention "teams" or "participation" anymore without someone's labeling them "Japanese-style management." In the first place, this is faintly ridiculous, because it is just as American to use teams, and when American companies do it, they are doing it as Americans, out of their own organizational priorities and images. But if we stopped there, with the idea of participation, we might indeed be missing a distinctively American strength: the initiative of individuals. Innovating companies emphasize teamwork, but they also reward individuals, and they give internal entrepreneurs free rein to pioneer—as long as they can also work with the team. So "American-style" participation does not and should not mean the dominance of committees over individuals, the submergence of the individual in the group, or the swallowing of the person by the team, but rather *the mechanism for giving more people at more levels a piece of the entrepreneurial action.*

Thus, companies need to be encouraged to *invest* in people rather than paying them off—that is, to channel more of their "rewards" into budgets for projects or new ventures and less into after-the-fact bonuses for executives. Tax incentives could help; e.g., by a combination of deferred compensation for individuals and write-offs for the company, pools of working capital could be made available to support innovative projects inside a corporation.

Indeed, Harry Olson, a former senior executive of American Express, has proposed that severance pay for personnel laid off during a recession could be treated this way, as "venture capital" rather than a payoff to the individuals terminated. He argued that the company could use the same amount of money that would otherwise be paid directly to the individuals to set each of them up in a business; this in turn might create long-term gain for the company. Even if only a small proportion of the businesses paid off, the company would be no worse off than it is by the present system of cash payments that are not reinvested for its benefit. A similar reasoning is behind the efforts of companies like 3M and Levi-Strauss to set up internal venture-capital banks to fund new ventures developed by internal entrepreneurs.

Investment in internal human-resource systems is also related to the encouragement of corporate entrepreneurship in the interests of productivity and innovation. To the extent that innovation and change can save jobs through better company performance, and to the extent that an investment in human resources creates a better labor pool not only for the current company but also for the society at large should people change companies, then it is in the public interest as well as the company's interest to support investments in these areas. There could be tax credits, for example, for the development of training programs or other internal educational efforts meeting certain standards—in part a way of acknowledging the important educational role increasingly played by major corporations in today's society.

While not all aspects of a company's human-resource practices are reducible to concrete manifestations that could be supported in this fashion, a surprising number would lend themselves to this: e.g., the start-up costs of improving labor-management cooperation or beginning a joint labor-management committee; the R&D costs of a program to encourage more innovation in manufacturing methods; the retraining costs of shifting workers from one manufacturing sector to another or giving them skills to be more adaptable in the face of the changing technical environment. (It is striking to contemplate what kinds of changes could be encouraged by this method. For the most part, government interventions in the human-resource realm, e.g., safety and affirmative action, have been negative—threatening companies with punishment if they do not comply with regulations but not providing any rewards for quick compliance and creative change.)

If more companies are encouraged to increase their investment in their people, following the lead of the innovating companies I have described by replacing segmentation with integration, then this could in fact turn out to be a transforming era—one that might even be termed an American "corporate Renaissance" because of its humanistic as well as economic benefits.

In an American corporate Renaissance, we could see the reawakening of a dormant spirit of enterprise at all levels of organizations, among all kinds of workers. Entrepreneurship and initiative would be rewarded in large as well as small companies, and there would be a sense of shared purpose—almost a missionary zeal—with which people approached their work. The humanistic thrust inherent in the idea of a Renaissance would be manifested in corporate attention toward ending the "miseries" of earlier corporate work systems, integrating quality-of-life concerns with productivity.

The potential for doing this already exists, in countless offices and factories all over America beginning to see the virtues of a more participative workplace.

The Renaissance analogy suggests a growing "intellectuality" surrounding the American corporation. No longer the mindless machine, the corporation could be the instrument for meaningful intellectual exploration. There is already a growing trend toward self-conscious corporate examination of purpose and philosophy, sometimes expressed in the narrower and more technical idea of a corporate "mission" but increasingly being expressed in more philosophical statements of operating principles that stress human concerns. I have participated in the drafting of several such statements by groups of executives, including phrases such as "work life and home life have interacting needs that will be recognized." What is striking is not the mechanics of producing such statements (talk is cheap, after all) but the self-reflective discussions that take place among corporate leaders who are now participating in perhaps their first chance to examine their own and others' values. I am not suggesting that such statements of philosophy always result in immediate action or solve all of the workplace problems to which they are addressed, but they are an important starting point for a corporate Renaissance.

This growing concern with corporate purpose and long-term responsibilities would be aided by a quest for leaders characterized by long-range, integrative thinking. And by their encouragement to operate, once in executive office, toward long-term objectives. Perhaps it would even be possible to build these encouragements into corporate charters, the very framework for the corporation itself.

Corporate governance—e.g., the shape and composition of the Board of Directors and the officer group—has been much discussed over the last few decades as a means to ensure that the public interest, as well as that of key groups such as employees, is reflected in corporate decisions. But there has been practically no discussion of the use of such mechanisms to encourage investment in the long term, including in its traditional form of R&D expenditures. But what if, for example, there were tax incentives, or even requirements, for publicly chartered firms to withhold a proportion of an officer's or a director's compensation until five years after retirement? Would this decrease the tendency to manage against stock price or quarterly income statements and encourage investment in activities that might not pay off until after the executive's term? Would this encourage more careful succession planning and development of successors?

At this point, of course, such suggestions seem fanciful, and we can only speculate about their likely results. But consider that entrepreneurs who *found* companies often have to be willing to wait years for a return, trying to build a long-term capacity rather than just make a quick killing. Why shouldn't corporate executives too have to wait for rewards until the ultimate results of their actions are known?

A long-term view and concern with corporate philosophy and mission is only one part of a Renaissance-style intellectual awakening. The intellectual dimension of a possible corporate Renaissance is represented in more mundane ways by the increasing numbers of "knowledge workers" whose task performance is linked to the quality of their intellect, a rapid growth in "intellectual" staff functions such as planning departments, and a general increase in the amount of education carried on by and within corporations. There are even some companies hiring historians and cultural anthropologists to help them grasp ineffable dimensions of the corporate experience.

We could see a potential Renaissance in the flowering of literature highlighting the drama and excitement of activities within the new-style innovating corporations. Most great literature about business in the past seemed to fall into one of two camps: muckraking treatments of the corporation as oppressor of the human spirit, or cynical accounts of how someone beat the system. But otherwise, great art has not come out of the corporate sector; only dull monotony and Babbitry. (The great executive/poet Wallace Stevens did not write about his insurance company.) Between Horatio Alger and the recent past, we have only Willy Loman and the man in the gray flannel suit—and stories about the smothering of creativity not unlike my accounts in Chapter 3. But today business stories are beginning to be told for their dramatic qualities as well as their immediate news value. The corporation is being seen as a human arena, and thus one out of which great tragedy and great comedy might be crafted—or gripping adventure stories like *The Soul of a New Machine*, about the design of a new computer.

There is drama in innovation and change that does not exist in a segmentalist environment. Out of the new high-tech companies in the Silicon Valley and Route 128, populated by the generation that gave us beads and plumage, has come a more colorful and expressive kind of existence, full of Friday parties-by-the-company-pool, tales of legendary heroes who found important companies but occupy the smallest office, and rituals like a "boot camp" to teach new managers about company culture. Thus, business life in an innovating company may be seen not only as a necessity, but as *interesting*—a life through which people can express themselves.

This Renaissance could also be signaled by the beginning of an end of a "Dark Ages" of insularity, closed boundaries, and chauvinism of all kinds. The potential for this is clear, though we still have a distance to go. With awareness and acknowledgment of the successes of foreign competition, American companies have become less smug and insular, willing to learn from other countries and other companies, integrating their overseas operations into the domestic mainstream—but respecting the differences of other cultures, and not automatically assuming American superiority. A greater sense of community and social responsibilities would also bring about a corporate Renaissance. Indeed, I nominate Minneapolis as the capital city for the corporate Renaissance because of the pioneering efforts of companies like Dayton-Hudson, General Mills, Honeywell, Control Data, and others to break down boundaries between company and community, behave responsibly, develop new work systems, and join together to promote these values.

Other forms of chauvinism and insularity would have to be overcome to warrant the Renaissance label, of course: assumptions of managerial superiority, male superiority, white superiority. But companies *could* do it—we have models of successful work systems that are more integrative environments—if they chose to put a commitment into the effort. Models exist.

Finally the potential for an American corporate Renaissance would be enhanced by the kinds of people developed and rewarded in leading-edge innovating companies: broader-gauged, more able to move across specialist boundaries, comfortable working in teams that may include many disciplines, knowledgeable about how to manage ambiguous assignments and webs of interdependencies. In short, Renaissance people—men and women of skill and cultivation who could function simultaneously in several organizational worlds.

The style of thought and problem-solving capacity associated with such Renaissance people are encouraged by a strong, affordable educational system that combats narrow vocationalism and permits people the luxury of studying a variety of fields before becoming too specialized. *Affordable* is the key word. When a liberal-arts education is not only priced out of the reach of most middle-income families but also appears to be a frill in a job-hungry society where there is no public assistance for either job finding or translation of a general education into a specific entry credential, then we encourage single-skilled people unable to function on the kinds of cross-disciplinary teams that produce innovation—and less adaptable when circumstances change. Thus, the potential for a corporate Renaissance would be enhanced by public—and really, federal—financing of higher education, particularly in the liberal arts.

Clearly we will always require a large number of specialists, particularly technically trained personnel skilled in the newest technologies. But if their education is balanced by a general education giving them a broader view and an ability to make intellectual and interpersonal connections with people in other fields, then the potential innovative capacity of the organizations that employ them is expanded. Some of those who become general managers in the most innovative high-technology firms have minimal technical competence but are well educated in an integrative discipline—including lawyers or personnel experts I met who had risen to head divisions in engineering-based companies. Purely technical experts are often unable to put all the pieces together to manage a business in a demanding, rapidly changing environment; Renaissance people are required.

If we were to have a corporate Renaissance, the organization itself would be the arena in which its great achievements would take place: new products, markets, policies, structures, methods, and philosophies. The excitement of change, the drama of invention captures the imagination in a way that routine, everyday work in a defined job does not. Being part of a team designing a new program for the company can give people a heightened sense of importance and involvement, an experience of creation that punctuates the rest of their ongoing work experience. Changing a part of an organization, inventing its shape can be fun, can be uplifting. And thus, some of the more deadening aspects of work in segmented systems could be alleviated by the opportunity to move beyond or outside of the job to innovate.

Of course, the organization itself can be the arena for innovation only if corporate leaders are focused on their own operations as the realm for investment, rather than seeking financial gains by manipulating assets—merging, acquiring, and divesting bundles of capacity rather than putting resources into increasing or redirecting that capacity itself.

As the recent Bendix debacle has made clear, attempts at mergers and acquisitions that serve no productive purposes are made possible in part by the ability of companies to write off certain of the costs involved against their taxes—in essence, a public subsidy of such activity. But it would be more clearly in the public interest to encourage companies like Bendix to reinvest their profits (or gains from the sale of assets) in the development of their own businesses, as Edgar Bronfman, chairman of Seagram's and himself a player in a large takeover battle with Du Pont, suggested in a column in *The New York Times*. To the extent that the marriage metaphor applies to a merger—whether it is a "shotgun romance" or a "courtship"—we can also see that our present system encourages companies to increase the "divorce" rate by "trading in spouses" rather than working on improving the quality of existing marriages. Under these circumstances, less attention is paid to internal innovation and fewer resources are made available to invest in it. But internal investment is what creates the climate for the innovations allowing companies to stay ahead in a changing environment.

In a corporate Renaissance, in short, companies would be more like "families" making long-term commitments to the development, health, and prosperity of each of their members, and looking to all of them for productive new ideas.

The potential exists for an American corporate Renaissance, with its implied return to greatness. Because recent economic conditions have been so unfavorable for American business, leaders should be motivated to search for new solutions—and to engage their entire work force in the search. I argue that innovation is the key. Individuals can make a difference, but they need the tools and the opportunity to use them. They need to work in settings where they are valued and supported, their intelligence given a chance to blossom. They need to have the power to be able to take the initiative to innovate.

Whether the promise of this corporate Renaissance is fulfilled depends on how fully corporate leaders understand this need and decide to act on it. It depends on whether we can come to embrace change, to see it as an opportunity, and thus to stimulate the people in our organizations to take action to master it.

As a nation, we can no longer afford to do otherwise.

Appendix:
The Core Companies
and Research Methods

THE RESEARCH that led to this book took place over a period of five years and included six focused research studies on specific topics in the general area of innovation, change, and corporate responsiveness to new environmental demands. More than 100 companies were involved in some fashion in one or more studies, including more than 40 that I visited personally. Thus, when I draw conclusions based on the 10 core companies that I examined in depth, I can be reasonably sure that my findings accurately reflect corporate America.

Two of the studies constituted background: changes in the cultural environment for American companies over the last twenty years, and an attempt to define the most "progressive" companies for these times and their characteristics, including financial performance. The other four provided in-depth analysis of patterns and behaviors in 10 major companies which served as the core group for my analysis.

BACKGROUND STUDIES

I. Content Analysis:
The Changing American Corporate Cultural Environment, 1960–1980

I began with the conviction, based on my earlier research, that the period from 1960 into the 1980s was a time of major transformation. Now I wanted to find ways to document shifts in the cultural context for the American corporation: assumptions about how a company should behave, norms and values about appropriate actions. This is an amorphous subject. I had already collected a large number of statistics on changes in the American labor force, in attitudes toward work, and in job content (see especially my article "Work in a New America," *Daedalus,* Winter 1978), and so I sought ways to capture, in addition, some more elusive concepts.

The strategy ultimately chosen was to look at the changes that showed up in the business press and in the speeches of key executives. But because this investigation was designed to provide background and was thus subsidiary to my main purposes, I also wanted to keep it manageable. My first idea was a content analysis of *The Wall Street Journal, Fortune, Forbes,* and selected industry trade magazines over the period from 1960 to the 1980s, but an initial attempt at coding the index for just one of those publications for just one year made it clear that this would take much too much time for the gains derived. The solution was to code information at five-year intervals in two major summary sources: the *Business Periodicals Index,* which catalogues articles in major business publications, and *Vital Speeches,* which collects major addresses by public figures, including corporate executives. Thus, David Summers, research director for this activity, conducted a content analysis of *BPI* in 1959–1961, 1964–1966, 1969–1971, 1974–1976, and 1979–1980 and of *Vital Speeches* in 1960, 1965, 1970, 1975, and 1980.

For the *BPI* study, the first step was to develop a list of topics of interest to us around both questions of internal organization and management and the events that had impinged on companies in the last twenty years, especially labor-force changes and strategic challenges such as increasing foreign competition and the energy crisis. A simple count was made of the number of articles devoted to subtopics on each of these issues, and an indexed figure was prepared that looked at this number as a function of the total number of articles reviewed. We then also looked at the topics receiving the most total space in the business press as measured by inches in the index, in each cluster of years, and watched how these changed. While both of these were relatively simple measures, they did provide some empirical confirmation for what I was already considering to be the major changes since 1960.

The content analysis of *Vital Speeches* was more straightforward. Summaries were prepared of the major speeches by corporate executives covered by the magazine, and these speeches were then examined together, by year, for a qualitative analysis of themes reflecting assumptions about the environment and assumptions about how organizations should be managed.

Chapter 2 reports on the findings.

II. The Progressive Companies Study

I next wanted to gather information on the reasons companies lead in change (innovating early in human-resource areas) and the benefits they derived from this. It is impossible to find an objective set of indicators akin to *Fortune*'s or *Forbes*'s financial indicators to generate a list of the "best" or "most innovative" companies with respect to their treatment of employees. Thus, we decided that the best initial source for such a list was the views of corporate experts in human-resource fields. The senior human-resource executive (generally the vice-president), or designated staff, from 65 companies participated in this phase of the study, out of 246 of the Fortune 1000 industrials and 250 nonindustrials originally approached (a 27-percent participation rate). We received 58 completed nomination forms, several of which represented the consensus of a group of human-resource staff from that company, and oral comments from another 7 executives, for a total of 65 nominators. Forty of the nominators were also interviewed by phone for 45 to 90 minutes to discuss the reasons for their nominations.

Nominations were sought in five categories:

- The most innovative companies in overall human-resource systems—the pioneers/developers (4 nominations sought).
- The most innovative companies in work alternatives such as QWL and job design (4).
- The most innovative companies in affirmative action and EEO (4).
- The presently most progressive companies in overall human resource systems—the best adapters, whether or not they were the developers (6 nominations sought).
- The companies most likely to serve as a model for others in human-resource areas (6).

Companies could be nominated in more than one category, and respondents could nominate their own companies.

A total of 909 nominations were received (out of a theoretically possible total of 1,950), based on *at least* the following degree of acquaintance with a nominated company other than one's own:

115 based on personal acquaintance with key staff (13 percent)

87 based on a contact with the company seeking
information (10 percent)

17 based on previous employment in the
nominated company (2 percent)

(I have good reason to believe that these figures vastly understate the degree of acquaintanceship simply because many respondents known by me to be in contact with certain companies left this question blank, and others nominated their own companies.) Non–U.S.-owned companies were excluded from the count, as were other types of organizations such as universities and consulting firms. There were 100 companies nominated at least once in any category (including companies nominating themselves); 47 were nominated twice or more; 33 were nominated three times or more and in more than one category; and 5 companies were mentioned the most often in single categories as well as overall, receiving 20 or more nominations in any one category: (AT&T, General Electric, General Motors, IBM, and Xerox).

The 33 top-ranking companies were:

Aetna	General Foods
AT&T	General Motors
Arco	Hewlett-Packard
Bank of America	Honeywell
Celanese	IBM
Citicorp	Johnson & Johnson
CBS	Kodak
Control Data	Levi-Strauss
Corning Glass	3M
Cummins Engine	J. C. Penney
Dana	Polaroid
Digital Equipment	Procter & Gamble
Donnelly Mirrors	Sears
Du Pont	Texas Instruments
Exxon	TRW
Ford	Xerox
General Electric	

The 14 also receiving two nominations were:

Connecticut General	Merck
Dayton-Hudson	Motorola
Delta Airlines	NCR
Disneyland/World	Saga
Eaton	Shell
ITT	Smith Kline
Mead	Tektronix

These 47 were then used as the basis for further analysis.

The 47 "progressive" companies were first analyzed in terms of the sources of the "progressive" reputation: number of nominations; categories/kinds of HR innovation; reasons for nomination; and degree of nominator's acquaintance with firm. Data on firm size as of 1981 were collected, including net sales, assets, and number of employees. Industry characteristics were also examined, using Dunn and Bradstreet, *Forbes* financial indicators, and other major sources. A clippings file of major news accounts of each company was also compiled.

Then, for each "progressive" company, a nonnominated match was made from the same primary industry and closest to the progressive on three 1980-size measures: net sales, assets, number of employees. For three very large and two medium-sized companies, no match was possible. Owing to industry clustering of progressives, two companies served as the match for more than one progressive. Thus, there was a total of 41 "nonprogressive" matches. Overall, 23 different industries were represented in the "progressive"/"nonprogressive" match-up.

The "progressive" companies and matches were then compared in terms of:

- Headquarters-city characteristics
- Growth patterns over 20 years (1960–1980)

- Profitability—for 20 years, at 5-year intervals, on such measures as return on sales (ROS), return on equity (ROE), and return on assets (ROA)
- Growth/profitability compared with industry norms in primary and secondary industry categories using similar financial indicators
- Composition of board of directors and senior executives
  - inside/outside
  - functions represented
  - age
  - education
  - company service
- Characteristics of highest human-resource official

The first round of analysis was based on these data.

Analysis of data from this study is still in progress as I complete this book (e.g., fuller statistical analysis, comparison of 1981 reputation with 1983 reputation, updating of nominations, and description of each company's practices against a checklist of ideal ones). But among the key findings to date that I draw on here are these:

Except for a few very large companies, perceived leadership in one HR category is unrelated to leadership in others; areas of HR innovation appear uncorrelated.

Though the "progressive" companies were generally very large, size alone accounts for only 41 percent of the variance in the number of nominations a company received. Furthermore, number of nominations is largely *uncorrelated* with company growth or profitability. Thus a reputation for leadership in HR systems is not defined by size or success alone; our nominators were not simply giving us the *biggest* or most glamorous U.S. companies.

The industries contributing the largest number of progressive companies are computers and electronics.

"Progressive" companies have been more profitable and faster-growing over 20 years than their matches ("nonprogressives") on each of three measures of profit and two measures of growth —even in fast-growing industries.

Progressive companies' senior executives, compared with the matches, are:
- more likely to have risen through the ranks;
- no different in age;
- less likely to have attended college or professional school;
- more likely, if they did go to graduate school, to have gone in a field *other* than law, business, or engineering;
- more likely to have worked in *no* other company and much more likely to have at least 20 years of service, including much more than 50 percent of their career;

• no different in likelihood of sitting on their own company's or others' boards of directors.

In short, they are more likely to have intimate knowledge of their company gained from in-depth experience and less likely to be "professional managers."

DEPTH-RESEARCH PROJECTS

As I explored issues of innovation and change with American companies over the five years of my investigations, I focused on a number of questions to examine in depth. In 1977 I began to collect information on cases of employee involvement in problem solving. Then in 1979 and 1980 I began to formulate specific research questions which resulted in the analysis here, and I solicited the participation of major companies. All the companies involved were aware of and willing to contribute to the research, in particular after I promised to maintain confidences and to use pseudonyms in place of the actual company names where the company preferred this or where use of the name would serve no constructive purpose. (Thus, none of the low-innovation companies are identified, and many details are disguised.) In a few cases, the research grew out of consulting projects with the companies, but formal permission was received to document the projects; in the majority of the cases, the companies were approached directly about participating in the research as an independent activity.

A total of 37 companies were involved in one or another of the four depth projects, and 10 of them took part in more than one. It is these 10 which constitute the "core group" for in-depth analysis in this book, and it is out of their experiences that I developed most of my interpretations and conclusions. All are prominent Fortune 500 companies, though they are scattered all over the size and age range, with one of them a fairly recent entrant to this list. The 10 core companies include 4 that come out of an earlier American corporate era:

> General Motors—auto manufacturer
> "Meridian Telephone"—a pseudonym for a communications company
> "Petrocorp"—a pseudonym for a raw-materials processor and manufacturer
> "Southern Insurance"—a pseudonym for an insurance and financial-services company

and 4 that are newer, "high technology" firms founded since 1940:

> "Chipco"—a pseudonym for a publicity-shy computer manufacturer
> Hewlett-Packard—electronic-equipment manufacturer
> Polaroid—photographic-equipment/optical-products manufacturer

> Wang Laboratories—manufacturer of word-processing equipment

and 2 that are "high technology" sectors of older manufacturing firms:

> General Electric (Medical Systems Division)—manufacturer of X-ray machines, CT scanners, and other medical diagnostic equipment
>
> Honeywell (Defense and Marine Systems Group)—manufacturer of electronic equipment for military and commercial markets.

The choices of companies and the studies in which they participated were a matter of *access* (where my research team was welcome), *fortuity* (where a change effort was under way), and *theory* (how to ensure relatively complete coverage of major sectors).

In addition to findings from the research projects on each of the companies, my research team and I collected financial data and growth histories of the companies from public sources and maintained a file of clippings from the business press about the activities of these companies since 1960. (The clippings files focused particularly on the years 1977 through 1980.)

Table I indicates the kinds of participation of each of the 10 companies in the four depth-research projects.

I. The Innovation Study: Managerial Accomplishments

This was a large-scale quantitative and qualitative empirical comparison of the factors encouraging initiative and innovation in organizations. It began with the question of how managers acquire and use power to move beyond the boundaries of their job to contribute to innovation. The initial assumption, confirmed by the findings, was that formal authority would be less important to managers attempting an innovation than power and influence which they exercised beyond the formal mandates of their organizational position. It was also assumed that in "new style" companies which encouraged innovative and entrepreneurial activity to begin with, especially those high-technology firms which came to maturity in the 1960s, authority would generally be loosened enough so that managers had to rely on whatever power they could gather, rather than on formal title and position. On the basis of my theory of organizational power, I developed a set of questions about the events and behaviors involved in carrying out a significant accomplishment and in a pilot study tested those questions on 26 middle managers from 18 different companies, some of which later participated in the main study. The pilot-study findings (reported in my article, "Power and Entrepreneurship in Action: Corporate Middle Managers," in *Varieties of Work*, edited by P. L. Stewart and M. G. Cantor, Beverly Hills: Sage, 1982) were used to develop a final interview guide.

Then I reviewed the previous research literature on innovation, especially the conditions associated with receptiveness to innovation in a wide variety of organizations, but also accounts of the development of prominent product innovations in American companies. This search (which resulted in "Ninety-Nine Propositions About Innovation from the Research Literature," in *Stimulating Innovation in Middle Management*, Cambridge, Mass.: Goodmeasure, Inc., 1982) was used to develop hypotheses about the differences in company environments accounting for their degree of innovation, to be tested in the study.

In the main study, 208 managers from six companies ("Chipco," General Electric, Honeywell, "Meridian," Polaroid, and "Southern") were asked to describe their most significant job-related accomplishment of the last two years; this brought the total number of interviews, including the pilot study, to 234. Later 30 interviews were conducted with managers at Wang and Hewlett-Packard to check the validity and generalizability of the findings; these were more informal and not included in the quantitative analysis.

This method uncovered 115 innovations, including among them a number of quite important developments with major financial, strategic, or organizational implications. Table II provides an illustrative list of some of the 115 innovations, indicating their range and significance. The position of the person reporting this as a key accomplishment is also described in Table II; in about six large-scale innovations several participating managers reported on the same event.

The wave of interviews in each company was preceded by conversations with informants to become clear about the nature of the company. The interviewees were selected by the companies, to represent "effective" managers whose positions were below officer or department-head level and above first-line supervisor. Interestingly enough, the range of levels represented (the distance between the highest- and the lowest-level interviewee) was greater in the less innovating companies, "Southern" and "Meridian," which were characterized by tall hierarchies; in the other companies we managed to get a more homogeneous "middle" population.

The 234 interviews were carried out by a research team and took a minimum of an hour and, in some cases, up to two and a half hours to complete. The interview covered:

- Identification of up to four significant accomplishments, with one serving as the interview focus;
- A chronology of the accomplishment from initiation through completion or to the present time;
- Conditions of initiation, including the existence of models and precedents, the nature and typicality of the assignment, the source of ideas;
- Acquisition and use of resources, including magnitude, kinds, number of sources (e.g., inside/outside the work area, organizational level), typicality of sources;
- Acquisition and use of information, including kind, sources, typicality;

- Acquisition and use of support, including sources, timing, exchanges involved, and typicality;
- Opposition, criticism, and roadblocks encountered and strategies used to overcome them;
- Results and payoff from the accomplishment, to self and organization, including costs as well as gains;
- The ways "success" or "achievement" was handled, including rewards and sharing of rewards;
- Changes the respondent would make the next time around;
- Description of the job, an organization chart for the department or function, history with the company, age, sex/race, self-description of management style, and discussion of any changes taking place around the department.

The interviewers tried to capture the respondents' own words as much as possible as well as checking appropriate categories on the interview guide. There were no formal tests of the reliability of the respondents' accounts (e.g., checking with other key actors), but we did have several instances of managers involved in the same major innovation, and the overall consistency of their accounts (though from different angles) suggests that the method was valid and reliable.

The interviews were then analyzed both qualitatively and quantitatively. The qualitative analysis consisted of reading and sorting into categories as well as preparing case studies describing the findings for each company. The interviews were also coded on 65 variables and analyzed by computer, using relevant statistical tests.

The most important cut through the data involved the coding of the accomplishments themselves. They were sorted into two overall categories, with several subdivisions each. Accomplishments were considered either:

BASIC—involving nothing more than carrying out the basic job-as-given, though perhaps with increments of improvement in performance; or

INNOVATIVE—adding to the organization's future potential or capacity by going beyond the job-as-given and "leaving a trace."

Basic accomplishments included three subcategories: simply doing the *basic job*; *impacting on individuals* only, with no organizational consequences (e.g., raising morale, getting a subordinate promoted, or managing one's own career); and *incremental advances* in performance (e.g., raising sales). Innovative accomplishments included: *new policies, new structures, new methods or technological processes,* and *new products or market opportunities.* Intercoder reliability in categorizing accomplishments was 88 percent; it was generally high on all the other variables as well. Where there was disagreement among research-team members involved in the coding, adjustments were made based on discussions among the coders about the reasons for their choices.

The "innovation index" was a simple score: the percentage of the accomplishments studied that were rated as "innovative" rather than "basic." This rather crude approximation was not used as an exact rating but rather merely to create categories, to differentiate the higher-innovation companies from the lower-innovation ones. And there was a significant gap between them. The validity of this score was then checked against other ratings of the accomplishments: the degree of self-initiation, risk, and significance. All three measures were highly correlated with the simple innovation code. On the basis of these distinctions, four organizations were considered high innovators ("Chipco," General Electric Medical, Honeywell DMSG and Polaroid) and two low innovators ("Meridian" and "Southern"). The differences between these groups of companies in terms of culture and structure, as well as supports and roadblocks for innovation, were then analyzed using both qualitative material from the interviews and statistical comparisons on the 65 variables.

A summary of some representative facts about these companies is found in Tables III and IV.

II. Histories of Change Projects Involving Employee Problem-Solving Teams

The six case histories developed in this study were straightforward. My team and I documented and analyzed all the steps involved in carrying out an organizational change at the "grass-roots" (nonmanagerial) level through participative methods. Such documentation was carried out in six companies (General Motors, "Chipco," General Electric, Honeywell, "Petrocorp," and "Southern"). In two of these (GM and "Southern") the documentation was arm's length, consisting of occasional snapshots or retrospective descriptions from informants. But in the other four cases, the researchers were involved at least weekly in direct observation of activities from start-up.

The material that I analyzed in putting together the accounts used in this book included:

- My own field notes and those of my colleagues;
- Internal memos and minutes of meetings;
- Surveys of the employees involved at various points in the project's history;
- Formal interviews with managers at various points in the project's history, especially detailed at the beginning and the end;
- Informal conversations with participants and others in the companies;
- Any documents or publications from the companies relating to projects. (In several of the cases, the managers or project's initiator have themselves gone on record with cases or publications about the projects.)

Notes and documents were sorted in two ways: in chronological order, to provide a history; and by participant or key event, to gain the actor's perspective. Since in several of the cases my colleagues and I were involved as consultants and advisers, these cases also constituted "tests" of the validity of our hypotheses about change technologies and their impacts. These are described in greater detail in the chapters on the "Chipco" Chestnut Ridge plant and the "Petrocorp" Marketing Services Department, as well as in Chapter 9, "Dilemmas of Participation."

Early versions of the case histories were reviewed by internal parties in the companies for accuracy and modified accordingly.

III. "Whole" Company Cases: Structure, Culture, and Change Strategies

I included seven companies in my analysis of how companies under a variety of conditions handle change and develop strategies to cope with it. General Motors was involved most extensively, but I also included analysis of "Chipco," Hewlett-Packard, Honeywell, "Meridian," "Petrocorp," and Wang.

In these cases, I was looking for clues as to the determinants of how a company responds to change. The analysis was highly qualitative and interpretive but based on extensive contact with the company and detailed interviews with informants. I sought the basic facts about the company's structure and leadership, the major changes it had gone through in the last 10 years, and how these were handled, especially for those which impacted on organizational structure or the management of people. (For example, I was more interested in changes in product or technology that resulted in reorganizing or handling the work differently than I was merely in changes in product that had no such organizational ramifications.) Changes in size, the addition or subtraction of functions, changes in structure, or shifts in policy and strategy were of greatest interest, but I let the questions and the issues be determined by those selected by the company itself as important—as reflected in public statements, annual reports, and the views of my high-level informants.

General Motors

Because of the key role of General Motors in the economy and the public impact of many of its precedents, I knew that I would want to provide a fuller account of that company's attempts to change in the light of the changing environment, and so my contact there was the most extensive. Data sources included:

- Thirty high-level corporate and UAW informants, including seven present and former officers, with whom I held onsite meetings in their offices or other company facilities, and in a few cases a series of meetings at several-month intervals;

- Review of statistics and reports of internal studies provided by corporate informants;
- About 50 interviews at eight facilities involving field trips of a day or more—including union leaders, personnel staff, and first-, second-, and third-line supervisors of both blue-collar and white-collar workers;
- Review of public documents such as annual reports, United Auto Workers contracts since 1967, personnel manuals and policy statements, Public Interest Reports, and executive biographies and speeches;
- Casual information such as conversations with other GM employees and officers at formal meetings both inside and outside the company and the experiences of students who had worked in GM plants with QWL programs;
- Presentations at Yale by GM officers and staff and UAW officials, including two sessions of my course and a conference on the auto industry held in November 1981;
- The opportunity to listen to high-level GM and UAW officials speak at programs where I was also appearing.

Because many others have also been interested in General Motors, I benefited from reading books, articles, and case studies about the company. A recent case on quality of work life at General Motors by Burt Spector, working under Paul Lawrence, and one on downsizing by James Brian Quinn were especially helpful, as were older business-school cases about General Motors that appeared under pseudonyms. (These were given to me and identified to me by personnel staff as concerning General Motors but I was not to reveal the case names here.) Especially useful articles included Robert Guest's account of the Tarrytown QWL efforts in the *Harvard Business Review* and Joseph Kraft's account of the downsizing decision in *The New Yorker*, as well as a series of articles by Charles Burck in *Fortune*. Books read for background material included scholarly treatises as well as anti-GM polemics, from Alfred Chandler on General Motors' history and William Abernathy on productivity in the auto industry to Emma Rothschild's *Paradise Lost*. In total, I scanned a large number of books and about a foot of recent business-press clippings, and my files on General Motors made a pile a yard thick.

As a final check, I sent early drafts of my chapter on General Motors to several of my most knowledgeable informants—people with a firsthand relationship to the events described. I also sent it to several consultants who had worked with GM. They read and critiqued the chapter, adjusting facts, suggesting additional data sources, disputing my emphases. In four cases I received extensive written commentary from those close to the events. Not surprisingly, their views differed. While one former line executive congratulated me for "an outstanding job of presenting the subject," another expressed disappointment that the chapter "fails to do justice to the corporation" in the conclusions I reach at the end. (The split in their reactions reflects the split I describe in the company itself. But I did try to take the comments into account and adjust the emphasis of my analysis.) One commentator warned me that including comments attributed to unidentified individuals would make it possible for people inside the company to write off some of my conclusions as just a "collation of gossip." What I told him in reply was that (a) I had promised everyone I would treat the interview material judiciously, and so I had decided not to cite a source where the items involved were sensitive (but I made sure I had at least one other source, and ideally, many, agreeing with the informant's interpretation—so no "single source" idea is ever used unless it is a person explaining his own behavior); and (b) I *was* given a great deal of what I consider "gossip," some of it quite provocative, but *none of it was used in the book*.

Thus, I feel reasonably confident that I am presenting a fair account of General Motors' achievements—and some of its "liabilities"—at least through early to mid-1982. I am aware that there are a number of important initiatives in the company that occurred too late for me to include them, and that others are still being contemplated. I also recognize that the chapter tries to compress a great deal of complexity into a short space, thus leaving little room for related events or facts that I knew about but could not include. Finally, none of my informants is responsible for the conclusions I drew, which are not a matter of fact alone but of an outsider's interpretation of the relative importance of particular facts (or events or people) as against others.

Cases on "Chipco," Hewlett-Packard, Honeywell, "Meridian," "Petrocorp," and Wang

The analyses of structure, culture, and change strategy in the other six companies were based on the following:

- at least 20 informants from the upper and middle management ranks;
- a minimum of a week onsite;
- public documents and any reports of internal studies made available by informants;
- an extensive clippings file on each company.

In five of the cases, the above was supplemented by participation in or observation at training programs for a variety of staff, including women managers, first-line supervisors, manufacturing managers, or sales managers. I maintained regular contact with corporate personnel staff over a period of at least a year and checked observations with them. The Wang account, which does not figure prominently in this book, was pulled together to test the hypothesis that had been emerging about the issues of management of change in companies where change is a routine event. The case was prepared after an in-depth plunge over a one-month period.

IV. The Changing Personnel Department:
   Internal Responses to New Demands

The fourth research project is not reported on directly in this book but should be described briefly because it offered an additional source of information on innovation and change in many of the core companies. Under a grant from the William H. Donner Foundation in 1979–1980, my research team and I investigated the ways in which the personnel departments of 32 major companies had changed in response to new demands such as the growth of the equal-employment-opportunity function. Thus, we were in effect documenting the way in which a company adjusted internally to externally mandated change. This involved a response to a government requirement that in many cases was supported by other external pressures, from lawsuits through activist attacks on the company.

The study used structured forms and open-ended interviews with from three to five key informants in each company (including the head of the personnel or human-resource function) to understand:

- the structure and activities of the human-resource department or departments;
- the history of events in the introduction and evolution of the company's EEO response;
- the development and history of a formal EEO function, the key committees involved in it, and the power base of the function and its staff;
- the patterns of support of and opposition to the changes involved in introducing this new activity;
- the relative proactivity or reactivity of the personnel function in responding to a variety of environmental pressures on the company;
- the functioning of corporate staff as "change agents" for the organization or as buffers to protect the line-operating core from disruption by external demands.

Five of the core group of companies for this book were among the 32 companies studied: General Motors, "Chipco," Polaroid, General Electric, and "Meridian Telephone."

A CONCLUDING NOTE

I am providing this information for those interested in the data base behind my findings and conclusions. But I do not rest my case for my analysis on any one study or any one way of examining American companies. Instead, I drew on everything I knew, everything I saw, to develop the ideas in this book: from statistical data to personal observations and conversations. Thus, this book is not intended to be merely a "report of research"; nor is the "whole" that it represents merely a sum of the specific studies that went into it. Indeed, in making choices about what material to use to express and illuminate my ideas, I leaned toward rendering those dramas of life in the corporation which would make my conclusions come alive, which would cause readers to believe me not because of my numbers but because of the echoes of my ideas in their own experience.

TABLE I

Core-Company Participation in Depth Research Studies

	INNOVATION STUDY (MANAGERIAL ACCOMPLISHMENTS)	CHANGE/EMPLOYEE INVOLVEMENT PROJECT HISTORIES	"WHOLE" COMPANY CASES	CHANGING PERSONNEL DEPARTMENT STUDY
"Chipco"	X	X	X	X
General Electric (Medical Systems)	X	X		X
General Motors		X [a]	X	X
Hewlett-Packard (parts)	X [b]		X	
Honeywell (Defense and Marine)	X	X	X	
"Meridian Telephone"	X		X	
"Petrocorp"		X	X	
Polaroid	X			X
"Southern Insurance"	X	X [a]		
Wang Laboratories	X [b]		X	

[a] Material gathered from retrospective accounts rather than documentation from start-up and over the course of the effort.
[b] Informal participation only; data used as check but not included in quantitative analysis.

TABLE II

Innovation Study:
Illustrative Examples of the 115 Innovations Examined

INNOVATION	REPORTING INNOVATOR AND COMPANY
New product application: monitor for a patient's heart cycles, successfully conceived and introduced to market	Marketing manager, GE Med
Reorganization of X-ray manufacturing plant, with improvements in performance (e.g., from meeting 75% to 95% of production schedules)	Shop operations manager, GE Med
First functional test station for marine sensing device	Design engineering manager, Honeywell, DMSG
New product package adding X-ray equipment to a vendor's complex product—rapidly created and marketed	Special products design engineering manager, GE Med
New markets for CT scanner equipment (during time of "market overcapacity," quadrupling both the revenues and the percentage return on sales)	Business area manager, GE Med
Inventory reduction plan (goal was saving $1.5 million; achieved $2.5 million saving)	Production planning manager, GE Med
Unique new compensation system for sales force	Product sales manager, GE Med
Unique new safety program	Administration and technical support manager, Polaroid
Ten-year manufacturing strategic plan (the group's first)	Planning manager, Honeywell DMSG
New system to minimize unit costs through integrated process management and control	Division data processing manager, Polaroid

Innovation Study:
Illustrative Examples of the 115 Innovations Examined
(Continued)

INNOVATION	REPORTING INNOVATOR AND COMPANY
Reorganization of maintenance function and new methods for repairing machines	Technical support manager, Polaroid
New method for looking at processes and coating yields, saving $720,000 annually	Product manager, Polaroid
New product components: development of 25 new emulsions in 2 years (previous record: 2 per year)	Product manager, Polaroid
New, more competitive recognition and compensation policy	Group personnel manager, "Chipco"
New budget process to integrate input with output dollars in manufacturing, revealing and correcting $200 million imbalance	Asset and tax manager, manufacturing "Chipco"
Computerized library of data from engineering to manufacturing	Product description systems manager, "Chipco"
New market for old products and reorganization of sales force to product basis, doubling market share in region	Regional operations, manager, "Chipco"
New pricing committee to integrate 8 product lines	Marketing manager, technical products, "Chipco"
50% staff reduction in a major facility without layoffs or unilateral transfers	Plant personnel manager, "Chipco"
Five-year computer-aided design plan for the division	Engineering manager, Honeywell DMSG
New state-of-the-art compensation system (developed through participative methods)	Operations employee relations manager, Honeywell DMSG

Innovation Study:
Illustrative Examples of the 115 Innovations Examined
(Continued)

INNOVATION	REPORTING INNOVATOR AND COMPANY
Separation of sales and service responsibilities, with major performance improvements	District staff manager, business service center, "Meridian"
Decentralization of corporate purchasing of all direct ("raw") material inputs for manufacturing, speeding the process	Corporate purchasing manager, manufacturing, "Chipco"
New long-range financial planning and budgeting system for the sales organizations, including charge-backs of sales expenses to the product groups	Sales and international finance manager, "Chipco"
New business creation, in the educational-services market area	Business manager, custom services, "Chipco"
Reorganization of dispatch and service record keeping and financial reporting, including establishment of a dispatch and coordination desk, vastly improving service	District service manager, GE Med
Reorganization to free production engineers from maintenance work, improving technical performance	Plant engineering manager, Polaroid
New automatic recording densitometer to measure the density, intensity, and trueness of color on film	Division technical manager, Polaroid
New organization structure: transformation of service office into district sales/service office, giving supervisors profit responsibility	District manager, "Southern"
New product project to design a digital radiographer (comparable to a CAT scanner)	Projects engineering manager, GE Med

Innovation Study:
Illustrative Examples of the 115 Innovations Examined
(Continued)

INNOVATION	REPORTING INNOVATOR AND COMPANY
New plant start-up, beating all timetables and cost estimates for producing small motors	Manufacturing engineering manager, GE Med
Division-wide word processing system	Administration manager, Honeywell DMSG
New applied mechanics unit with a $250,000 laboratory	Senior project engineer, GE Med
New market: opening of Japanese market for GE equipment, especially CT scanner—the first time a single country was defined as a foreign market area —beating Japanese competition in Japan	Operations manager for Japan, GE Med

TABLE III

Innovation Study:
Summary of Representative Six-Company Comparisons

	"CHIPCO" N=34	GE N=42	HONEYWELL N=44	"MERIDIAN" N=39	POLAROID N=29	"SOUTHERN" N=20
INNOVATION-INDEX SCORE *(% innovative accomplishments)*	71	64	61	36	62	45
	N(%)	N(%)	N(%)	N(%)	N(%)	N(%)
BASIC ACCOMPLISHMENTS: *Basic Job*	8 (24%)	6 (14%)	7 (16%)	8 (21%)	4 (14%)	5 (25%)
Individual impact	2 (6%)	2 (5%)	– 0 –	8 (21%)	1 (4%)	– 0 –
Incremental advance	– 0 –	7 (17%)	10 (23%)	9 (23%)	6 (21%)	6 (30%)
INNOVATIVE ACCOMPLISHMENTS: *New policy*	3 (9%)	1 (2%)	5 (11%)	2 (5%)	1 (4%)	– 0 –
New product/ market opportunity	5 (15%)	7 (17%)	5 (11%)	– 0 –	4 (14%)	2 (10%)
New structure	8 (24%)	11 (26%)	3 (7%)	8 (21%)	5 (17%)	7 (35%)
New method	8 (24%)	8 (19%)	14 (32%)	4 (10%)	8 (28%)	– 0 –
MEAN DURATION OF PROJECTS *(months)*	33	24	25	30	32	20
MEDIAN AGE OF MANAGERS	37	43	46	38	42	39
MEDIAN YEARS OF COMPANY SERVICE	5	13	19	15	13	4

(N = Number)

TABLE IV

General Characteristics of the Six Companies Bearing on Support for Innovation

	THE FOUR MORE HIGHLY INNOVATING				THE TWO LOWER-INNOVATING	
	"CHIPCO"	POLAROID	GENERAL ELECTRIC MED SYSTEMS	HONEYWELL DMSG	"SOUTHERN INSURANCE"	"MERIDIAN TELEPHONE"
CURRENT ECONOMIC CLIMATE	steadily up	trend up but currently down	steadily up	steadily up	slipping	down
CURRENT CHANGE ISSUES	change "normal"; constant change in product generations; proliferating staff and units	change "normal" in products, technologies; recent changeover to 2nd management generation with new focus	reorganized about 3–4 years ago to install matrix; "normal" product-technology changes	change "normal" in product, organization—esp. in 2 of 3 operations	change a "shock"; new top-management group from outside reorganizing and trying to add competitive market posture	change a "shock"; undergoing reorganization to install matrix and add competitive market posture while reducing staff
ORGANIZATION STRUCTURE	*matrix*	*matrix* in some areas; product lines act as quasi-divisions	*matrix* in some areas	*matrix* in some areas	*divisional;* unitary hierarchy within divisions, some central services	*functional* organizational currently overlying a matrix of regions and markets
	decentralized	mixed	mixed	decentralized	centralized	centralized
INFORMATION FLOW	free	free	moderately free	moderately free	constricted	constricted
COMMUNICATION EMPHASIS	horizontal	horizontal	horizontal	horizontal in matrixed areas; vertical in functions	vertical	vertical

CULTURE	*clear*, consistent; favors individual initiative	*clear*, though in transition from emphasis on invention to emphasis on routinization and systems	*clear*; pride in company, belief that talent rewarded	*clear*, consistent, participative management and good treatment of people emphasized	*idiosyncratic*; depends on boss and area	*clear* but top management would like to change it; favors security, maintenance, protection
CURRENT "EMOTIONAL" CLIMATE	pride in company, team feeling, some "burnout"	uncertainty about changes	pride in company, team feeling	pride in company	low trust, high uncertainty	high uncertainty, confusion
REWARDS	*abundant* include visibility, chance to do more challenging work in the future, get bigger budget for projects	*abundant* include visibility, chance to do more challenging work in the future, get bigger budget for projects	*moderately abundant* conventional	*abundant* in technical areas, somewhat less so in others; many formal awards and ceremonies.	*scarce* primarily monetary	*scarce* promotion, salary freeze; recognition by peers grudging

Notes

CHAPTER 1

1. The Appendix discusses in detail how the "progressive companies" and their matches were identified. Here are the figures for the five-year period to 1980:

| 1980 | PROGRESSIVE | | "NONPROGRESSIVE" | | T-TEST[a] |
	Mean	(N)	Mean	(N)	
Return on equity (5-year average)[b]	17.8	(44)	15.4	(38)	2.40
Return on equity (1-year average)[b]	17.8	(41)	15.2	(38)	2.02
Return on total capital (5-year average)[c]	14.4	(44)	12.2	(38)	2.25
Return on total capital (1-year average)[c]	14.2	(42)	12.2	(38)	1.73
Debt/Equity ratio[d]	0.25	(44)	0.39	(41)	2.31
1-year Net Profit margin[e]	7.0	(41)	5.4	(38)	2.15
5-year average growth in sales (%)	15.0	(44)	13.4	(37)	1.55
5-year average growth in earnings per share (%)	16.5	(44)	13.4	(37)	1.82

[a] The t-test is a measure of the significance of the difference between the means for the two groups. For this sample size, which was approximately 44 for the "progressive" companies and 38 for the others, a t of 1.66 yields a 95% probability that the difference is significant. A t of 2.40 means the probability is greater than 99%.

[b] The return on equity is the proportion of stockholders' investment that is returned in profit.

ᶜ The return on total capital is the return on stockholders' investment and debt together.

ᵈ The debt/equity ratio is a measure of the proportion of total capital that is debt rather than stockholders' equity. Because it is desirable to minimize debt relative to equity, the lower this ratio the better.

ᵉ The net profit margin is the proportion of sales that constitutes a profit after expenses and taxes.

The figures from 1960 to 1975 show a similar pattern.

    2. Peter Drucker, "Japan Gets Ready for Tougher Times," *Fortune*, November 3, 1980, pp. 108–14.

    3. My definition of innovation is based on Victor Thompson, "Bureaucracy and Innovation," *Administrative Science Quarterly*, 10 (1965): 1–20. The term "innovation" is used in several different ways in the literature: confined to original inventions, defined as implying something new for *this* organization but perhaps not original, or synonymous with any kind of change; Lloyd A. Rowe and William B. Boice, "Organizational Innovation: Current Research and Evolving Concepts," *Public Administration Review*, 34 (1974): 284–93. To confuse the matter further, much of the investigation of "innovation" has focused on the *acceptance* of new ideas or the willingness to be influenced to change, relatively passive processes, rather than on the active process of pushing for a change. See, for example, Floyd Rogers and Everett Shoemaker, *The Communication of Innovations*, New York: Free Press, 1971. And some analysts call the *idea* the innovation ("Why don't we offer widgets in three colors?"), separating this from implementation (actually getting three colors of widgets into production).

    On the association of innovation with internal entrepreneurs: it is common among experts on R&D, such as Edward Roberts of MIT, to use the term "entrepreneur" to describe the people behind an innovation, those who pick up an idea and drive it toward support and use within an organization. This challenges the conventional confinement of the term "entrepreneur" to those who take the risk of founding an organization, who bet themselves and their savings on an enterprise they create. But I find it useful to extend the term, appropriately modified, to those inside large organizations who in effect bet their *jobs* on the outcome of an innovation.

    4. William Abernathy and Edward Roberts, as discussed in Michael Wolff, "What Do We Really Know About R&D? A Talk with Edward B. Roberts, *Research Management*, 21 (November 1978): 6–11; H. Igor Ansoff and John M. Steward, "Strategies for a Technology-Based Business," *Harvard Business Review*, 44 (November-December 1967): 71–83; John R. Kimberly, "Managerial Innovations," in W. Starbuck, *Handbook of Organizational Design*, New York: Oxford, 1981.

    5. For a review of the evidence on invention and innovation as a function of enterprise size, see Barry A. Stein, *Size, Scale, and Community Enterprise*, Cambridge, Mass.: Center for Community Economic Development, 1973. Sharon Oster also found that innovations in steelmaking were adopted more rapidly by smaller firms, in "The Diffusion of Innovation Among Steel Firms: The Basic Oxygen Furnace," *Bell Journal of Economics*, 13 (Spring 1982): 45–56.

    6. A related idea was suggested in two classic works: Tom Burns and G. M. Stalker, *The Management of Innovation*, London: Tavistock, 1968; and Paul R. Lawrence and Jay Lorsch, *Organization and Environment*, Boston: Harvard Business School, 1967.

7. "Ninety-nine Propositions about Innovation from the Research Literature," *Stimulating Innovation in Middle Management,* Cambridge, Mass.: Goodmeasure, Inc., 1982.

8. My concepts of segmentalism/segmentation versus integrative approaches are not the same as "differentiation" and "integration" as used in Lawrence and Lorsch, *Organization and Environment,* and elsewhere. The division of labor and the recognition of differences is just as important, in my view, in an integrative as in a segmentalist organization, but the critical issue is whether the differences become hardened as segments with virtually uncrossable boundaries and limited cross-segment cooperation. The "magic rings" discussion in Chapter 12 also addresses this point.

9. The idea of segmentalist versus integrative structures and cultures also differs from two other concepts in current use by organizational analysts: tight versus loose coupling [Karl Weick, "Educational Organizations as Loosely Coupled Systems," *Administrative Science Quarterly,* 21 (1976): 1–19, and *The Social Psychology of Organizing,* second edition, Reading, Mass.: Addison Wesley, 1979], and mechanistic versus organic organizations (Burns and Stalker, *Management of Innovation*). The tight/loose coupling distinction cuts across mine in that highly segmented organizations may have their parts only loosely attached to each other, in terms of communication and exchange of information, personnel, resources, etc., but still be tightly coupled to the demands/decisions of centralized authorities. The mechanistic/organic distinction is closer to the one I am making here, e.g.:

MECHANISTIC	ORGANIC
• problems/tasks broken down into specialist roles	• problems not broken down/divided
• each sees tasks as distinct from tasks of whole, as if each were a subcontractor	• individuals have to perform specialized tasks in light of knowledge of tasks of whole
• precise definition of technical methods, duties, powers in each functional role	• jobs lose formal definition in terms of methods, duties, powers—continually redefined through interaction
• vertical interaction within management	• interaction lateral as much as vertical

10. Thompson, "Bureaucracy and Innovation," pp. 8, 9.

11. Richard M. Cyert and James G. March, *A Behavioral Theory of the Firm,* Englewood Cliffs, N.J.: Prentice-Hall, 1966, p. 117.

12. Cyert and March, *ibid.,* propose that the search process is *motivated* (driven by both the perception of problems and the eagerness to apply pet solutions), *simpleminded* (focused on perceived symptoms and locally available alternatives), and *biased* by the training, experiences, and political interests of those searching.

13. Harold Wilensky, *Organizational Intelligence,* New York: Basic Books, 1967.

14. Michael D. Cohen, James G. March, and Johan P. Olsen, "A Garbage Can Model of Organizational Choice," *Administrative Science Quarterly,* 17 (March 1972): 1–25.

15. The psychic-economy principle is why Cyert and March called searches "simpleminded."

16. Five of the six high-innovation companies out of my ten core companies were on the list of progressive companies; Wang Labs' reputation seems to be largely local, in the Boston area, but it is in that arena well regarded and considered a progressive employer. General Motors, my mixed-innovation case, is also on the list, but largely for its quality-of-work-life programs.

CHAPTER 2

1. Daniel Nelson, *Managers and Workers: Origins of the New Factory System in the United States 1880–1920*, Madison, Wis.: University of Wisconsin Press, 1975.

2. *Fortune*, January 25, 1982, and *Wall Street Journal*, December 8, 1981.

3. Gene Bylinsky, "The Japanese Score on a U.S. Fumble," *Fortune*, June 1, 1981, pp. 68–72.

4. *Ibid.*

5. *Statistical Annual of the United States*, 1981.

6. National Science Foundation and Stanford Research Institute figures; reported in Hunter Lewis and Donald Allison, *The Real World War*, New York: Coward, McCann & Geoghegan, 1982.

7. UPI release, June 24, 1982.

8. Thomas S. Kuhn, *The Structure of Scientific Revolutions*, Chicago: University of Chicago Press, 1962.

9. Nelson, *Managers and Workers*.

10. Rosabeth Moss Kanter, "Work in a New America," *Daedalus: Journal of the American Academy of Arts and Sciences*, 107 (Winter 1978): 47–78.

11. Michael R. Cooper *et al.*, "Changing Employee Values: Deepening Discontent?" *Harvard Business Review*, 56 (January-February 1979); Daniel Yankelovich, "We Need New Motivational Tools," *Industry Week*, August 6, 1979; D. Quinn Mills, "Human Relations in the 1980s," *Harvard Business Review*, 56 (July-August 1979).

12. *The New York Times*, December 27, 1981.

13. Increases in technical knowledge embedded in tasks means that managers cannot simply concentrate on getting the work out, because they are often less knowledgeable about the work process than those with whom they interact, including their subordinates. As Victor Thompson wrote: "Authority is centralized, but ability is inherently decentralized, because it comes from practice and training rather than from definition. Whereas the boss retains his full *rights* to make all decisions, he has less and less *ability* to do so because of the advance of science and technology"; *Modern Organizations*, New York: Knopf, 1961, p. 47. Thus, managers are increasingly less able to exercise the authority of command, and it is increasingly less appropriate to what their organizations need. They need instead to have "political" skills such as identifying issues, persuading, building coalitions, campaigning for points of view, and servicing constituencies, including subordinates. Chapter 8 identifies some of these skills in action.

14. Rosabeth Moss Kanter, "Power Failure in Management Circuits," *Harvard Business Review*, 57 (July-August 1979): 65–75.

15. Paul W. MacAvoy, "The Business Lobby's Wrong Business," *The New York Times*, December 20, 1981.

16. R. Kelly Hancock, "The Social Life of the Modern Corporation: Changing Resources and Forms," *Journal of Applied Behavioral Science*, 16 (July 1980): 279–98. Also H. E. Meyer, "Remodeling the Executive for the Corporate Climb," *Fortune*, July 16, 1979, pp. 82–92.

17. "The Corporate Image: PR to the Rescue," *Business Week*, January 22, 1979.

18. Heidrick and Struggles survey, reported in *Wall Street Journal*, December 8, 1981.

19. Robert H. Hayes and William J. Abernathy, "Managing Our Way to Economic Decline," *Harvard Business Review*, 58 (July-August 1980): 67–77.

20. Hayes and Abernathy, *ibid.*, are leading exponents of this view.

21. Kanter, "Work in a New America."

22. Rosabeth Moss Kanter, *Men and Women of the Corporation*, New York: Basic Books, 1977, Chapter 7.

23. This is a focus barely realized in the organizational literature, with a few exceptions: C. R. Hinings. D. J. Hickson, J. M. Pennings, and R. E. Schneck, "Structural Conditions of Intraorganizational Power," *Administrative Science Quarterly*, 19 (1974): 22–44; Andrew M. Pettigrew, *The Politics of Organizational Decision-Making*, London: Tavistock, 1973; Henry Mintzberg, *The Structuring of Organizations*, Englewood Cliffs, N.J.: Prentice-Hall, 1979.

24. On adversity's binding organization members together: Rosabeth Moss Kanter, *Commitment and Community*, Cambridge, Mass.: Harvard University Press, 1972. On resource sharing in lean environments: Howard Aldrich, *Organizations and Environments*, Englewood Cliffs, N.J.: Prentice-Hall, 1979. On differentiation of organizational parts creating different outlooks: Paul R. Lawrence and Jay Lorsch, *Organization and Environment*, Boston: Harvard Business School, 1967.

25. Barry A. Stein and Rosabeth Moss Kanter, "Building the Parallel Organization: Toward New Mechanisms for Permanent Quality of Work Life," *Journal of Applied Behavioral Science*, 16 (July 1980): 371–88. Mintzberg, *Structuring*, uses the less specific term "adhocracy" in his lengthy discussion of this form, a term coined by Warren Bennis and subsequently picked up by Alvin Toffler.

26. Kanter, "Work in a New America."

27. Richard Edwards, *Contested Terrain: The Transformation of the Workplace in the Twentieth Century*, New York: Basic Books, 1979.

28. Kanter, "Work in a New America."

29. Allan R. Cohen and Herman Gadon, *Alternative Work Schedules*, Reading, Mass.: Addison-Wesley, 1978. Alan Westin, ed., *Individual Rights in the Corporation*, New York: Pantheon, 1980. David Ewing, *Freedom Inside the Organization*, New York: Dutton, 1977.

30. Rosabeth Moss Kanter, "Power and Change: Toward New Intellectual Directions for Organizational Analysis," Plenary Address, American Sociological Association Annual Meeting, 1980; and "Contemporary Organizations," in *Common Learning*, Washington: Carnegie Foundation for the Advancement of Teaching, 1981, pp. 75–94. Credit for first recognizing the shift goes to a number of analysts, including Arthur Stinchcombe, James Thompson, Paul Lawrence and Jay Lorsch, William Evan, Eric Trist, and others who wrote major papers or books in the 1970s. I summarize the shift of models as follows:

Old Model Assumptions	New ("Political") Model Assumptions

• Organizations and their participants have:
  choice
  freedom of contract
  limits set only by own abilities and capacities
	• Organizations and their participants face:
	  environmental constraints
	  resource limits
	  conflict and unequal power

• Organizations as tending toward "closed system" (rational focus and economic models)	• Organization as tending toward "open system" ("institutional" focus and political-economy models)
• Organizations as having limited purposes (and therefore able to stay bounded because they produce bounded and identifiable outputs)	• Organizations as having multiple activities and impacts ("uses") any one of which is subject to scrutiny by other groups; bargaining by stakeholders to set organizations' "official goals"

• Key management problems:
  control (internal and external)
  coordination of isolated segments
  reducing friction around the work process
	• Key management problems:
	  "strategic decisions"
	  issue management
	  external political relations

• Internal, micro-focus primacy of leadership and interpersonal issues	• External, macro-focus
• Need to study static or relatively invariant properties of the organization—e.g., how size or formal structure affects "success"	• Need to study bargaining, competition, and mutual adjustment
• Organizational effectiveness as a technical matter, based on objective standards and relatively universal human and organizational requirements	• Organizational effectiveness as a political matter, based on standards set by an organization's "dominant coalition" after bargaining among constituencies

31. Paul R. Lawrence, "How to Deal with Resistance to Change," *Harvard Business Review*, 46 (January–February 1969): 4–13.

32. Of course, government actions and public policies play a role too; I am confining myself to the actions companies themselves can take. For some public-policy suggestions, see Chapter 12.

Chapter 3

1. Among other things, as my colleague Walter Powell has pointed out, the ways an organization finds to reduce uncertainty in its environment can lead to an inappropriate repetition of past patterns.

2. Reported in *The Boston Globe*, "Corporate Overhaul at John Hancock," February 9, 1981.

3. Rosabeth Moss Kanter, *Men and Women of the Corporation,* New York: Basic Books, 1977, Chapter 7.

4. Harold Wilensky, *Organizational Intelligence,* New York: Basic Books, 1967. On low motivation and the creation of mutual-comfort groups as a result of limited opportunity, see Kanter, *Men and Women of the Corporation,* Chapter 6.

5. Melvin L. Kohn, "Bureaucratic Man: A Portrait and an Interpretation," *American Sociological Review,* 36 (June 1971): 461–74.

6. Donald Schon, *Technology and Change,* New York: Delacorte, 1967.

7. *Ibid.*

8. *Ibid.*; Charles Perrow, *Complex Organizations,* Glenview, Ill.: Scott, Foresman, 1979.

9. Schon, *Technology and Change,* pp. 71–73.

10. Tracy Kidder, *The Soul of a New Machine.* Boston: Atlantic–Little, Brown, 1981.

11. Frederick Bergerson, *The Army Gets an Air Force,* Baltimore: Johns Hopkins, 1980. Ken Farbstein was one of my research assistants.

12. T. A. Wise, "The Rocky Road to the Marketplace," *Fortune,* October 1966.

CHAPTER 4

1. *Wall Street Journal,* February 5, 1982.

2. David A. Whetten, "Sources, Responses, and Effects of Organizational Decline," in J. R. Kimberly and R. H. Miles, eds., *The Organizational Life Cycle,* San Francisco: Jossey-Bass, 1981, pp. 342–74. See also W. H. Starbuck and B. Hedberg, "Saving an Organization from a Stagnating Environment," in H. Thorelli, ed., *Strategy + Structure = Performance,* Bloomington: Indiana University Press, 1977.

3. Linda Frank and Richard Hackman, "A Failure of Job Enrichment: The Case of the Change That Wasn't," *Journal of Applied Behavioral Science,* 11 (October 1975), pp. 413–36.

4. Richard Walton, "The Diffusion of New Work Structures: Explaining Why Success Didn't Take," *Organizational Dynamics,* 4 (Winter 1975): 3–21.

5. Richard Walton, "Establishing and Maintaining High Commitment Work Systems," in J. R. Kimberly and R. H. Miles, eds., *The Organizational Life Cycle,* San Francisco: Jossey-Bass, 1980, p. 285.

6. On Tom West: Tracy Kidder, *The Soul of a New Machine,* Boston: Atlantic–Little, Brown, 1981.

7. Karl Weick makes the excellent point that an innovation (or "variation," in evolutionary-theory terms) must *persist* in order to be selected, even if one buys quasi-Darwinian evolutionary theory. Despite great long-run adaptive potential, then, an innovation must get over its threshold of vulnerability when it may be too early for a selective mechanism in the environment to *know* if the innovation will be a superior adaptation. Weick, *The Social Psychology of Organizing,* second edition, Reading, Mass.: Addison-Wesley, 1979, p. 129.

8. Walton, in "Diffusion," offers a variety of insightful political explanations for the failure of diffusion of successful workplace innovations: threatened obsolescence of some people if it succeeds; bureaucratic barriers such as organizationwide rules; envy of "stars"; rewards to the first

creators/adopters implying fewer rewards left for the next ones; the innovators' communicated feelings of specialness leading others to reject the new system as unique; and signs that those who fight the rules to get a change through hurt their careers (or as one company I know puts it, "make a career-limiting move").

CHAPTER 5

1. On the shared characteristics of rapidly growing organizations, see Rosabeth Moss Kanter and Barry A. Stein, "Growing Pains," *Life in Organizations,* New York: Basic Books, 1979; and Kanter, Myron Kellner-Rogers, and Janis Bowersox, "Strategies for Management of Growth in Two Rapidly Growing High Technology Firms," in *Managing Growth,* Cambridge, Mass.: Goodmeasure, Inc., 1982.

2. See "General Electric: The Financial Wizard's Switch Back to Technology," *Business Week,* March 16, 1981; "Piercing Future Fog in the Executive Suite," *Business Week,* April 28, 1975.

3. Jerald Hage and Michael Aiken, "Program Change and Organizational Properties," *American Journal of Sociology,* 72 (1967): 503–19.

4. Interdependence as a stimulus to innovation is a common research finding. For example, John L. Pierce and André Delbecq, "Organization Structure, Individual Attitude, and Innovation," *Academy of Management Review* (January 1977): 27–37.

5. See Stanley Davis and Paul R. Lawrence, *Matrix,* Reading, Mass.: Addison-Wesley, 1977; Kenneth Knight, "Matrix Organization: A Review," *Journal of Management Studies,* 13 (1976): 111–30.

6. Thomas J. Peters, "Putting Excellence into Management," *Business Week,* July 24, 1980; and *Three Yards and a Cloud of Dust,* New York: McKinsey and Co., 1980.

7. On forms lateral relations can take, see Jay Galbraith, *Designing Complex Organizations,* Reading, Mass.: Addison-Wesley, 1979. On the matrix as a transition, see Davis and Lawrence, *Matrix.*

8. Hage and Aiken, "Program Change."

9. Everett M. Rogers and F. Floyd Shoemaker, *Communication of Innovations: A Cross-Cultural Approach,* New York: Free Press, 1971. On the role of elite values: Jerald Hage and Robert Dewar, "Elite Values vs. Organizational Structure in Predicting Innovation," *Administrative Science Quarterly* (September 1973): 279–90.

10. In my progressive-company study, 46 percent of the 1980 senior executives of the progressive companies had been with the company their entire career, and 60 percent since 1960, compared with 36 percent and 49 percent in the nonprogressive matched companies in each category. These comparisons are statistically significant at the .03 and .05 levels respectively.

11. Barry M. Staw, "Attribution of the Causes of Performance," *Organizational Behavior and Human Performance,* 13 (1975): 414–32; Camille B. Wortman and Joan A. W. Linsenmeier, "Interpersonal Attraction and Techniques of Ingratiation in Organizational Settings," in B. M. Staw and G. Salancik, eds., *New Directions in Organizational Behavior,* Chicago: St. Clair Press, 1977.

12. Research by Alex Bavelas cited in Wortman and Linsenmeier, *ibid.*

13. See Chapter 10 of this book for an elaboration.

14. John Brooks, "The Money Machine," *The New Yorker,* January 1, 1981, pp. 43, 53.

15. Myron Magnet, "Managing by Mystique at Tandem Computers," *Fortune,* June 28, 1982.

16. Tracy Kidder, *The Soul of a New Machine,* Boston: Atlantic–Little, Brown, 1981.

CHAPTER 6

1. Rosabeth Moss Kanter, *Men and Women of the Corporation,* New York: Basic Books, 1977, Chapter 7.

2. Ken Farbstein, "Achieving at Chipco: Supporting Managerial Enterprise in a High Technology Firm," in *Stimulating Innovation in Middle Management,* Cambridge, Mass.: Goodmeasure, Inc., 1982.

3. I am indebted to Ken Farbstein for much of the discussion of "buy-in" at "Chipco," as indicated in the text.

4. The overall determination of how much power anybody has in the organization is twofold: first, how much he or she gets of the three basic commodities (information, resources, and support), and second, what he or she manages to use of them. Organizational power is in this sense *transactional:* power exists as potential until someone makes a bid for it and then invests it in activities and people that will produce results. Three variables on both the acquisition and the investment side further determine relative power: the amount of each of these commodities, the number of suppliers of investments, and the certainty of transactions.

5. This has echoes of the practice of "mortification," or public display of weakness, that serves to build commitment in strong communities; see Rosabeth Moss Kanter, *Commitment and Community,* Cambridge, Mass.: Harvard University Press, 1972.

6. Organizations with more career opportunity appear to be more innovative in general; J. Victor Baldridge and Robert A. Burnham, "Organizational Innovation: Individual, Organizational, and Environmental Impacts," *Administrative Science Quarterly,* 20 (1975): 165–76. This is perhaps because of the motivational aspects of opportunity; Kanter, *Men and Women of the Corporation,* Chapter 6. Furthermore, those with upward-mobility aspirations are likely to be more innovative; Floyd Rogers and Everett Shoemaker, *Communication of Innovations,* New York: Free Press, 1971. But mobility patterns have more than an incentive value; a major contribution to innovation may lie in the information and support networks they create. The research literature is very clear on this point: there is more innovation where there is more "communication integration" —closer interpersonal contact or interconnections via interpersonal-communication channels. And innovators are more likely to have more exposure to such channels. Innovation may be facilitated, furthermore, to the extent that frequent mobility breaks up project groups and adds new blood. Ralph Katz's studies of R&D teams show that increasing group longevity (especially beyond five years) may lead to innovation-stifling outcomes (more behavioral stability, more selective exposure to information, greater group homogeneity, and reduced communication within and outside the project group—all forms of what I call "segmentalism"). Katz, "Project Communication and Performance: An Investigation into the Effects of Group Longevity," *Administrative Science Quarterly,* in press.

7. In this present analysis, I am viewing mobility in terms of its organizational structure-forming impacts and not in terms of its effect on people—as it is usually considered, and as I did in *Men and Women of the Corporation*, Chapter 6.

8. Melvin L. Kohn, "Bureaucratic Man: A Portrait and an Interpretation," *American Sociological Review*, 38 (February 1973): 461–474.

9. "Teamwork Pays Off at Penney's," *Business Week*, April 12, 1982. See also Charles Burck, "The Intricate 'Politics' of the Corporation," *Fortune*, April 1975.

10. There is a great deal of agreement among researchers that specialization and functional differentiation in organizations aids innovation, even though specialization is a phenomenon we usually associate with bureaucracy. It is not so much that specialists themselves innovate—they may, indeed, be subject to a narrowness of focus—but that the existence of many different kinds of internal experts and professionals enlarges, in my terms, the marketplace of ideas. Some of the related characteristics that researchers claim are innovation-enhancing include: a larger number of professional specialties, greater specialization in jobs, and more administrative components. These coalesce into multiple demands, pushing for new ideas and practices to suit their professional interests, and in general, pushing for change. See Rogers and Shoemaker, *Communication of Innovations*; Baldridge and Burnham, "Organizational Innovation"; John R. Kimberly, "Managerial Innovations," in W. Starbuck, ed., *Handbook of Organization Design*, New York: Oxford, 1981; Gerald Zaltman, Robert Duncan, and Jonny Holbek, *Innovations and Organizations*, New York: Wiley, 1973. By placing a high value on knowledge, these groups also ensure that there is a diversity of sources of information to bring to bear on problems. Potential innovators are thus likely to scan more of the internal environment for ideas, and benefit, in turn, from the diversity of approaches which ensures that more of the external environment will have been scanned, with better ideas resulting.

"Chipco," Polaroid, Honeywell, and GE fit this model. There are large numbers of scientists and professionals, working on a range of distinctive products and technologies, divided into decentralized local units with their own administrative apparatus.

11. In political theory, the crosscutting ties of a pluralist system have long been held to prevent strong ideological divisions in electoral politics in the United States and to reinforce a consensus-seeking centrist orientation. If crosscutting ties aid political democracy, then perhaps they may aid "organizational democracy" too.

12. John Kimberly has argued that organization structure and process innovations are more likely when an organization is in distress, cutting back or looking for ways to improve internal operations. This gets weak support from the pattern at "Meridian Telephone," in Chapter 3; a high proportion of its innovations concerned the reorganization and reduction-in-force. But still, overall my findings indicate that even structure/method innovations are more common in the richer, less-distressed companies. Kimberly, "Managerial Innovations."

13. Frederick C. Klein, "Some Firms Fight Ills of Bigness by Keeping Employee Units Small," *Wall Street Journal*, February 5, 1982.

14. In my emerging theory of power-in-use, there are limits on how much power can be expanded, on how widely it can circulate. These limits include:

- *The size of the resource pool.* Several scholars argue that jockeying for advantage in organizations occurs when the resource pool is rather small to begin with, but when it is rather large and looks expandable, there is more cooperative and less oppositional activity.
- *The transformability or transferability of power commodities.* There are certain things that cannot really be passed on to other people, that cannot be transformed and used beyond the original intent residing in the object. The commodity can be spent in only certain places and certain ways: like a gift certificate. Investment of certain kinds of highly technical information in nontechnical parts of the organization, for example, could be wasted.
- *Too much counterproductive circulation of some of the commodities.* Organizational structure is one way we have of channeling the circulation of resources, information, and support so that organizational purposes are served. Not all directions are equally possible, and there is not equal access to these commodities from all parts of the system.
- *The capacity of components of an organization to process, or act upon, supplies.* Information overload, too much noise in the system, can also be dysfunctional. Attempting to expand power by circulating all information to everybody can make the system break down.
- *Differential time span involved in acquiring certain kinds of supplies.* Resources, information, and support come in a variety of forms, some of which take longer to produce than others. Technical knowledge as a kind of information, for example, often takes a great deal of time to acquire, and the replacement costs of producing it again can be very high. Experts can often gain more power in organizations from the fact that they have already spent the time to get the knowledge, and for other people to get equivalent amounts would take much too long. So acquisition time for cumulative knowledge—professional expertise—places a limit on power expansion.
- *The capacity of the organization to handle new activities.* Even the largest organization cannot do everything at once, and this places limits on the expansion of power via the circulation of such commodities as support. The boundary condition for legitimacy of support seems to be the capacity of the organization to handle all the things people would do with the support they get. Each of us may have a proposal for a project we want our function to carry out, but the ultimate limit on how many people can get legitimate authority to do theirs may have more to do with organizational capacity than with the desires of the individuals involved to limit power. (The implication for entrepreneurial activity is clear; an organization has limits on the amount of such enterprise it can realistically incorporate.)

15. Zaltman, Duncan, Holbek, *Innovations.* Decentralization/centralization measures show some of the same kind of fluctuation with phase, but the findings are less clear-cut. There seems to be more initiation of innovation in decentralized systems, but much less adoption; John L. Pierce and André Delbecq, "Organization Structure, Individual Attitude, and Innovation," *Academy of Management Review* (January 1977): 27–37. There is evidence that more "participation" favors innovation—a positive relationship between participation in decision making and rate of organizational change in core activities, and a negative relationship between power concentration in a hierarchy and change rate; Hage and Aiken, "Program Change." Kimberly argued that formalization and centralization in a company reduce the probabilities of adoption of innovation, holding that there

is greater receptivity to innovation in organic organizations with minimum procedural specification, minimum routinization of behavior, and widespread internal communication—just the conditions that fit my integrative model; Kimberly, "Managerial Innovations." But more decentralization can also raise the rate of conflict and disagreement, which interferes with implementation; Zaltman, Duncan, Holbek, *Innovations*. This is perhaps why one study finds more innovativeness in organizations with conflict-prevention committees; Baldridge and Burnham, "Organizational Innovation." By and large, then, the evidence for the effects of decentralization versus centralization is mixed—if we can even make the distinction between two ends of the same process anyway. Centralized power concentration seems to help in the adoption of innovations that require the organization, as opposed to individual units only, to change, as I am arguing here.

16. This is an interesting example of the institutionalization process: persistent use of a label with important symbolic meaning even where it clearly does not apply.

17. Gene Bylinsky, *The Innovation Millionaires*, New York: Scribner's, 1976. For a theoretical account of this process, see Jack W. Brittain and John H. Freeman, "Organizational Proliferation and Entity Dependent Selection," in J. R. Kimberly and R. H. Miles, eds., *The Organizational Life Cycle*, San Francisco: Jossey-Bass, 1980, pp. 291–338.

CHAPTER 7

1. It can be argued that similarities of technology make all jobs look more similar at the bottom of the wage and supervision hierarchy in any organization engaged in mass production; indeed, the "bottom" is generally the most routinized end of any organization and as such tends to be subject to rules of efficiency that can homogenize job designs. At the bottom, the impact of organizational culture is muted by the more dominant role of technology (unlike the middle and above, which tend to be more affected by cultural factors). I would argue that there is greater technological determinism as one moves down the organization to production tasks, as well as greater desire for managerial control over the jobs at the bottom. (See my article with Barry Stein, "Life at the Bottom," in our *Life in Organizations*, New York: Basic Books, 1979.)

Thus, while the design engineers or marketing staffs at a computer firm would operate in worlds very different from that of their counterparts in an auto manufacturer, the workers in assembly on the shop floor might have much more in common across industries. Joan Woodward's great contribution to organizational theory was to document the kinds of organizational structure likely to develop around three kinds of technology (small batch or craft work, as in glass blowing; mass production, as in assembly operations; and process technology, as in chemical refineries), though she assumed that the organization surrounding production (the direct labor at the "bottom") would define the entire system. Neglected was the fact that other functions operate in different "micro-environments," as Mariann Jelinek has also pointed out. (See Woodward, *Industrial Organization: Theory and Practice*, London: Oxford, 1965.)

In short, a company like "Chipco" can contain variations on its basic entrepreneurial culture depending on the specific tasks and pressures facing different parts of the organization. The people at the bottom may not experience the "Chipco" culture in the same way as those at higher levels do; a "logic of production efficiency" may cut across the culture.

2. Motorola has an advertising campaign based on its Participative Management Program, as Ford does for its employee-involvement groups ("Quality is Job No. 1"). Motorola says it launched its PMP more than ten years ago, and in 1981 about a third of its 45,000 U.S. employees operated under it, with problem-solving teams and a suggestion system the cornerstones.

3. All organizations, even the most innovating ones, must also maintain the momentum of ongoing routines while seeking change. And even in the most routinized, maintenance-oriented positions, there are still opportunities for involvement in solving unique problems not yet encountered by the organization and not yet built into routines—or inventing better routines to improve performance. The issue is not efficiency *versus* innovation, but rather, how to build both into the system.

4. Rosabeth Moss Kanter, *Men and Women of the Corporation,* New York: Basic Books, 1977, Chapters 6, 7. Among the conclusions: low opportunity leads to "passive resistance" or foot-dragging from the sidelines; a feeling of powerlessness leads to inappropriately tight control and turf-mindedness.

5. Howard Carlson, in Ernest Miller, "The Parallel Organization Structure at General Motors," *Personnel,* September-October 1978, pp. 64–69. The term was popularized by Barry Stein and me in "Building the Parallel Organization: Toward Mechanisms for Permanent Quality of Work Life," *Journal of Applied Behavioral Science,* 16 (Summer 1980), a report on the "Chestnut Ridge" project.

6. The report was based on extensive data collected by the pilot group using their own survey-and-interview design. The other groups produced: (1) an employee intake and development package made up of modules that could be combined in various ways to suit the needs of users. These included performance-appraisal training for supervisors, orientation programs for new hire and transfers, and career-path planning; (2) a set of recommendations for modifying supervisory training, and for making the new activities available when supervisors themselves saw the need, designed in collaboration with the plant training and personnel departments.

7. There are cycles of disadvantage and cycles of advantage around opportunity and power. Low opportunity breeds low motivation, low ambition, and low commitment, which reinforces low opportunity; and low power breeds overcontrol, rules focus rather than results focus, and low morale, which in turn keeps power low. But the positive cycles of advantage involve upward spirals: opportunity producing higher motivation producing more opportunity, and power producing better management behavior producing more power. Kanter, *Men and Women of the Corporation,* Chapters 6, 7, 9.

8. On external visibility's keeping a system in place, see John Meyer and Brian Rowan, "Institutionalized Organizations: Formal Structure as Myth and Ceremony," *American Journal of Sociology,* 83 (July 1977): 340–63. Thus, publicity not only functions as a source of reinforcement for those involved but also establishes expectations on the part of external constituencies. One of the companies to which I consult, which has been getting a great deal of positive publicity for its parallel organization, was recently faced with the problem of how to maintain the steering committee and project teams in the light of a major system reorganization. In stressing the importance of keeping them going, I said, only half in jest, "The eyes of corporate America are upon you." Management found this persuasive. Thus, external visibility serves as an important "retention"

mechanism in organizational evolution; see Howard Aldrich, *Organizations and Environments,* Englewood Cliffs, N.J.: Prentice-Hall, 1979.

9. Victor Turner, *The Ritual Process,* Chicago: Aldine, 1969.

10. Pehr Gyllenhammer, *People at Work,* Reading, Mass.: Addison-Wesley, 1978. See Rosabeth Moss Kanter and Barry Stein, "Where Leaders Can Lead and Followers Have Power," *Wharton Magazine,* 3 (Summer 1979): 66–69.

11. The differences between the routine hierarchy for maintenance and the parallel organization for change can be summarized as follows:

MAINTENANCE ORGANIZATION	PARALLEL ORGANIZATION
• routine operation—low uncertainty	• problem solving—high uncertainty
• focused primarily on "production"	• focused primarily on "organization"
• limited "opportunities" (e.g., promotion)	• expandible "opportunities" (e.g., participation in a task force)
• fixed job assignments	• flexible, rotational assignments
• competency established before assignment	• developmental assignments
• long chain of command	• short chain of command
• objectives usually top-down	• objectives also bottom-up

• rewards:
    pay/benefits
	• rewards:
	    learning
	    recognition/visibility
	    different contribution
	    bonus possibility
	    new contacts

• functionally specialized	• diagonal slices—mixed functions
• leadership a function of level	• leadership drawn from any level

12. The two kinds of involvement are not entirely independent, of course, and we can expect that the existence of a parallel participative organization may start to modify relationships in the hierarchy. At "Chestnut Ridge," potential dissatisfaction with the routine job after tasting the excitement of innovation was prevented by the fact that even regular jobs began to be more team-oriented in the course of the project, and many people received promotions. We can expect some "spillover"—positive and negative—from the coexistence of two organizational modes, simply because they share the same people.

CHAPTER 8

1. Tracy Kidder, *The Soul of a New Machine,* Boston: Atlantic–Little, Brown, 1981.

2. Richard Peterson makes a similar point in "Entrepreneurship and Organization," in W. H. Starbuck, *Handbook of Organization Design,* New York: Oxford, 1981. This conclusion matches Anthony Oberschall's findings after looking at entrepreneurial activity in Africa: where barriers to

innovation were low, entrepreneurs came from all sections of the population, but if barriers were high, they came only from special groups; "African Traders and Small Businessmen in Lukasa," *African Social Research,* 16 (1973): 474–502.

3. Statistical comparisons showed higher dollar payoffs and/or wider organizational significance to innovative accomplishments, but it was not always possible to make these judgments, since not all innovations can be cast in financial terms.

4. Those variables which most discriminated between innovative and basic accomplishments, at a statistical significance level of .05 or better, in descending order of sharpness of differences, were:

   a. style: managers with innovations more collaborative.

   b. risk: higher perceived risk for the innovations.

   c. selection of goals: innovations more likely to be self-initiated rather than just understood as part of the job.

   d. higher-level support: innovations need more.

   e. number of sources of resources, information, support: innovations draw on more.

   f. amount of money: innovations require more beyond budget.

   g. opposition from peers in work unit: innovations encounter more, more frequently.

   h. source of technical support: innovations use more from both inside and outside work unit.

   i. opposition: innovations more likely to encounter some.

   j. peer support: innovations more likely to get this.

   k. reward of a promotion: more likely after innovations.

   l. location of key actors: innovations more likely to involve actors from outside the immediate work unit.

   m. space: innovations need more.

5. For example, James G. March, "The Business Firm as a Political Coalition," *The Journal of Politics,* 24 (1962): 662–78.

6. There are many variations, of course, depending on the kind of task and how it is initiated. A clear assignment from higher officials may dramatically shorten the definition phase, whereas carrying out a complex technical project may lengthen the action phase. But despite these kinds of variations, I have found striking universality in the overall rhythm of innovative accomplishments.

7. Donald G. Marquis and Sumner Myers, *Successful Industrial Innovations,* Washington: National Science Foundation, 1969.

8. Gerald Zaltman, Robert Duncan, and Jonny Holbek, *Innovations and Organizations,* New York: Wiley, 1973.

9. James Thompson, *Organizations in Action,* New York: McGraw-Hill, 1967.

10. Allan Cohen and David Bradford suggested the conductor image to me. They are using it in their forthcoming book, *Managing for Excellence.*

11. Kidder, *Soul,* p. 227.

12. *Ibid.,* p. 275.

13. *Ibid.,* p. 83.

14. The same conclusions are reached by a more statistically sophisticated comparison:

   a. *Mean level of resources,* based on a scale of 1 (low) to 3 (high) dividing resource needs across all accomplishments:

	Mean	Standard Deviation	Range
Basic accomplishments (88)	1.60*	1.87	1–2
Innovative accomplishments (120)	2.34*	1.75	1–3

b. *Mean level of support,* based on combination of subordinate, peer, and higher-management support, on a scale of 1 (low) to 10 (high):

	Mean	Standard Deviation	Range
Basic accomplishments (88)	4.12*	1.89	1–8
Innovative accomplishments (120)	5.31*	2.21	1–10

c. *Mean level of activity outside work unit,* based on combination of location of key actors, amount of peer support, use of staff from outside work unit, on a scale of 1 (low) to 10 (high):

	Mean	Standard Deviation	Range
Basic accomplishments (88)	4.93*	2.06	1–10
Innovative accomplishments (120)	5.66*	2.27	1–10

* Differences in means significant at $p < .01$ on t-test

15. Kidder, *Soul,* p. 272.

CHAPTER 9

1. This has sometimes been labeled a "Hawthorne Effect," after the classic studies at the Western Electric Hawthorne Works in the 1930s. In one of the Hawthorne experiments, a team of women workers were given a separate work area where their production would be measured while a variety of environmental conditions such as lighting and rest breaks were varied. Productivity tended to go up regardless of the changes in physical conditions. One conclusion was that being singled out to be in a high-visibility experiment was highly motivating in and of itself; calling this the "Hawthorne Effect" was in part a way of dismissing the claims made by new "human relations" programs, arguing instead that any change involving increased management attention and special treatment would have positive effects for a little while. But of course, as I have pointed out elsewhere, there was more going on at Hawthorne than just the effects of an experiment. The women had their own separate territory. They were freed from supervision and instead had a research coach, so they became a team (an early example of a self-managed team inside a conventional factory?) helping one another do the work. A few who did not get along were replaced, helping the team coalesce. There was constant feedback on performance. An ambitious young woman saw this as her chance to catch the eye of top management and advance—and indeed, these workers *were* more visible to the top. There were special celebrations, such as ice cream parties after regular medical exams. Hawthorne, then, was one of our early experiments in increasing opportunity and power through participation and attention to "quality of work life." (One of the original reports of Hawthorne is F. J. Roethlisberger and William J. Dickson, *Management and the Worker,* Cambridge, Mass.: Harvard University Press, 1939.)

2. I am using the term participation to mean involvement in a team with joint responsibility for a product—which might be a set of recommen-

dations, a plan, a decision, a solution to a work-area problem, or the output of the work area itself. For the purposes of this analysis, I am equating participation with teamwork, and participative management with the building and nurturing of a collaborative team that is more fully consulted, is more fully informed, and shares responsibility for planning and reaching outcomes. But it is important to note that a wide range of policies and practices not involving teamwork are currently in use in companies under the *label* of participation, including employee-opinion surveys, suggestion systems, and job enrichment. See Rosabeth Moss Kanter, "Participation," *National Forum*, Spring 1982.

3. Victor H. Vroom and Philip W. Yetton, *Leadership and Decision-Making*, Pittsburgh: University of Pittsburgh Press, 1973.

4. Thus, there are times when autonomy and individual responsibility are more important than participation and team responsibility. Invention and innovation are often not "democratic" processes, and they may sometimes be best pursued by individuals who care passionately about an issue and build their *own* team of supporters and workers, as we saw in the previous chapter. Participation and autonomy are two different—and occasionally contradictory—forms of empowerment. In two book-publishing firms, one a "collective" and the other a conventional hierarchy, the distinction was clear. Some employees were happier in the low-participation hierarchy than in the high-participation collective, because while those in the collective had a greater voice in decisions and a greater involvement in tasks outside their job, they also had more intrusion from others who had some right to a say in these same activities—and in their jobs. (Frederic Engelstad, Norwegian sociologist, personal communication on work in progress.)

5. Daniel Zwerdling, "At IGP, It's Not Business as Usual," in R. M. Kanter and B. A. Stein, eds., *Life in Organizations*, New York: Basic Books, 1979, pp. 349–63.

6. Victor H. Vroom, *Some Personality Determinants of the Effects of Participation*, Englewood Cliffs, N.J.: Prentice-Hall, 1960.

7. J. Timothy McMahon, "Participative and Power-Equalized Organizational Systems," *Human Relations*, 29 (1976): 203–14.

8. John F. Witte, *Democracy, Authority, and Alienation in Work*, Chicago: University of Chicago Press, 1980, pp. 122–23.

9. *Ibid.*

10. *Ibid.*, p. 146.

11. *Ibid.*, p. 32.

12. Richard Hackman's research on job design supports this. The ideal design involves: experienced meaningfulness (task significance); task identity or responsibility for a defined chunk; and knowledge of results. Hackman and Greg R. Oldham, *Job Design*, Reading, Mass.: Addison-Wesley, 1980.

13. Bertil Gardell, "Automony and Participation at Work," *Human Relations*, 30 (1977): 513–33.

14. R. M. Powell and J. L. Schlacter, "Participative Management: A Panacea?" *Academy of Management Journal*, 14 (1971): 165–173.

15. Richard J. Long, "The Relative Effects of Share Ownership vs. Control on Job Attitudes in an Employee-Owned Company," *Human Relations*, 31 (1978): 753–63.

16. Mauk Mulder, "Power Equalization through Participation?" *Administrative Science Quarterly*, 16 (1971): 31–40.

17. Jane Mansbridge, "Time, Emotion, and Inequality: Three Problems of Participative Groups," *Journal of Applied Behavioral Science*, 9 (1973): 351–77.

18. Vroom, *Personality Determinants*; Thomas L. Ruble, "Effects of One's Locus of Control and the Opportunity to Participate in Planning," *Organizational Behavior and Human Performance*, 16 (June 1976): 63–73.

19. E. Linden Hilgendorf and Barrie L. Irving, "Workers' Experience of Participation: The Case of British Rail," *Human Relations*, 29 (1976): 471–505. Furthermore, it has also been found that the greater the autonomy and skill level in the job, the greater the interest in and demand for participation; see Gardell, "Autonomy and Participation."

20. On "outsiders" or people who are "different" in a work group: Rosabeth Moss Kanter, *Men and Women of the Corporation*, New York: Basic Books, 1977, Chapter 8. On the "competence multiplier": S. S. Weiner, "Participation, Deadlines, and Choice," in J. G. March and J. P. Olsen, *Ambiguity and Choice in Organizations*, Bergen, Norway: Universitetsforlaget, 1976, pp. 225–50.

21. Richard Walton, "Establishing and Maintaining High Commitment Work Systems," in J. R. Kimberly and R. H. Miles, *The Organizational Life Cycle*, San Francisco: Jossey-Bass, 1980, p. 227.

22. Muzafer Sherif *et al.*, *Intergroup Conflict and Cooperation: The Robbers Cave Experiment*, Norman, Okla.: University Book Exchange, 1961.

23. Hilgendorf and Irving, "Workers' Experience of Participation."

24. Walton, "High Commitment Work Systems," p. 237.

25. *Ibid.*, p. 287.

26. Witte, *Democracy, Authority*, p. 65.

27. Some critics argue that most participation is a subtle extension of management control anyway, but in a sophisticated new form.

28. Kanter, *Men and Women of the Corporation*, Chapter 7; "Power Failure in Management Circuits," *Harvard Business Review*, 56 (July-August 1979): 65–75.

29. Gardell, "Autonomy and Participation."

30. On productivity and satisfaction: A. C. Filley and R. J. House, *Managerial Process and Organizational Behavior*, Glenview, Ill.: Scott, Foresman, 1969; Suresh Srivastva *et al., Job Satisfaction and Productivity*, Cleveland: Case Western Reserve School of Management, 1975. On the experiment: Ruble, "Locus of Control." On saturation effects for those already involved in many decisions: Joseph A. Alutto and James A. Belasco, "A Typology for Participation in Organizational Decision Making," *Administrative Science Quarterly*, 17 (1972): 117–25.

31. Zwerdling, "IGP."

32. Rosabeth Moss Kanter and Louis A. Zurcher, "Evaluating Alternatives and Alternative Valuing," *Journal of Applied Behavioral Science*, 9 (1973): 381–97.

33. Walton, "High Commitment Work Systems."

34. Michele Hoyman, "Leadership Responsiveness in Local Unions and Title VII Compliance: Does More Democracy Mean More Representation for Blacks and Women?" *Proceedings of the Industrial Relations Research Association*, December 1979, pp. 29–42.

35. Rosabeth Moss Kanter, *Work and Family in the United States*, New York: Russell Sage Foundation, 1977.

36. Victor Turner, *The Ritual Process*, Chicago: Aldine, 1969.

CHAPTER 10

1. As the noted anthropologist Clifford Geertz commented, "What we call our data are really our own constructions of other people's constructions of what they or their compatriots are up to . . . Explanation often consists of substituting complex pictures for simple ones while striving somehow to retain the persuasive clarity that went with the simple ones." Geertz, *The Interpretation of Cultures,* New York: Oxford, 1975. For a similar perspective in a larger context see Peter Berger and Thomas Luckmann, *The Social Construction of Reality,* Garden City, N.Y.: Anchor Books, 1967.

2. Mayer N. Zald and M. A. Berger, "Social Movements in Organizations," *American Journal of Sociology,* 83 (1978): 823–61.

3. William H. Starbuck, "Organizations and Their Environments," in M. D. Dunnette, ed., *Handbook of Industrial and Organizational Psychology,* Chicago: Rand McNally, 1976, pp. 1069–1123.

4. Karl Weick, *The Social Psychology of Organizing,* second edition, Reading, Mass.: Addison Wesley, 1979, especially page 178.

5. Robert J. Litschert and T. W. Bonham, "Strategic Responses to Different Perceived Strategic Challenges," *Journal of Management,* 5 (1979): 91–105. Robert H. Hayes and William J. Abernathy, "Managing Our Way to Economic Decline," *Harvard Business Review,* 58 (July-August 1980): 67–77.

6. James Brian Quinn, *Strategies for Change: Logical Incrementalism,* Homewood, Il.: Richard D. Irwin, 1980.

7. Howard Aldrich called my attention to research by Baruch Fischoff on reinterpreting the past. He shows, through experiments, that people cannot disregard what they already know about something, when it comes to constructing an explanation about why something happened in the past. That is, once they know the outcome, people build stories which lead, inevitably, to that outcome. The researchers investigated this by altering historical outcomes, using cases most people don't know much about. They took real historical data, and simply changed the outcome of some series of events. When people were asked to estimate the probability with which they could have successfully predicted the outcome of the events, given knowledge only of the past, they constantly overestimated their ability to successfully predict. They also, in writing up stories about the justification of their prediction, were able to put together a very coherent and compelling story line. Of course, they were historically wrong! See Fischoff, "For Those Condemned to Study the Past . . ." in D. Kahneman, P. Slovic, and A. Tversky, eds., *Judgment Under Uncertainty,* New York: Cambridge University Press, 1982.

8. Tracy Kidder, *The Soul of a New Machine,* Boston: Atlantic–Little Brown, 1981.

9. Weick, *The Social Psychology of Organizing,* pp. 158, 165. The philosopher Abraham Kaplan is associated with this position.

10. For arguments about the symbolic functions of various aspects of organizational structure, see John W. Meyer and Brian Rowan, "Institutionalized Organizations: Formal Structure as Myth and Ceremony," *American Journal of Sociology,* 83 (July 1977): 340–63.

11. It has become fashionable to urge the management of organizational myths and stories. See Ian I. Mitroff and Ralph H. Kilmann, "Stories Managers Tell: A New Tool for Organizational Problem Solving," *Management Review,* July 1975; David M. Boje, Donald B. Fedor, and Kendrith M.

Rowland, "Myth Making: A Qualitative Step in OD Interventions," *Journal of Applied Behavioral Science*, 18 (1982): 17–28.

12. Quinn, *Strategies*, p. 58.

13. Weick, *Social Psychology of Organizing*.

14. The idea of random or planned deviance setting a change cycle in motion is consistent with the idea that organizational, like human, evolution begins with "variations," as captured in what has become known as the "population ecology" model. This is well explicated in Howard Aldrich, *Organizations and Environments*, Englewood Cliffs, N.J.: Prentice-Hall, 1979. I am suggesting that the evolutionary view is useful but requires a cognitive model of conscious and directed human action to flesh it out.

15. Michael D. Cohen, James March, and Johan P. Olsen, "A Garbage Can Model of Organizational Choice," *Administrative Science Quarterly*, 17 (1972): 1–25.

16. Eleanor Westney, personal communication, based on her research for her forthcoming book, *Organization Development in Meiji Japan*.

17. For a similar idea see Richard D. Beckhard and Reuben Harris, *Organizational Transitions: Managing Complex Change*, Reading, Mass.: Addison-Wesley, 1977.

18. Weick, *Social Psychology of Organizing*.

19. See Modesto A. Mardique, "Entrepreneurs, Champions, and Technological Innovation," *Sloan Management Review*, 21 (Winter 1980). Edward Roberts of the Sloan School at MIT is one of the first innovation researchers to identify the importance of champions; there is also a similar idea in Quinn, *Strategies*.

20. Thomas J. Peters, "Symbols, Patterns, and Settings: An Optimistic Case for Getting Things Done," in H. J. Leavitt, L. R. Pondy, and D. M. Boje, *Readings in Managerial Psychology*, 3rd ed., Chicago: University of Chicago Press, 1980.

21. See Peters, *ibid.*, for other examples.

22. For an account of one of the most elaborate attempts at institutionalizing a system for organizational strategy and change by tying it to every unit—Texas Instruments' OST system—see Mariann Jelinek, *Institutionalizing Innovation*, New York: Praeger, 1979.

23. In a recent study of a large number of employee-involvement programs, one of the strongest predictors of new-program adoption was the staff's expectation of benefits. Philip H. Mirvis, "Assessing Factors Influencing Success and Failure in Organizational Change Programs," in S. Seashore, E. Lawler, P. Mirvis, and C. Cammann, eds., *Observing and Measuring Organizational Change*, New York: Wiley, in press.

24. In evolutionary theory ("population ecology") this is a "failure of retention." See Aldrich, *Organizations and Environments*.

25. Richard Walton, "The Diffusion of New Work Structures: Explaining Why Success Didn't Take," *Organizational Dynamics*, 4 (Winter 1975): 3–21.

26. Richard Pascale and Anthony Athos, *The Art of Japanese Management*, New York: Simon and Schuster, 1980. Warren Bennis, *More Power to You*, New York: Doubleday, forthcoming. Quinn, *Strategies*. Charles E. Lindblom, "The Science of Muddling Through," *Public Administration Review*, 19 (1959): 78–88. Cohen, March, and Olsen, "Garbage Can Model."

27. "Conversation with Charles L. Brown," *Organizational Dynamics*, 11 (Summer 1982): 28–36.

28. David Nadler includes politics and anxiety as two of his three key tasks of change management. The third is control: ensuring ongoing

organizational maintenance while change is still in process. His solution to
politics is the use of teams and visionary leaders; to anxiety, a clear vision;
and to control, breaking the change down into a series of smaller, shorter,
and thus more manageable transitions. Nadler, "Managing Transitions to
Uncertain Future States," *Organizational Dynamics*, 11 (Summer 1982):
37–45. Noel Tichy has provided a full picture of the political-management
tasks in change efforts in his *Managing Strategic Change*, New York: Wiley,
1983.

CHAPTER 11

1. According to personnel staff, the three universities most heavily
represented in GM's U.S. salaried work force are distinguished but still
regionally based: Michigan State, the University of Michigan, and Purdue.

2. The Commerce Department study findings were reported in *The
Wall Street Journal*, December 7, 1981.

3. United Press International release, "Analysts Say J-Cars May
Become Edsel," *The Boston Globe*, April 27, 1982. Also *Consumer Reports*
in January 1982 made this recommendation (p. 11): "Should you buy a J-
car? Our advice: Not yet. Maybe not at all." GM's experience with offering
free cars was reported in *Time*, December 7, 1981, and the company's explanation of the financial advantage to those who took cash came from two
unrelated high-level informants.

4. E. Mary L. Balbaky, "General Motors and Ford: Decision Making in Turbulent Times," Harvard Business School Case, 1981.

5. Bart Spector, "General Motors and The United Auto Workers,"
Harvard Business School Case, 1981.

6. Charles Burck, "How General Motors Stays Ahead," *Fortune*,
March 9, 1981, p. 100.

7. Anne Armstrong, GM's other woman director, also served on the
committee. The Public Policy Committee was a major factor in the establishment of the European Advisory Council, the General Motors Institute
Visiting Committee, and the Science Advisory Committee, created in 1971,
which brought together distinguished outside academic scientists to discuss
a variety of technological matters.

8. Some executives, already looking at the threat of Japanese competition in the early 1970s, were long eager for a GM world strategy. The
general manager of the assembly division commented in 1972 that "in my
judgment, our division doesn't compare with the Japanese in productivity."
Since his unit was the one hit most by the labor unrest at Lordstown, it was
natural that he be concerned. But the question of GM's role in the world
auto market began to be raised. The quote is from Emma Rothschild, *Paradise Lost: The Decline of the Auto-Industrial Age*, New York: Random
House, 1973.

9. Described in greater detail in *General Motors' 1979 Public Interest Report*, pp. 20, 32. The European Advisory Council and the Austra-
lian Advisory Council are made up of local representatives of business,
academia, and the financial community.

10. Furthermore, the language of most host countries is now the
language at work, a phenomenon some insiders call "revolutionary." There
are emergency "affirmative action" plans overseas, to get sufficient numbers
of host-country nationals involved. Situations like that in Mexico—decades
of operations there with no Mexican general manager in that period—are

not to be repeated, according to my executive informants. Also, William MacKinnon, the current Vice-President of Personnel Administration and Development, indicated that GM's personnel data base (part of the HRM system described later in the chapter) now includes language competence, to support the international strategy. But building an effective international production and marketing base takes time, and in 1982 there were still pieces of reorganization occurring (e.g., merging three American divisions making and selling light-duty trucks and vans into one unit of the international truck and bus group) and concerns about whether GM's commitment to its international business would continue, given losses abroad; see *New York Times*, April 20, 1982, and John Koten, "GM's Overseas Drive Continues to Suffer After Three-Year Push," *Wall Street Journal*, July 19, 1982.

11. Charles Burck, "How G.M. Turned Itself Around," *Fortune*, January 16, 1978. Also these remarks by Elliott M. Estes while still GM President: Speech to the Detroit Auto Show Industry Dinner, January 13, 1981; news briefing on July 9, 1980; and transcript of interview on the Phil Donahue program, September 27, 1979.

12. Burck, *ibid.*, p. 89.

13. *Ibid.*

14. James Brian Quinn, "General Motors Corporation: The Downsizing Decision," Amos Tuck School of Business Administration, Dartmouth College, case, 1978. See also Joseph Kraft, "The Downsizing Decision," *The New Yorker*, 56 (May 5, 1980): 134–62.

15. Company publications about the HRM system say that it was established to "maintain the high quality of the General Motors salaried work force, assure an adequate future supply of management talent, and provide maximum opportunities for all employees, including minorities and women. . . . HRM committees in each employing unit administer the system's seven elements: inventory; appraisal; forecasting; collection; acquisition; career planning; and employee development. Each committee is reponsible for fully implementing standards and procedures for each element of the system within its employing unit."

16. The divisional and plant controllers report directly to their division general managers or plant managers. The sole exception is Chevrolet, where, for historical reasons, plant controllers report directly to Chevrolet's divisional controller. Then the divisional controllers also have an indirect reporting relationship to the corporate controllers. Hence, the top financial person at Buick, for example, has two titles: "Comptroller of Buick Motor Division" and "Assistant Comptroller of General Motors Corporation for Buick Motor Division."

17. Fuller himself traveled widely, visiting all the divisions, developing friendships with the general managers. "In Detroit they used to say, 'Is he out of town again?' Around here, if you're not in the office you are not working." His staff also began to visit divisions and made sure they were working on the division's problems, not corporate ones, by enlisting the assistance of divisions in defining and prioritizing problems. The corporate staff identified a contact person for each division, who could talk to his or her divisional counterpart six or eight times a day, if necessary; the staff could take any contemplated act and learn instantly how the divisions felt. Eventually, they held regional conferences four times a year to ask, What's troublesome? How's our new policy?

18. Fuller did not hesitate to work with the general managers and the Executive Committee to back or block appointments on the basis of professional criteria. The old image of personnel as a "dumping ground"

was reinforced, for example, by one personnel director who came to work only about four days a month, or another who was so intimidated by the general manager that he would not come out of his office all day if the general manager was around. Fuller insisted that people like this be replaced, and when he insisted, the general managers responded—perhaps because he had just arrived and was an officer, or perhaps later because of their growing respect for his staff and the clear backing from Estes, who was then Fuller's boss.

19. General Motors and AT&T were among the earliest companies to invest heavily in employee-attitude surveys.

20. William F. Dowling, "At General Motors: System 4 Builds Performance and Profits," *Organizational Dynamics*, 3 (Winter 1975): 23–38.

21. General Motors is not only a highly unionized employer; it is also one of the nation's leading firms in terms of setting a pattern for industrial bargaining and for the relations between American corporations and American unions. No program intended to impact on the corporation's workers could go very far without the active involvement of the unions, especially the United Auto Workers. Although the history of collective bargaining in the United States does in fact include a very significant period in which labor–management cooperation was common (after the First World War and through the first years of the 1920s), an extremely strong adversarial relationship, characterized by much violence and hostility, had prevailed between the two sets of institutions well through the 1960s.

22. Spector, "GM and the UAW."

23. Important support for QWL developed within the corporation, including that of Joe Godfrey, vice-president of GM's assembly division (thus responsible for all the assembly plants, including Tarrytown), Terrell, and eventually Gerstenberg himself. The 1971 split of the corporate labor-relations staff from the personnel staff concerned with human resources and development provided more potential support and legitimacy for a cooperative stance toward personnel matters, as opposed to the traditional adversarial one represented by labor relations, and Stephen Fuller became critical to the emerging commitment to QWL.

24. Bluestone continued to push and to work at handling opposition and interference even as QWL took hold, just like the entrepreneurs in Chapter 8. For example, Tarrytown became a demonstration site for him. He arranged for a major conference of union officials to be held at Tarrytown in 1978, partly to convince the doubters that QWL was both feasible and effective, which allowed for presentations by local union representatives and GM assembly-plant management—a pivotal event in turning things around. (For a sensitive portrait of Bluestone's leadership style, see Michael Maccoby, *The Leader*, New York: Simon and Schuster, 1981, pp. 93–121.)

25. Spector, "GM and the UAW." A turning point occurred when a group of militant blacks with a history of disciplinary problems and alienation became part of a joint labor-management committee aimed at undercutting QWL, and instead became converts. Thus, QWL also helped improve race relations, decreasing segmentation by category of worker.

26. Robert Guest, "Quality of Work Life: Learning from Tarrytown," *Harvard Business Review*, 56 (July-August 1979): 76–87. However, the QWL experiment was nearly terminated because of concerns about both its costs and its potential, given troubled financial circumstances. This was the point at which people began to reflect on the successes and failures and to realize that, hopes for instant transformation aside, the process did take

time. And eventual surveys showed that 95 percent of the work force wanted to participate.

27. A postscript on Tarrytown. By the spring of 1981, I was told, Tarrytown ranked last among the sixteen GM assembly plants in a standard monthly measure of quality and productivity posted on charts in the factories; the Lakewood plant was also near the bottom of the list. What happened? It's not entirely clear, but at Lakewood the plant manager and architect of the program, Frank Schotters, had departed for corporate headquarters. At Tarrytown, the introduction of the "X" car had necessitated a major overhaul in the plant itself; along with that, perhaps, went some potentially divisive negotiations between the UAW and GM, as a part of the 1979 agreements. But meanwhile, the Fleetwood plant in Detroit, about to be described, was at the top of the spring 1981 list with a score of over 120 against a norm of 100—far ahead of Tarrytown's 80.

28. The program is supported by a two-day QWL workshop which over 70 percent of the plant had attended by the time of our visit. The group leadership has been given a great deal of flexibility and voice beyond the role it normally plays, and new positions—e.g., QWL coordinator—have been created and occupied by plant personnel.

29. Spector, "GM and the UAW."

30. There was an interesting example of this ignorance in a *New York Times* update (May 25, 1982) on Lordstown ten years after the famous strike. Several of the workers interviewed expressed little knowledge of Lordstown's QWL activities. (At the same time, it was also clear that there were major changes for the better in the climate at Lordstown.)

31. I am reporting this the way it looked to outsiders. Insiders have, not surprisingly, a different view. Because the matter, owing to salaried union organizing drives, is sensitive, I was asked not to quote anyone on this subject, nor to reveal what I was told.

32. The GM *Public Interest Report* for 1978 justified the traditional divisional-autonomy stance and its slow response time—the long time it takes to incorporate changes into GM cars: "Design changes are first tried in one or two divisional product lines before being adapted to others, and it takes a lot of lead time to incorporate changes into every GM vehicle. Some consider this overcautious, but General Motors' record—the industry's best —proves it is wiser to get it right than to get it first." Even those executives who speak positively about the project-center idea do not wish to counter divisional autonomy, as one of them told me: "The project center is basically a good concept. There are some wrinkles—for example, the question of who's responsible. We don't want to remove a marketing tool from the divisions." The current thrust at GM is to reinforce the divisions' independence, an idea Chairman Roger Smith explained to me at length in an interview in March 1981. See also "General Motors: The Next Move Is to Restore Each Division's Identity," *Business Week*, October 4, 1982; "General Motors: Are the Old Ways Still the Best Ways?" *Forbes*, September 27, 1982. The cover of *Forbes* for the latter story showed a portrait of "Roger Smith Contemplating the Bust of Alfred P. Sloan," satirizing the famous Old Master painting.

33. Thomas Murphy made clear to me that it is simply not true that the chairman and CEO of GM always was and inevitably is a financial person. Most recently, James Roche was CEO from 1967 to 1972 and had never had a financial job in his career. Both Charles Wilson and Harlow Curtice were operating people (Curtice held the titles of both chairman and

president in the 1950s), and Frederic Donner, who became chairman in 1958, was the first primarily financial person. Still, in the recent history of GM and other companies, financial executives have begun to predominate. See Robert H. Hayes and William J. Abernathy, "Managing Our Way to Economic Decline," *Harvard Business Review*, 58 (July-August 1980): 67–77.

34. For those in the salaried ranks, beginning to implement participative efforts was often countercultural, as I was told again and again. For example, one first-line supervisor I interviewed, from a divisional accounts-payable department, an area of 35 people and 3 supervisors, initiated a problem-solving meeting like a quality circle—he called it a "little training program with my people." He had some trouble at first with his boss, who said he liked the idea but wanted the supervisor to cut down the time; there was some implication that the general supervisor thought they were just shooting the breeze. The meetings took about four hours—an afternoon. The boss was concerned: "Why can't you cut it down to an hour?" But the boss was too busy to come to a meeting, to see for himself.

35. The 1979 GM–UAW negotiations were something of a problem. The primary issue was what the UAW regarded as GM's "Southern Strategy," a device that was thought to represent the corporation's effort to shift its manufacturing base from the heavily unionized North and Midwest to the thinly unionized and still relatively antagonistic to unions Southern Tier. GM insisted that the motivation had much more to do with the cost structure in the South, including energy as well as labor, and a desire to broaden its political and institutional influence by extending its direct operations beyond its traditional areas. Then GM's new Oklahoma City plant, designed to be run by worker teams, voted to join the union in what the media jumped on as a vote "against QWL methods." (But QWL plants were often the first to sign the agreement.) Ultimately, and in part because Bluestone decided that the issue needed to be settled "once and for all," GM eventually agreed to automatically recognize the union at *all* new plants opened after 1979. Presumably that settled the issue of collective bargaining and representation in GM plants.

36. Thomas C. Hayes, "Behind GM's Labor Troubles," *New York Times*, February 26, 1982.

37. John Holusha, "Union View: Fair Sacrifice," *New York Times*, April 29, 1982; Robert L. Simison, "GM's New Alliance with UAW's Starting to Look Rather Shaky," *Wall Street Journal*, June 2, 1982. See also Associated Press release, "White Collar Workers at GM Want to Join Up with UAW," *Boston Globe*, May 15, 1982; Donald Woutat, "The Isuzu Decision by GM," *Boston Globe*, June 2, 1982; Kevin Tottis, "Malfunctions Plague Fancy New GM Buses," *Wall Street Journal*, June 2, 1982.

38. One executive provided me with this summary of why he concluded in 1981 that there was still too much short-range thinking in the company, as of that time:

"There is no long-range planning function in the company still, and the only planning groups we do have are product-related. We have a new staff, worldwide product planning, as of about five years ago. This was the first planning function at all, and we still have no coordinated planning function. There are four or five different groups in the corporation that do all product planning for trucks. Then there are fifteen or twenty people on the research staff that do this, involved in 'societal analyses.'

"There is no planning here because most long-range planning is influenced by the external environment and changes in it. This company has been regional, insular. We don't communicate much with IBM, GE, Bechtel. We don't go to meetings, thinking of this as a waste of time. We act self-satisfied. We think we know more about things than anyone, so how can they teach us? And we measure ourselves against Chrysler and Ford. Our overseas operations were largely bought, not built. We have no experience in *developing* the overseas market.

"The best evidence of our short-range planning is around plant and equipment. For twenty years, despite investments in plant and equipment, we didn't do enough. The reason we haven't invested is because return on investment is our primary measure for executives when it comes to the bonus system. . . . But now we need attention to planning, because our problems are of such magnitude that they can't be dealt with in one year. For example, you can't invest forty billion dollars in one year."

But among the signs of change in 1982: Under Chairman Roger Smith, there was a consolidation of four planning groups that had formerly been assigned to different functional or product areas.

39. William P. Safire, "Accent on the Caribbean," *New York Times Magazine*, March 21, 1982, p. 13.

40. Some of the examples of young senior executives: When William MacKinnon was appointed to replace Stephen Fuller on April 1, 1982, he was then, at 43, GM's youngest vice-president. But he held that "honor" for only one month, as Robert Eaton was appointed engineering-staff vice-president on May 1 at age 42. GM's current controller, its treasurer, and the vice-presidents/general managers of Buick and Pontiac are all in their early or mid 40s.

Index